T0231286

Luigi Balugani's Drawings of African Plants

James Bruce of Kinnaird, aged 32, by Pompeo Batoni. Courtesy Scottish National Portrait Gallery, Edinburgh.

LUIGI

From the collection made by James Bruce of Kinnaird

BALUGANI'S

on his travels

DRAWINGS

to discover the source of the Nile 1767–1773

OF

By Paul Hulton · F. Nigel Hepper · Ib Friis

AFRICAN

Yale Center for British Art · New Haven · A.A. Balkema · Rotterdam · 1991

PLANTS

CRC Press is an imprint of the
Taylor & Francis Group, an **informa** business

Paul Hulton was formerly Deputy Keeper of Prints and Drawings at the British Museum

F. Nigel Hepper is a Principal Scientific Officer in the Herbarium, Royal Botanic Gardens, Kew

Dr Ib Friis is Senior Lecturer in the Department of Botany, Copenhagen University

CRC Press
Taylor & Francis Group
6000 Broken Sound Parkway NW, Suite 300
Boca Raton, FL 33487-2742

First issued in hardback 2017

CRC Press is an imprint of Taylor & Francis Group, an Informa business

Published by
A. A. Balkema, P.O. Box 1675, Rotterdam, Netherlands

ISBN 13: 978-1-138-40765-7 (hbk)
ISBN 13: 978-90-6191-779-3 (pbk)

Contents

Foreword *and* Acknowledgments

In 1968 and 1977, Paul Mellon acquired from the collection of Lord Elgin of Broomhall the drawings, watercolors, journals, letters and other records relating to the expedition which James Bruce of Kinnaird undertook in search of the source of the Nile. Nine years later, he gave the entire archive to the Yale Center for British Art and encouraged the proposal, which was put forward shortly afterwards by Nigel Hepper of the Herbarium of the Royal Botanic Gardens at Kew, to publish from it the drawings of African plants. This volume exists thanks to Mr. Mellon's abiding interest in the publication and to the generous support he has given to it.

In 1981, Patrick Noon, Curator of Prints and Drawings, took charge of the project. He has worked closely with Paul Hulton, formerly Deputy Keeper of Prints and Drawings at the British Museum, whose art historical treatment of the material is matched by the scientific knowledge of his two co-authors, Nigel Hepper and Ib Friis of the Department of Botany, Copenhagen University. They wish to stress that, although separate chapters bear the names of individual authors, the book is the product of close-knit cooperation.

Jointly, they thank the Director of the Royal Botanic Gardens, Kew, for many facilities during the preparation of this work. They have benefitted from the expertise of several colleagues at Kew, especially the following: D. V. Field (Asclepiadaceae), C. Jeffrey (Compositae, Cucurbitaceae), Susan Carter (Euphorbia), M. G. Gilbert (Asclepiadaceae, Labiatae, Loranthaceae), and W. Marais (Liliaceae), also N. K. B. Robson (Hypericaceae) of the British Museum (Nat. Hist.). The staff of the Yale Center for British Art, New Haven, have been very helpful, especially Patrick Noon and his former assistant Paula Joslin, who compiled the first basic list of the plant drawings. Constance Clement has patiently attended to the many production details. Photography was carried out by Joseph Szasfai and Michael Marsland. In London, Brian Allen and the staff of the Paul Mellon Centre for Studies in British Art, and the Department of Prints and Drawings at the British Museum, kindly provided facilities. They are grateful for the interest shown by colleagues at the University of Copenhagen and to Dr. Kirsten Grubb-Jensen for help with the reading of copies of the Italian letters from the Bruce archives. Lord Elgin of Broomhall kindly allowed access to his collection of Bruceana and the photography of the drawings remaining in his collection. The staff of the Biblioteca dell'Archiginnasio in Bologna provided access to manuscripts and books relating to Balugani. Dr. Jacques Mercier of CNRS Paris kindly commented on the vernacular names. The maps were drawn by Tim Aspden.

In conclusion, the Center acknowledges Lawrence Kenney's contribution as copy editor of the complex typescript, and the help which Greer Allen provided in the design of the book. We thank A.T. Balkema for taking full responsibility for its printing and for his readiness to co-publish the volume.

Duncan Robinson
Director
Yale Center for British Art

Introduction

While James Bruce was exploring in the Red Sea and in Ethiopia Captain James Cook was navigating the *Endeavour* in the South Pacific. The two expeditions would seem to have few similarities. Cook's was a well-prepared scientific enterprise carried out by professionals and for which responsibility was shared by the Admiralty and the Royal Society. Besides Cook, the navigator, were Sir Joseph Banks, the naturalist (who had equipped the vessel at his own expense), two other naturalists, and two artists. Cook's primary aim was to explore the ocean and to chart his discoveries, and Banks's was to observe and record the human and natural life they encountered.

By contrast Bruce's expedition was a private venture, without governmental or institutional backing, and financed solely from his own resources (though he did receive some help from France in replacing expensive equipment lost at sea). Bruce, as an amateur artist himself, took only one assistant, Luigi Balugani, the young Bolognese architect and draughtsman who was not, so far as we know, a professional botanical artist. Bruce first set out to find and record Greek and Roman architectural remains in North Africa, hence Balugani's recruitment. Only when this work was completed did he turn toward Ethiopia, where he aimed to discover the source of the Nile, then widely believed to rise there. Balugani remained with him but now spent much of his time drawing the zoological and botanical life they found en route. Bruce was motivated by ambition and curiosity. He was an example of the all-round amateur, well endowed with the mental and physical attributes every explorer needs to carry him through. In Ethiopia, Sennar, and the Nubian Desert he was at the mercy of man and climate even more than was Cook on his long sea voyages and briefer landfalls.

Yet if we look at the essential routines carried out by the two expeditions they are remarkably similar. Bruce calculated his position by instrumental observation, whatever the conditions, and made maps, as did Cook. Bruce recorded temperature, barometric readings, wind, and weather conditions generally, as did Cook. Bruce and Balugani observed and drew animals, and Balugani (alone probably) plants, as did Sydney Parkinson on the Cook expedition after the early death of his fellow artist. Though Bruce's plant descriptions are erudite and classically colored but, botanically speaking, largely unscientific (characteristic of the cultivated amateur of the time), Balugani's drawings have scientific value greater perhaps than those of Parkinson, who was overworked and too busy to make finished watercolors. He sketched hundreds of outlines with color notes to be finished in England by Frederick Nodder. And on both expeditions logs and journals were kept showing to what degree their achievement depended on observable and measurable fact.

The primary purpose of this book is to publish all the plant drawings Bruce brought out of Africa. Inseparable from this is the need to identify and proclaim their author, Luigi Balugani, a name hitherto virtually unknown as a botanical artist and consistently obscured by Bruce–whether from vanity or jealousy is not altogether clear. As Balugani died in Ethiopia he himself could not redress the balance. This volume aims to do so with essays on his life and on the scientific and artistic value of his drawings. As my co-authors Nigel Hepper and Ib Friis, botanists with a special interest in African plants, were the first to identify the drawn plants and to publish articles on them, it was decided that the catalogue should be arranged in the conventional botanical order and that the entries should cater both for the reader with botanical interests as well as for the reader of art history. But it is hoped that the catalogue and essays together will also appeal to the general reader.

Bruce has been the subject of many studies and biographies since his death in 1794, and his *Travels to Discover the source of the Nile* (1790) has appeared in many editions and languages. Yet the definitive biography remains to be written, and a full, up-to-date edition of the *Travels* still awaits an editor. In recent years there has been fresh interest in Bruce, and several recent books have aimed to make him more accessible to the general public. The best are C.F. Beckingham's abridged edition of the narrative of the *Travels* of 1964, with a brief introduction on his life, and J.M. Reid's *Traveller Extraordinary* of 1968. Both give excellent accounts of the man and his travels. The former allows Bruce to speak for himself at length, the latter gives fuller biographical details with ample, though briefer, quotations from his book. As for any new life of Bruce or new edition of the *Travels*, Alexander Murray's editions with his essay on the life of Bruce (1804/5, 1813) will remain the basis for all new research. Here the first publication in quantity of drawings from Bruce's collection requires a detailed outline of his life, placing more emphasis on this graphic aspect of his achievement than previous writers have done.

Though Bruce unhappily exaggerated his own part in their making, the drawings at the Yale Center for British Art and those still in the possession of his descendant Lord Elgin at Broomhall remain very much his creation. They were made as the result of his efforts and leadership, of his ambition to get them drawn and finished in watercolor. They became even more "his" after Balugani's death in 1771, and when, against all odds, he got them home intact, even after having been forced to abandon them temporarily in the desert sands near Aswan. Bruce's achievement was by any standards remarkable, but his lengthy account of his travels is not easily accessible. The chapter on his life and journeys is given not only to provide background information, but in the hope of stimulating the reader to turn, if he is able, to Bruce's own narrative, many passages of which must rank among the most lively and entertaining in English travel literature.

· I ·

An outline of the Life of James Bruce *of* Kinnaird *with Particular Emphasis on the Drawings He Brought out of Africa*

James Bruce was born at Kinnaird, the family home in Stirlingshire, Scotland, on 14 December 1730.[1] He was a son of David Bruce, whose mother was Helen, heiress of the Bruces of Kinnaird. By a condition of her marriage contract with David Hay of Woodcockdale, the Bruce name and estate passed to their issue. James was intensely proud of the name Bruce, for he claimed descent from Robert the Bruce among other noble forebears. His mother died when he was an infant and his father married again. To provide for his numerous offspring he realized the necessity for James to make his own way in a profession. As the family sympathies were Hanoverian, it was decided to send him to England for his education. He spent four years at Harrow (and was there during the '45 Jacobite rebellion), where he had the traditional English teaching in the classics. By his contemporaries he was held to be a quick learner who rapidly became proficient in Latin and Greek. His interest in classical studies led later in life to an interest in classical antiquities and particularly architecture. Some of his contemporaries at Harrow, among them Daines Barrington and William Graham of Airth, his maternal uncle, were to remain lifelong friends. He left Harrow in 1746, an unusually tall schoolboy and, as he was far from robust, was suspected of having outgrown his strength. He then joined an academy in London run by a Mr. Gordon and continued his classical education while extending his studies to French and mathematics. Surprisingly, he briefly entertained the idea of going into the Anglican Church but quickly changed his mind and, no doubt as a result of his father's persuasion, decided to study law. He had in mind to become an advocate at the Scottish bar, so, in May 1747, he returned to Scotland.

Before starting his university career at Edinburgh, he developed a taste for and considerable skill in riding and shooting which began to strengthen his physique and was to enhance his reputation later, on his travels.[2] In November 1747 he took up his studies in Edinburgh. It soon became clear that the law was not for him, and, as his health remained delicate, he decided to return to Kinnaird. If he had not learned much law he had begun to study Italian, for his textbooks were said to have been scribbled over with Italian verses. He certainly had no wish to return to Edinburgh and continue with his law studies. After some time his imagination fixed on India. He was by now too old to join the East India Company as an underwriter, and he decided to travel to London and apply to the directors for permission to become a free trader under the company. He renewed his old friendships and as a result was introduced to the widow of a prosperous wine merchant, Mrs. Allan. He fell in love with her daughter, Adriana, and with her mother's approval married her on 3 February 1754. All ideas of trading in India now faded. The marriage arrangements included a share in the wine business. He seems to have had something of a flair for the trade and helped to manage it with every prospect of success. But his wife inherited a delicate constitution and was soon realized to be suffering from tubercolosis, the illness from which his own mother had died. It was decided that she should spend the winter in Provence. She and her mother left for France in early September and Bruce joined them in Boulogne. They got no further than Paris and Adriana died there about 9 October.

Bruce's attitude toward Roman Catholicism was unquestionably colored by his experiences during the last days of her life and by the circumstances of her burial. The Catholic clergy had, in spite of Bruce's objections, gained access to his dying wife. They would also have prevented her burial in consecrated ground had not the British ambassador intervened. To escape public notice she was buried at midnight in the graveyard of the embassy. Tormented with grief, Bruce took horse and rode through appalling weather to Boulogne and so returned to England. His account of those days is given in a moving letter to his father.[3] The picture he gives of his wife's last moments, of her burial, and of his journey through wind and rain on horseback, without food or rest, is consistent with other accounts of critical events in his life. Though his feelings were unquestionably genuine, the style takes on the quality of a romantic novel with actions and the central figure appearing larger than life.

Bruce now retired from active partnership in the wine business, though he was still associated with it, and began to study Spanish and Portuguese. He had an aptitude for languages which his knowledge of the classics had sharpened. At the same time he began to practice drawing and took lessons from an artist named Bonneau recommended to him by a friend, Robert Strange, the Scottish engraver. According to Murray, he had previously taken up drawing but only in connection with the design of fortifications in which he had become interested. We do not know for certain who his new master was, but he is likely to have been Jacob Bonneau, son of a French engraver, who was well known as a drawing master and was exhibiting watercolor landscapes at the Royal Academy and the Incorporated Society of Artists.[4]

In July 1757 Bruce traveled to Spain and spent the rest of the year there and in Portugal. We must suppose that he was much more concerned with enlarging his knowledge of the arts and architecture of these countries than to trade in wine, though he was still representing his firm. His journeys in the southern provinces of Spain unquestionably stimulated an already existing interest in Moorish language and culture. He was, in Murray's words, "in the character of a merchant, with the taste and science of a scholar."[5] But his scholarly aims were to be frustrated. He was strongly prejudiced against the Portuguese, not only on religious grounds, but more perhaps because of what he considered to be their general obstructiveness and inefficiency. The Spaniards, he felt, were not much more helpful. Even with the influence of Don Ricardo Wall, of English extraction, and minister to his Catholic Majesty, the king of Spain, he was not allowed access to Moorish manuscripts in the Escorial. The subject now closest to his heart was the history of the establishment of Mohammedan power in Spain, and with it the opportunity of extending his small knowledge of Arabic.

Bruce left Madrid at the end of December and returned to England by way of France and the Low Countries. His warm reception in France, particularly by the merchants of Bordeaux, contrasted with the rather cool treatment he had received in Spain, though this may have been because of his comparative lack of Spanish and his wine interests in Bordeaux. Whatever the reason he seems to have held the French in high esteem, whereas the Spanish and Portuguese he continued to regard as bigoted, ignorant—their clergy in particular—secretive, and indifferent to the enquiries of the intelligent outsider.

Before returning to England, he was anxious to see for himself some of the many examples of military architecture in the Low Countries. Two events there in which he involved himself tell us more about his character and tastes. He was an impetuous man and, seeing an acquaintance slighted, took up his cause and fought a duel, wounding his adversary twice. The wounds were not mortal, and he was not obliged to leave the country in haste. He then moved on to Rotterdam. From here he went with Dutch friends to visit the theater of war beyond the Dutch frontier in north Germany. He was just in time to observe the action between the French and the British and Hanoverian forces and the victory of the allies at Crefeld on 23 June 1758. The experience seems to have excited his imagination and aroused latent military ambitions.[6] Even more significant for his future life was the acquisition of books in Holland on oriental literature, in particular Arabic studies and the work of Job Ludolf on Ethiopic languages and literature. Perhaps at this time he first conceived of the idea of going to Ethiopia, if not of exploring for the source of the Nile. (He was never to believe, or admit, that the river which rose there was the secondary stream of the Blue Nile, not the main stream of the White Nile.)

At this juncture a letter arrived telling him of his father's death in Edinburgh the previous May. He decided to return home but stayed some time in London to see to his business affairs and continue his studies. Though now Laird of Kinnaird there was no certain prospect of his becoming financially independent, as he would need to be if he continued to travel. But on 4 November 1760 he made a contract with the Carron Company for the supply of coal from small mines on the Kinnaird estate. This company was a new venture, the largest of its kind in Europe.[7] With coke from a certain type of coal it was now found possible to smelt iron ore. The site chosen for the new works was near the mouth of the Carron Water, two miles south of Kinnaird. The contract meant a substantial income for Bruce and the beginning of heavy industry in Scotland. It meant too that he had no need to stay on his estate as an indigent laird. Nor did he need the wine business and he withdrew from it in August 1761.

When traveling in northwest Spain Bruce had visited the naval port of Ferrol. Through curiosity and obliging Spanish naval friends he had gathered considerable information about the dockyards and defenses of the harbor. Spain was at this time neutral but its entry into war against England was felt to be imminent. Bruce realized that if this happened information about Ferrol might soon be turned to good account. He prepared a plan for the capture of Ferrol by a British naval squadron as a prelude to the invasion of Spain.[8] At the end of 1760 he told his friend Robert Wood, undersecretary of state, about his scheme, who laid it before Pitt. Though Bruce was not yet known in government circles, the plan seems to have aroused interest though at the time it was filed away. Not long afterward Pitt was thought to have reconsidered an assault on Ferrol and wished to discuss the project with Bruce. However, quite a different plan involving the invasion of France from Bordeaux was joined with it, much to Bruce's disapproval, which he expressed to Pitt and about which he felt strongly enough to write a paper. This was well received by Pitt, who showed it to the king with the approval of Lord Halifax, with the result that his plan alone now received new attention. However, on the intervention of the Portuguese ambassador, it was withdrawn. But it had the merit, for Bruce, of making his name known in the highest places as a man of action and originality, and alerted Lord Halifax, secretary of state, to some of Bruce's more obvious virtues.

Immediately after the Ferrol scheme had been abandoned, Halifax saw fit to offer Bruce the post of consul-general at Algiers. Bruce, having certainly told him of his interest in classical antiquities, formed the impression that the main purpose of the appointment was to allow him the opportunity of traveling in North Africa and making drawings of classical remains.[9] These would particularly gratify George III and would obtain for him some fitting recompense—promotion in the diplomatic service or perhaps a baronetcy and a pension. It was pointed out to him that the work of consul at Algiers could be carried out by a deputy (though this was far from the truth, as he was later to learn). Bruce was flattered and accepted the offer. During ensuing conversations with Lord Halifax and Robert Wood, other ideas equally appealing to Bruce were mooted—exploration in Africa and an expedition to discover the source of the Nile, then widely believed to rise in Ethiopia.[10]

The taking up of his post could not be arranged immediately. In late 1761 Bruce returned to Scotland. He came to London again the following spring and was introduced to the king. He was believed to have promised him drawings of all the classical remains discoverable in North Africa, recorded as accurately and completely as possible.[11] Until the consulship could be taken up, Wood secured for Bruce the opportunity of traveling in France and Italy with sufficient time to allow him to study something of their arts and antiquities, in preparation for his intended work in North Africa. Bruce, who also had astronomical interests, looked ahead and hoped to be able to travel in Armenia to observe in 1769 the projected transit of Venus over the sun—the objective of many other explorers and scientists, including Captain James Cook.

In June 1762 he left for France and traveled by way of Lyons to Italy and then through Turin and Parma to Bologna. His leisurely journey allowed him the opportunity to study ancient sculpture and architecture and the paintings of later periods. He made a careful and extensive catalogue of much of this material which included his own perceptive comments. He was received in Bologna by the Marchese Ranuzzi, patron of his future assistant, though there is no mention that on this occasion he met Luigi Balugani. He reached Rome early in August and was accorded a good deal of attention by the English residents there and by the Roman nobility. Through them he was able to see much of interest in churches, palaces, and smaller private collections. It seems that at this time Bruce intended to write a dissertation on ancient and modern Rome.[12] More relative to his future plans, he met Andrew Lumisden, secretary to the exiled Jacobite court and Scottish antiquary who was later to publish a work on the antiquities of Rome.[13] His lasting friendship with him (in spite of Bruce's strong Hanoverian loyalties) no doubt focused his attention on plans to record the classical remains of North Africa and the need to find a qualified assistant to help with the work.

Bruce then made contact with the British consul at Leghorn, James Mann, to whom orders about his posting to Algiers would be sent by the government in London. After another visit to Rome he spent three months in Florence. Letters referred to by Murray show that he spent much of his time practicing and improving his drawing and perhaps taking drawing lessons. He also purchased there drawings of the Greek and Roman remains at Paestum in the Kingdom of Naples by an unnamed Spanish officer.[14] His interest in them grew, and he conceived the idea of having them engraved and published with his own account of Paestum. He was soon to have the opportunity of visiting the site, for in January 1763 he was instructed to travel to Naples to await further orders on his voyage to Algiers. After his arrival he got in touch with the British ambassador, Sir James Gray, and told him of his plans to publish the Paestum drawings. Gray was a collector of old drawings and medals and himself something of an authority on classical architecture. He encouraged Bruce in his project and advised him to travel to the site and make his own drawings, promising him all necessary assistance. This was, so far as we know, Bruce's first experience in drawing and describing classical architectural remains. The work he did there was unaided and considerable, consisting of plans and elevations of the main buildings and the position of the fortified walls. The Paestum portfolio survives in the Bruce collection at Yale (see Ill. 1). Though he was not the first to record these remains, his aims were to make a more accurate and detailed record than any in existence. He does not yet seem to have made use of the camera obscura, the optical instrument now coming into fashion for projecting images of distant objects onto paper, there to be traced.

Since no instructions had arrived about traveling to Algiers he returned to Rome and Florence. Here he seems to have got his work on Paestum in order, found an Italian artist by the name of Zocchi[15] (perhaps the Florentine, Giuseppe Zocchi, 1711–67) to make a frontispiece and "landscape" some of the drawings, and arranged for his friend Robert Strange, the brother-in-law of Lumisden, to engrave them, though there is no evidence that the work was ever carried out.[16] In February a ship destined for Algiers was ordered to put in at Leghorn and take Bruce on board. He arrived in Algiers on 20 March 1763.

In his new post he was not at first confronted by any major problems so that he had leisure to learn more Arabic and consider plans for visiting sites and making drawings of ancient architecture. Some of the earlier letters he received from the government show that his advice and actions made a very favorable impression. But difficulties soon arose. Relations between the British and Algerians were regulated by a treaty of 1682, periodically renewed and amended. Under this agreement British ships were allowed to trade in these waters if they carried a pass, customarily a printed form. Some of these forms were seized by the French when they took Minorca in 1756 and were sold to the enemies of Algeria, principally the Spaniards. To avoid confusion, governors of British ports wishing to trade in Algerian waters issued written certificates for British ships as passavants. These, out of ignorance or design, were not acceptable to Algerian officials. A group in the dey of Algeria's council pressed their ruler to order the seizure of all British ships carrying passavants. In spite of Bruce's vigorous protests, seizure of British shipping and piracy now became commonplace. The dey solicited the British government for the consul's removal. Worse, a messenger from the consul carrying dispatches to London was seized. Bruce managed to warn British ships of the danger they were in, but one ship putting into the harbor of Algiers was seized, its crew made slaves, and the vessel destroyed. The point was reached when Bruce himself received orders from the dey to leave the country within three days. But a peace party in his council forced the dey to change his mind, pointing out the ruinous consequences of a war with Britain. Moreover, the rumor of a British squadron assembling at Gibraltar destined for the Barbary Coast was enough to ease the situation.

Yet Bruce seemed to be losing the support of the ministry at home.[17] He was certainly quite unused to, and perhaps incapable of, exercising the gentle art of diplomacy in the way expected of him. There is no doubt that he made enemies among the British and other European residents in Algiers and that reports from them against him were being sent to London. For he was contemptuous of those who stood in his way and of all influence and corruption. His manner of administration was sometimes arrogant, cutting across accepted diplomatic convention. He dealt with the dey, not through an interpreter, but directly, in Arabic, which may not have endeared him to those who felt he should have kept his distance from one whom they considered not merely untrustworthy but tyrannical. But Bruce had a particularly dangerous enemy at the court of the dey, a Scotsman called Duncan. He sent reports to the dey from London that Bruce did not have the full support of his government, which was indifferent to the arrests and acts of piracy in the western Mediterranean against British shipping. Bruce, feeling his authority slipping, put his case strongly to Lord Halifax in a letter of 24 April 1765 and asked to be recalled, not for the first time. But his successor had already been appointed, and a naval captain was sent to Algiers as ambassador. The captain treated Bruce with studied coolness and then ignored him. Bruce's resignation prevented his dismissal.

Curiously, though the policy of the dey was to take as much advantage as possible of Bruce's lack of support from London, he had begun to admire Bruce's strong stand and straight dealing. No doubt the latter's use of Arabic helped to encourage the friendly intimacy which had grown up between them since the last crisis. Bruce made the best use of it during his travels in Algeria, and he found himself free to start on his journey within the next two months. This had been after all the chief reason for his accepting the consulship, and in his letters to the ministry there had been a repeated request, like a constant refrain, to be given leave of absence (at first he speaks of three months, later of two) to travel and make drawings of classical remains in Algeria and Tunis. These requests seem to have been ignored for a long time and he was, in any event, too heavily involved in consular affairs to go. Now he was free.

It seems probable that he had to rely largely on his own resources rather than on those of his office for the travel arrangements. He possessed considerable personal advantages—a good working knowledge of Arabic, the support of the dey, a deep interest in Greek and

Roman architecture, and a prudent and methodical approach to solving the difficulties he would invariably meet. In addition he now had the assistance of a highly qualified architectural draughtsman, Luigi Balugani, whom he had recruited to his service. He had for some time been searching for an assistant, for he realized that he could not achieve his purpose alone. The classical remains he intended to record were too numerous and often too extensive. He needed in particular the help of an artist who could offset his own weaknesses and supply convincing landscape settings with human figures in the approved fashion of the time. He had been in touch with Andrew Lumisden, who had let him know that the younger Nathaniel Dance was unwilling to travel as his assitant without his father's authority, which was not given.[18] As explained below, Andrew Lumisden had been on the lookout for an artist assistant for some time before Balugani came to his notice. There were doubtless many young artists competent to do the work Bruce required and eager to make themselves a reputation, but as many would be frightened off by the dangers of the journey, which were very real.

Bruce was able to test Balugani's abilities before they set off on their main travels, for there was a gap of about five months between the latter's arrival in Algiers and their departure toward the end of August 1765. In that period he and Balugani visited Cherchel, about seventy miles southwest of Algiers, where they drew the tomb of Juba II and the aqueduct of the old Roman city of Julia Caesarea.[19] A letter from Lumisden to Bruce makes it clear that the latter had already expressed doubts about Balugani's ability to draw figures and ornament.[20] The former weakness Balugani had readily admitted to Lumisden, but the latter criticism is difficult to accept as his ability to draw ornament in architecture was particularly praised by the judges who awarded him the gold medal at Parma (see p. 42), and his handling of the decorative elements of plants is demonstrably competent.

In addition to a personal assistant Bruce had a mechanical aid, the camera obscura. A good light and strong linear qualities in the object in view could provide effective results, and North Africa with its classical ruins no doubt proved ideal in both respects. In the introduction to his *Travels* Bruce writes, "By means of this instrument a person of moderate skill, habituated to the effect of it, can do more work, and in better taste, while executing views of ruined architecture in one hour than the readiest draughtsman, so unassisted, could do in seven."[21] Was he in reality referring to his own performance as against that of Balugani? Perhaps, though Balugani seems to have made use of an inferior model of the camera obscura at some stage of the journey.

Bruce was familiar with Thomas Shaw's *Travels or observations relating to several parts of Barbary and the Levant* (1738), a pioneering work and the foundation of his own new researches. Armed with letters of authority from the dey of Algiers, Bruce and his party set sail for Tunis on 25 August 1765, and before they reached it came ashore to examine the ruins of Utica and Carthage, which were well known.[22] At Tunis the bey received him well and he was welcomed by the consul, Mr. Gordon, a relative of his, as well as by the French consul. Both promised full assistance and he obtained from the bey permission to travel throughout his dominions.

Based in Tunis, Bruce made two separate excursions between September 1765 and

February 1766. He first followed Shaw's route southwest along the Medjerda valley toward Algeria by way of Tucca, Kef, and Hydra, where he and Balugani made careful drawings of the ruins of Greek architecture. They then made their way to Tebessa and to Constantine, the ancient capital of Algeria, in classical times known as Cirta, which they reached at the end of December. Here they stayed in the dey's palace for some days. They then traveled west to Setif, then south toward the Aures Mountains. They passed Medrassen, where the ancient kings of Numidia were buried, and the walls and gates of Lambessa. Thence they turned toward Tunis, traveling by way of Kasserine, Sbeitla, where Bruce "found magnificent ruins, either of temples, triumphal arches, or other public buildings, erected by the Romans, in the best ages of the Empire." Here they were surrounded by warlike tribesmen while Bruce had with him ten Tunisian spahis—"a fair match between coward and coward," he wrote—but they were rescued by a friendly tribe. They reached Tunis, passing and recording the ancient sites at Maktar and Dugga. Bruce claimed, "There is not either in the territories of Algiers or Tunis, a fragment of good taste of which I have not brought a drawing to Britain." An exaggeration, no doubt, but Bruce was unquestionably a pioneer of classical archaeology. His general accuracy is attested by Playfair, who followed his route rather more than a century later.[23] He gives detailed examples. There are also drawings of buildings which in that interval had disappeared or been considerably reduced. The folders of architectural drawings in the Bruce collection show much preliminary work and some more finished work by both Bruce and Balugani (see Ills. 2, 3). These architectural records must have taken much of their time, perhaps, at this stage, leaving too little to draw more than a comparatively few species of animals and plants. The balance would be restored later.

Bruce's second expedition was south to Thala, then east to Gafsa, Gabes, El Hamma and the great amphitheater of El Djem, and so to the seacoast and back to Tunis. Now Bruce intended to press on to Tripoli. But the bey of Tunis was hostile to the pasha of Tripoli, who, though promising an escort for the journey through the desert, where independent, predatory, and sometimes warring tribes made traveling dangerous, sent none. The help of the British consul at Tripoli was sought, but he was out of favor with the ruler so that any assistance would be limited.

Bruce grew tired of waiting and impetuously set off across the desert with a small retinue, traveling again through Gabes and so to the island of Jerba, under Tunisian control. Here they were fed and accommodated by the bey of Tunis in great comfort and again they waited a month for an escort from Tripoli. Again, none appeared and Bruce resumed his journey through the desert. The third night out they were attacked by armed horsemen, and though they put up a stubborn resistance they lost four men. Only by extreme resolution did they reach Tunis, where they had been given up for dead. Though they were warmly received by the British consul they were still without authority to travel in Tripoli and were forced to return to Tunis by the coastal route.

On 2 April 1766 Bruce wrote to Robert Wood of his journeys, of the danger and the kind of people they met, of cave dwellers near Jerba who lived with snakes, never raising their hand against them and allowing them like dogs to share their meals. And he describes at some length the harvest of their drawings, of the triumphal arches, Corinthian temples, aqueducts, the ruins of the three principal ancient cities of Africa, Julia Caesarea, Cirta, and Carthage, drawings enough to fill three large folio volumes. He speaks of his hopes, by his observations, of correcting and redrawing the map of Africa, of the hundreds of medals he has found, of vases and bronze statues, of copying about a thousand inscriptions, and finally mentions his natural history drawings: "I have not entirely neglected, but have made about thirty drawings of the rarest animals, insects, birds, and plants of this country, particularly the interior and remote parts of it, all in their natural colours."[24] As always, he writes in the first person without mention of Balugani.

He took the precaution of sending drawings, books, and some of his instruments from Tunis and Tripoli to Izmir (Smyrna), a wise move as it turned out, keeping with him only the essentials. He also left with Consul Gordon at Tunis some of his books and drawings. He then took ship across the Gulf of Sirte, landing at Benghazi. In this province there ruled a brother of the pasha of Tripoli. He found the city in a state of famine and obtained permission from the bey to journey into the interior. They made drawings of Tolmeita (Ptolemais), but Bruce's intention of moving eastward to Darnah and Alexandria was thwarted by hostile Arabs, famine, plague, and the unhelpful attitude of the pasha of Tripoli. Bruce decided to take ship to Crete. The vessel was in poor condition and was driven by storms to return to Benghazi. They were shipwrecked near Tolmeita. Bruce gives a graphic account of the disaster and of his escape.[25] Though the party survived intact, some drawings and some valuable astronomical instruments were lost.

It was two months before he made a fresh attempt to reach Crete. He boarded a French vessel and arrived at Khania on the north coast of the island after five days. He became ill with intermittent fever, perhaps malaria, but was, as so often, given the most generous help in this instance by the French Consul Amoureux. His illness forced him to remain for several months on the island. Finally he sailed to Rhodes, where his books and papers had been forwarded and from which he intended to search out classical remains in Asia Minor, or what today is southeast Turkey. But he felt too unwell to contemplate this and went on to Beirut and Sidon in modern Lebanon. On 29 July 1767, near the source of the Jordan, he observed and drew, or caused Balugani to draw, papyrus sedge (*Cyperus papyrus*, p. 107) and two days later, the sweet banana (*Musa* × *sapientum*, p. 113).[26]

Bruce was on the very best terms with the French community in Syria, particularly with their consul at Sidon, Clairembaut, and afterward looked upon this interlude as one of the happiest times of his travels. Though his fever returned for a time, by early August he was contemplating visits to Baalbek and Palmyra. These he accomplished, though he had no intention of trespassing on the antiquarian field of his friend Undersecretary Robert Wood (who had published work on both these sites) by publishing his own descriptions. He was at Baalbek by 19 September, and he and Balugani measured and drew the remains at their leisure. The task was even greater at Palmyra, where they worked a few weeks later.[27] They were astonished at their first sight of it. Bruce described it as "one of the most splendid works of human industry and genius, that ever had been abandoned to solitude, desolation and ruin."

Bruce arranged a division of labor for the two of them on the site, which he divided into six parts to bring the main monuments into prominence from different angles. In his account of the expeditions Bruce is critical of some of Wood's work and was sure that his own records were more thorough and more accurate.

After a short stay at (Syrian) Tripoli Bruce traveled to Aleppo, where he became dangerously ill. He had been in correspondence with Belville, a French merchant there and a knowledgeable collector of classical antiques. Here the Frenchman nursed him and called in an English doctor, Patrick Russell, to advise.[28] By now Bruce had decided on traveling to Egypt—he believed that Egyptian architecture could have supplied Greek architects with ideas which they developed with such success—and perhaps also to Ethiopia. He slowly responded to medical treatment and, perhaps with Egypt and possibly Ethiopia in mind, persuaded Dr. Russell to teach him the rudiments of tropical medicine. That specialist on the plague in fact gave him a very thorough idea of the treatment of many of the diseases he was likely to meet on his travels. He also supplied him with the medicines he would require. This was perhaps the asset which was to serve Bruce best later and earn him most respect.

He now received information that much of the astronomical and measuring equipment he had lost in the wreck could be provided from France. A quadrant from the military academy at Marseilles was ordered to be sent to him thanks to the efforts of the Comte de Buffon through the minister Choiseul and an appeal to Louis XV. Other material was coming from London and would be shipped to Alexandria.

In Belville's house Bruce found a room given over to Strange's engravings after Italian masters. He recounts how he presented his host with "a copy of your St. Cecilia in water-colours." This he thought might surprise Strange. Had Bruce seen Raphael's original painting of *St. Cecilia and Saints* in Italy and noted the colors or made a copy there, or had he, from his knowledge of Raphael's work, invented the colors in Aleppo? He discusses the state of the arts in England and laments the lack of royal patronage. In the letter written to Strange telling him about this he speaks of his two small landscapes by Albani, which he has left in Strange's care.[29] There is no doubt that Bruce was deeply and intelligently interested in the fine arts, as many of his letters and his record of collections seen in Italy testify.

He sailed from Sidon to Alexandria on 15 June 1768. Wearing Arab dress, as they had hitherto, his party were taken for Barbary Arabs. They traveled to Rosetta and there embarked on the Nile for Cairo, which they reached in early July. Here for the first time Bruce let it be known to the members of the French community that he intended to travel into Ethiopia.[30] They were apparently astonished at such an ambitious plan, for the interior of Ethiopia was hardly known, even in Egypt. Bruce with his instruments and medicines made a considerable impression on Ali Bey, the Egyptian ruler, and his Coptic astrologer, Maalem Risk. He also renewed his acquaintance with Father Christopher, a Coptic priest who had been his chaplain in Algiers. Father Christopher introduced him to the patriarch of Alexandria, who was also the head of the Ethiopian church. From him, Maalem Risk, and Ali Bey himself, Bruce received letters of recommendation which were addressed to subordinates in the countries through which he planned to travel and were considerably to ease his passage.[31] In particular Ali Bey provided him with letters to the sherif of Mecca, a descendant of the Prophet and governor of western Arabia, to the naib of Massawa—the port and surrounding land through which Bruce must pass in order to enter Ethiopia—and to the king of Sennar, through whose state Bruce may already have decided to return to Egypt. He already carried with him a firman of the sultan, which he had obtained through the ambassador at Constantinople, a fellow Scot. This described him as a noble Englishman, servant of the British king. In Turkish provinces this was, or should have been, the most powerful of all passports. He obtained other supporting letters to local rulers, all of which showed the extent of the trouble he took to give his journey full chance of success.

The party set sail from Cairo south up the Nile on 12 December 1768. He visited the site of Memphis, the ancient capital of Egypt, Dendra with its large Ptolemaic temple, and Thebes, where he saw the Valley of the Kings. Here he was tremendously impressed with the sheer quantity and solidity of the material he found, such as the "prodigious sarcophagus... sixteen feet high, and six broad, of one piece of red granite...," now identified as the tomb of Rameses III.[32] The wall paintings intrigued him, and he and Balugani copied the harp players among others.[33] He would have liked to have made many more copies had time allowed. A conventional classicist such as he could only express disappointment with the architecture, for there seemed no link with the development of the orders of Greek architecture.

Crossing the river, they visited the magnificent ruins of Karnac and sailed on up the Nile as far as the First Cataract of Aswan, which Bruce called by its classical name of Syene. Here he met Nimr, sheikh of the Ababde Arabs, who controlled the southern frontier of Egypt, and whom he provided with medicine—a prudent insurance against his return here from the desert years later. Bruce now turned round and sailed downriver to Qena, where he joined a caravan traveling overland to the Red Sea port of Quseir.

The Nile voyage was a restful interlude in an otherwise physically demanding expedition. Bruce describes the boat they used (in the *Travels* called a "canja" but now generally known as a "felucca"), which Balugani drew and which, with the Egyptian harpist, is illustrated in the *Travels* (see Ill. 4). It had a "mainsail yard near 120 English feet, which is larger than that of a 74 gun ship. With this prodigious pressure of sail we went with a very moderate wind eight or nine miles an hour, against one of the strongest currents in the world."[34] Though Bruce obviously enjoyed this Nile experience the strangeness of Egyptian culture seems to have blunted his appetite. He was, however, much impressed by the evidence he saw of the ancient Egyptians' advanced knowledge of mathematics and astronomy.

They left Qena on 16 February 1769 and crossed the barren and mountainous neck of land between the Nile and the Red Sea—a foretaste of Ethiopia—in six days. Bruce was based at Quseir until 5 April, but while he was waiting to enship for Arabia, and ultimately Ethiopia, he seized the opportunity of investigating the coasts and harbors of the northern parts of the Red Sea. He sailed up the west coast and across to Tor in the Sinai Peninsula, along the Arabian coast to Yanbu 'al Bahr, Rabigh, and Jidda. He was much concerned with the hydrography of these waters, and he and Balugani made their observations with the intention of mapping the area more accurately than in hitherto published maps. They made drawings of all the harbors they visited. Now they also had the chance to enlarge their records of natural history specimens, in particular, of fish.[35]

At Jidda Bruce was welcomed by the community of British merchants dealing in goods shipped from India. Yet Bruce's dress, which could now be taken for that of a Turkish sailor, "made no impression in his favour" and "he was driven from the gate of the [English] factory by one of his countrymen and relations who mistook him for a vagrant."[36] In spite of his wayward eccentricities the letters he carried made a great impression on the governor of Jidda. Great efforts were made by the local administration and the English to provide Bruce with the supplies he would need in Ethiopia. He had in particular the support of an influential minister of the sheikh of Mecca, Metical Aga, by origin an Ethiopian slave, who provided him with a trusted servant to accompany him to Massawa. Bruce filled in the time before the escort arrived by a further voyage as far south as the Straits of Bab el Mandeb. On his return he picked up his escort at Al Luhayya.

The Red Sea voyages and the journey to Quseir were the beginning of Bruce's explorations in the true sense. His task of recording ancient architectural remains had been all but completed. Besides his instrumental observations he was interested in the land's geological structure. Referring to the journey to Quseir, he writes,

> We found the quarries from which the ancients had extracted that prodigious quantity of porphyry and granite. After the second day's journey begin the porphyry and granite, then marbles of different colours, chiefly green, all of the finest kinds. We saw quantities of serpentine, and that called verde antico in less proportion; variegated marble, of different sorts; and in that last day's journey, before we arrived at the Red Sea, jasper in great plenty. The jasper does not grow like the marble in a mountain by itself, it runs in large veins in mountains of green marble. I have sent eleven specimens of different kinds of porphyry to Cairo.[37]

Observations of latitude and longitude had been made from about the time that Bruce reached Thebes on the Nile, and weather records, temperatures, and barometric pressures were recorded from his landing at Massawa in September 1769, all these normally the duties of Balugani.[38] The observation and drawn records of natural history specimens had now become perhaps the main task, though there are not very numerous references to sightings of animals and plants to be found in Balugani's journal, which he started at Quseir, or in Bruce's more general records.

Now that Bruce was on the threshold of Ethiopia how well equipped was he as an explorer in a largely unknown land? He had obvious physical assets, for as well as being a giant of a man he had great strength and stamina, enough to carry him through several serious illnesses. His prowess as a horseman and in the use of firearms was of more than usual advantage in a country where good horsemanship was much rarer than in Arab lands and where firearms were comparatively uncommon and old-fashioned. In character Bruce was ambitious and determined to achieve his ends and had the boundless curiosity of the natural explorer. Moreover, he was a natural leader. His medical knowledge and experience, though obviously limited, brought him an unlooked for reputation everywhere he traveled. Though not a scholar, he had scholarly inclinations, and his linguistic ability enabled him to acquire a working knowledge of Amharic, the most widely spoken Ethiopian language, quite quickly. He must have had a fair grasp of Geez, the literary language of Ethiopia, which enabled him to carry out considerable research into its history and literature. Both languages he would initially have learned from studying Ludolf.[39]

His knowledge of the country would have been derived largely from the writings of Portuguese Jesuits, whose accounts he affected to despise and which he dismissed as unreliable or even in parts invented.[40] He would have known Samuel Purchas's incomplete English version of the description of Ethiopia by Francisco Alvarez, chaplain to the first Portuguese mission to that country. He would also have known the English version of 1710 of an abridgement and revision of the *History of High Ethiopia* by Manoel de Almeida, published by Balthasar Tellez in 1660. There was, too, the best-known work on Ethiopia in English, Samuel Johnson's translation of Abbé Le Grand's French version of Jerónimo Lobo's *Itinerário* (at that time lost but rediscovered in 1947), which appeared in 1735, Johnson's first book. His *Voyage to Abyssinia, by Father Jerome Lobo a Portuguese Jesuit* would certainly have been familiar to Bruce. Lobo was once described by Bruce as "a grovelling, fanatic priest" and his work as "a heap of fables." C. F. Beckingham gives an example of the reckless prejudice shown toward Jesuits by Bruce, who found it strange that Tellez did not mention the execution in 1714 of some Capuchin missionaries in Ethiopia. "It would have been much stranger if he had done," Beckingham writes," His book was published in 1660."[41] We shall meet other examples of Bruce's prejudices in the course of his travels. But all the works known to Bruce were long out-of-date, and he had to rely on his own observations and the contemporary reports which he considered reliable to form any idea of the Ethiopia he would find.

To add to Bruce's difficulties it happened that the country was at this time going through a period of civil war and anarchy.[42] The Emperor Joas, claiming descent from Solomon and the queen of Sheba, was murdered in the spring of 1769 and was succeeded by Hannes II, who in turn was murdered soon after Bruce's arrival at Massawa in September 1769. He was replaced by his young son, Takla Haimonot, whom Bruce was to come to know and admire. He was, however, firmly under the control of Ras (a title approximately equivalent to a Turkish grand-vizier) Michael, who ruled the province of Tigré to the northeast. The real power was disputed by three main rivals ruling the different provinces of the country: the ablest of these, undoubtedly Ras Michael, who spoke for the emperor; Fasil, who controlled the provinces of Gojam and Damot contained within the loop of the Blue Nile south of Lake Tana and who had the support of the largely pagan Galla people to the west; and the governors of the allied provinces of Amhara and Begemder to the north and east of Lake Tana. Bruce's movements were later to be considerably restricted by the activities of the rival armies, as his destination, the source of the Blue Nile, and Gondar the capital, where for long periods he was based, were within the area of action.

Bruce's aim on landing was to make for Gondar by the usual route, that is, by way of Adowa and Axum, the ancient capital. His ship had anchored off Massawa on 19 September 1769. As soon as he disembarked he met difficulties placed in his way by the naib of Massawa, who disregarded all the letters of commendation he carried and would probably have murdered him had it not been for the intervention of the naib's nephew, Ahmed. Bruce gives a characteristically lively picture of the naib.

> All the procession was in the same style. The Naybe was dressed in an old shabby Turkish habit, much too short for him, and seemed to have been made about the time of Sultan Selim [1512–20]. He wore upon his head a Turkish cowke, or high cap, which scarcely admitted any part of his head. In this dress, which on him had a truly ridiculous appearance, he received the caftan, or investiture, of the island of Masuah [Massawa]. ...In the afternoon of that day I went to pay my respects to him, and found him sitting on a large wooden elbow-chair, at the head of two files of naked savages, who made an avenue from his chair to the door. He had nothing upon him but a coarse cotton shirt, so dirty, that, it seemed, all pains to clean it again would be thrown away; and so short, that it scarcely reached his knees. He was very tall and lean; his colour black; had a large mouth and nose; in place of a beard a very scanty tuft of hairs upon the point of his chin; large, dull, and heavy eyes; a kind of malicious, contemptuous, smile on his countenance; he was altogether of a most brutal appearance. His character perfectly corresponded with his figure, for he was a man of mean abilities, cruel to excess, avaricious, and a great drunkard.[43]

Another two months passed before Bruce was able to disentangle himself from the web of treachery in which the naib sought to entangle him. He even obtained from him, as a measure of good grace, a Christian guide, some bearers, and letters of authority for the journey. They left Arkeeko on 15 November and made their first march across the coastal plain. Here the travelers were caught up by Ahmed, who had helped to save Bruce from the naib's evil designs at Massawa. He replaced the four bearers provided by the naib and persuaded Bruce to alter course. Instead of taking the shortest route to a town under the naib's control he was persuaded to climb over Taranta, the highest mountain of the ridge overlooking Ethiopia, and make for Dixan (Digsa), part of which Ahmed himself controlled. There he would be welcomed in perfect safety.

The rains were just beginning but they were still able to move through dry water-courses. On the steep climb Bruce became worried about the safety of his instruments—the quadrant, in two pieces, carried by eight men, the telescopes, and the pendulum-controlled timekeeper—for the bearers were flagging. Bruce and Yasine, an Ethiopian Moslem who had shown much resourcefulness by helping Bruce to avoid shipwreck on the voyage to Massawa, took over carrying the quadrant-head themselves on the steepest slope. This so impressed the other bearers that they found new strength to struggle with the instruments up the mountain. Next day the party reached the summit and had their first sight of Ethiopia.

They reached Dixan a week after leaving Massawa and camped just beyond the frontier which separated the province of Tigré in Ethiopia from the territory of the naib of Massawa. For the first time since they reached the coast Bruce recovered his peace of mind. Before them lay the fantastic mountains of the Ethiopian landscape—"some flat, thin and square, in the shape of a hearthstone or slab that scarce would have been sufficient to resist the winds; some like pyramids, others like obelisks or prisms; and some, the extraordinary of all the rest, pyramids pitched upon their points with their base uppermost." On 25 November they crossed the riverbed which marked the boundary of Ethiopia proper.

Though the going was difficult these were cheerful days, but there can be no doubt that Bruce was already aware of the possibility that war and anarchy could thwart his aim to reach the source of the Nile and might even prevent him ever returning home. The party had been joined for protection by a number of Arab traders with laden donkeys as well as by the emperor's messengers. Yasine had assumed the duties of Bruce's deputy and hoped that the latter might help him to regain power in the province of Ras el Feel from which, as the governor's son-in-law, he had been exiled.

In these days they met their first notable Ethiopian, the bahrnagash, or sea king. He was the impoverished successor of a line of powerful officials who governed the broad band of country along the coast to the northeast of Ethiopia, now part of the region of Eritrea, but of late years his power and status had been much reduced. Bruce described him as "a little man of an olive complexion, or rather darker; his head was shaven close, with a cowl or covering upon it; he had a pair of short trousers; his feet and legs were bare." His retinue of seven horsemen and about a dozen men on foot were "all of beggarly appearance, and very ill-armed." He was not at first impressed, nor was Bruce likely to have impressed the bahrnagash with his ragged Arab dress and torn feet. But Bruce was soon to form a much more favorable view of the man. Always an admirer of the martial qualities and of direct dealing, he later found that he was a courageous soldier, frank, and with a becoming humility. Bruce bought a black horse from him, christened Mirza, and at first pretended he had never seen a horse before, perhaps to get a better bargain. When news leaked out through the servants that Bruce was a formidable horseman the bahrnagash thought his pretended ignorance a great joke. The horse was an excellent investment, and on it Bruce exhibited a riding skill that was to increase his reputation remarkably with Ethiopian leaders. The two men parted on the best of terms after a "carouse on the mead of the country," which gave Bruce "such a pain in the head that I could scarce raise it the whole day." The bahrnagash also supplied Bruce with a guard of twenty armed horsemen. In more than usual security the party moved on toward Adowa.

Occasionally there is a mention in the journals kept by Bruce and Balugani of the natural life they observed en route, though in periods of rest Balugani must have been busier than ever before drawing specimens, most of them plants. Approaching Taranta mountain, Bruce writes of a "caper-tree [*Capparis* species] as high as the tallest English elm, and its fruit, though not ripe, was fully as large as an apricot." Three days later on the mountain there are groves of "oxy cedrus, the Virginia or berry-bearing cedar," almost certainly *Juniperus procera,* to which he gives the vernacular name Arze. Leaving Dixan, he observes "nothing but the Kol-quall tree [*Euphorbia abyssinica,* p. 81] on the hill all around it." And on the same day there is mention of the daroo tree (*Ficus vasta,* p. 92). Balugani's journal is a mainly topographical record, a record of distances and times taken, but there are occasional mentions of plants. On 4 December, the day they crossed the river Angueah, approaching Adowa, he writes of a tree bearing small white flowers of five petals, with a most pleasing scent more delicate than jasmine, which was called Mapta (*Acokanthera schimperi,* p. 71). Very many of the plants drawn were first found on this journey to Gondar through country often mountainous but varying considerably in the nature of the terrain.

On 6 December they came in sight of the mountains of Adowa, "nothing resembling in shape those of Europe, nor, indeed, any other country. Their sides were all perpendicular rocks, high like steeples, or obelisks, and broken into a thousand different forms." Ras Michael was away on campaign against Fasil, but they were welcomed at Adowa by Janni, Michael's Greek steward. The care with which Bruce had obtained letters of authority nearly always paid off. The natural hospitality of the Greeks was enhanced in this instance by the letter of the Greek patriarch in Cairo. Bruce was provided with the pleasantest accommodation. Janni went further by sending friendly messages about him to the empress and her daughter at Gondar and to Ras Michael. No doubt Bruce's medical skill was one of the virtues he enlarged on. The smallpox which had been raging at Massawa had now reached Adowa. Bruce treated a few cases but was anxious not to become too involved or distracted from his main purpose, which was to push on to Gondar and the source of the Nile. He nevertheless spent more than six weeks there, and they did not resume their journey until 17 January 1770.

They reached Axum, the ancient capital of Ethiopia, the following day. It was by now largely ruinous but retained intact its small cathedral, where the emperors were still crowned. Bruce believed the city had been the trading metropolis of the Cushites, a people quite distinct from the aboriginal Ethiopians. He was impressed by the extensive remains and was particularly intrigued by the forty "obelisks," more accurately steles, fallen and still standing there which he considered to be Egyptian work carried out under the Ptolemies—a quite mistaken theory. He, or Balugani, drew one of them, which is illustrated in the *Travels* (see Ill. 5).[44]

Axum was the excuse for Bruce to embark on an absurd diatribe against the Jesuit traveler Lobo, whom he regarded as being "full of ignorance and presumption." He cites Lobo as sailing from India bound for Zeyla, Somalia (near Djibouti) on a ship going to Caxume, which he equates with Axum. But how, he says, can a ship sailing to Zeyla possibly choose to land three hundred miles beyond it? The explanation is that Caxume is not Axum but Qishn on the south coast of Arabia. Beckingham makes the point that only someone blinded by prejudice would have failed to search for a different location for Caxume.[45]

Soon after leaving Axum, Bruce observed an incident which was received with general disbelief in England but was remembered when other of his adventures were forgotten. They overtook three soldiers driving a cow before them, which they tripped and held down on the ground, while one of them cut "two pieces, thicker, and longer than our ordinary beef steaks... out of the higher part of the buttock of the beast." They then covered the wound with the flap of skin which they had left and which they secured with pins. Then they put a "cataplasm" of clay over it and drove the beast on for "a fuller meal when they should meet their companions in the evening. I could not but admire a device so truly soldier-like" was Bruce's ironical comment.

They journeyed on through the province of Sire to the valley of the river Tacazze. "It was shaded with fine lofty trees, its banks covered with bushes inferior in fragrance to no garden in the universe: its stream most limpid, its water excellent, and full of good fish of great variety." This arcadian vision is characteristic of Bruce, who is rarely specific. Balugani, if he had noted it at all, would probably have identified some of the plants. But soon afterward they were approaching wilder country, which showed signs of the devastation of war, and the last mountain barrier, Semien, before Gondar.

They had a brush with a local governor, but a display of Bruce's blunderbusses was enough to get the expedition by. They reached the village of Lamalmon at the foot of the pass on 8 February. At the customs post Bruce's letters allowed him and his party through without payment, but Yasine and the other Moslem traders faced extortionate charges. The son of the official, however, was a soldier and, seeing that Bruce had arms, insisted on a shooting match. This was much to Bruce's taste. He had a rifle which was far superior to the soldier's musket and, to the young man's astonishment, shot quails on the wing. Then he gave a riding display on Mirza with lance and javelin and shot a pigeon at gallop. The soldier was so impressed that even the Moslems were allowed through without payment.

Bruce now learned that Ras Michael had won a victory against Fasil, the Galla chief, who had been driven back to his country beyond the Blue Nile—comforting news for Bruce, who now saw his way to the river's fountains cleared. On 14 February, "after having suffered with infinite patience and perseverance the hardships and danger of the long and painful journey, at forty minutes past ten we were gratified at last with the sight of Gondar," about ten miles distant, according to Balugani, who allows himself to express no emotion when noting this. The towers of the emperor's palace were seen above the trees which masked the city. But when they reached the river Angareb, before the city, Bruce found no welcoming party to receive him. All his contacts were away with the army. As a friend of the Moslems, Bruce was in fact welcomed into the Moslem settlement, which was separate from the capital.

Bruce made first contact at Gondar with a highly placed official who was closely connected with the royal family, Ayto (Lord) Aylo. He was far from being an establishment figure. Though he was a relative of the dowager Empress Mentwab, and her adviser, and had been a soldier and horseman in his youth, he had come secretly to hate war. Secretly, too, he detested the Ethiopian priesthood and dreamed of leaving the life of a courtier to retire as a monk to the Convent of the Holy Sepulchre in Jerusalem. He saw himself as the protector in Ethiopia of Greeks and Franks (a term used by Ethiopians for European Roman Catholics or indiscriminately for Europeans generally). Aylo surprised Bruce sitting in his tent with his Ethiopian books and dictionaries. He had come to bring him to court, where his services as a doctor were urgently required. Seeing Bruce surrounded by his books, he was impressed to hear that he could speak Amharic and evidently had some knowledge of Tigrinia, though at other times Bruce confessed that his knowledge of the latter was very limited. Bruce was equally eager to show off his paces on horseback to a connoisseur of horsemanship.

The two men rode to the palace at Koscam nearby, where Bruce was presented to the Iteghe, the dowager empress. Though partly Portuguese and secretly sympathetic to Roman Catholicism, she seems to have accepted with relief Bruce's candid denial that he was a Frank (who could not be officially tolerated). She was eager for him to treat the smallpox patients, mainly children of her daughters. One of them died next day, and Bruce saw that his future hung on his ability to treat the disease at the palace. He put into operation a new régime. When he found the children languishing in an overheated, airless room, he "opened all the doors and windows, fumigating them with incense and myrrh in abundance, washed them with warm water and vinegar, and adhered strictly to the rules which my worthy and skilled friend, Doctor Russell, had given me at Aleppo." He was dramatically successful, and though others caught the disease all except one of the children recovered.

It was now, when Bruce was permanently in residence at the palace, that he developed a growing friendship with Ozoro (Princess) Esther, the empress's daughter and wife of Ras Michael. She was spirited, intelligent, and moreover beautiful, and there can be no doubt that Bruce developed a deep affection for her. His services to the family were such that his own security was assured insofar as those in power could assure it.

In contrast to the civilities at court Bruce was soon appallingly aware of the barbarities perpetrated by the emperor's returning army: the flaying of a Galla chief, the castration of the dead, the gouging out of eyes of captives. After some delay he was officially introduced to Ras Michael. Though when he returned from his campaign he had seemed at first old and lame, red-eyed and weary, Bruce saw him differently at his official reception. He now looked

> thoughtful, not displeased... his eyes quick and vivid... He seemed about six feet high, though his lameness made it difficult to judge with accuracy. His air was perfect, free from constraint, what the French call *dégagé*. In face and person he was liker my learned friend, the Count de Buffon than any two men I ever saw in the world. They must have been very poor physiognomists that did not discern his capacity and understanding by his very countenance.[46]

The ras explained that the emperor had agreed to appoint Bruce one of his baalomals, lords of the bedchamber, and also commander of his black horse-guards. These offices would give him authority and a certain freedom and respect which would protect him. Security was no theoretical matter. "You are," said Michael, "a man who makes it your business to wander in the fields in search after trees and grass in lonely places, and to sit up all night looking at the stars... These wretches here... would murder you for mere mischief"—or for all the gold Bruce was reported to possess, for he was believed to be immensely rich because he had refused payment by Esther for his medical services.

His introduction to the young emperor, which followed immediately, was a protracted affair in which Takla Haimonot deliberately tried Bruce's patience by keeping up a meaningless discussion until he was ready to drop with weariness. Bruce was later to form a high opinion of him as intelligent and wise beyond his years, though under the control of Ras Michael. After the emperor had finally dismissed him Bruce joined the Greeks and his fellow *baalomals* at a late supper. One of them, Gebra Maskal, commander of the ras's musketeers and his nephew, boasted in his cups of his knowledge of firearms. Someone suggested that he had something to learn from Bruce but he was contemptuous. Bruce, always quick to be provoked, declared, "the end of a tallow candle, in my gun, shall do more execution than an iron ball in yours." The Ethiopian called him a Frank and a liar. There was a violent quarrel. Gebra Maskal drew his knife and, though floored by Bruce, succeeded in wounding him superficially in the scalp. Bruce in a passion seized the knife and struck him, fortunately with the handle, violently on the face. At this point the company stopped the fight as the seriousness of the situation became clear to them. Lifting a hand in violence in the precincts of the palace could mean death.[47]

That the fault lay with the Ethiopian was clear to everyone. It was not so clear how the problem of his guilt could be resolved. Ayto Aylo suggested that Bruce should not revengefully press his case with the ras, who had put Gebra Maskal in chains. Michael was already jealous of him, suspecting his adultery with one of his wives in Tigré. But the soldier was enormously popular with his troops. His execution or severe punishment could well cause a mutiny. Bruce, as ever, was anxious to avoid making further enemies among people who often looked upon him with deep suspicion as a Frank. His policy was therefore to belittle the affair and to show that his wound, which had bled profusely, was a mere scratch. Ayto Aylo recounted the whole story to the ras, who was furious that the security he had provided for Bruce should have been so outrageously threatened. Bruce, with the support of Ozoro Esther, interceded with Michael for Gebra Maskal's release, which was finally obtained. The emperor took the matter as something of a slight to his authority and was anxious to make it up with Bruce. He took him aside and asked him if he was not drunk when he made the claim for the power of his gun. Bruce replied that he had been perfectly sober and meant exactly what he said: "Lying... is infamous... in my country, whatever it may be in this." So, to the astonishment of the emperor and the onlookers, he shot half a farthing candle through three shields from the length of a room, and with the other half pierced through a dining table of sycamore wood three-quarters of an inch thick. The priests dismissed it as magic but on the emperor it made a deep and lasting impression. As for Bruce, though he was more secure at court than before, his way to the source of the Nile was no clearer. In fact the emperor made it plain to him that there was only one thing he could not ask and that was to go home.[48]

Frustration at the unlikelihood of leaving Gondar and the need to tread warily as a sort of superior servant to the emperor had their effect on Bruce. He became depressed and ill. At about this time he makes his one reference to Balugani in Ethiopia. He speaks of the longing to return home accentuated by "the loss of the young man who accompanied me through Barbary." The question of Balugani's death is discussed at length below (pp. 48–50). It was not in fact to happen for nearly a year. This mistake was not a question of memory since Bruce was well aware that Balugani was with him when he reached the source of the Nile. Both Balugani's journal and a draft letter recording that achievement were in his possession and had Bruce's glosses on them. It seems to have been a way of "disposing" of his assistant at this stage of the story in order that he could appear to be alone, without a European rival, when he at last stood at the source of the Blue Nile.

His periods of despondency were relieved by events such as the royal wedding of Ayabdar, granddaughter of the empress and of Ras Michael, with Powussen, the new governor of Begemder province, intended to bind that land firmly to the cause of Michael and the emperor. There were bread and circuses for the people too or, more precisely, raw meat and wine. Bruce was also present at an Ethiopian feast. His detailed picture of a barbaric orgy must have seemed all the more shocking to his Western readers because of the restrained way it is described. But Bruce had had enough: "I found it necessary to quit this riot for a short time and get leave to breathe the fresh air of the country."

Two unforeseen events turned his thoughts to the Nile and home. First, the Agows of Damot, after being defeated by Fasil and pressed into his army, revolted. Damot was the very area where the infant river flows. Fasil then attacked the Agows' army. Though the rainy season was near, when campaigning would become impossible, Michael decided to march south against him. That, it was pointed out, would provide Bruce with the protection he needed to achieve his main ambition. Second, Bruce had been pressing Yasine's case with the emperor for the deputy governorship of Ras el Feel—the office had normally been filled by a Moslem—from which he had formerly been expelled. Not only would the post be a fitting reward for the help he had given Bruce on the voyage to Massawa and the journey to Gondar, but it would be in Bruce's interest as he intended returning to Egypt by way of the Sudanese kingdom of Sennar, which bordered on Ras el Feel.

Circumstances were soon to allow the emperor to send Yasine as deputy governor to Ras el Feel. More surprisingly, and largely through Esther's influence, Bruce was made governor. This promotion was a little way in the future. For the present, depressed and ill, Bruce was determined to escape to Emfras, a small town overlooking the eastern shores of Lake Tana. This was a place known for its tent making, and he planned a tent with a flap in its roof through which he could observe the sky with his telescope without interruption. Though the empress did her best to persuade him to stay at Koscam, Bruce insisted on going where he could make his observations and collect plants in peace.

Bruce left Gondar with Balugani (whose journal records the journey) on 4 April 1770 and reached Emfras the next day. During the following month, before the ras's army was ready to march, Bruce returned to his patients at Koscam at least twice, on one of which occasions his promotion was announced at court. He abandoned himself to joy. This short interlude, when he was free to do what he most desired, in at least temporary security, must have been one of the happiest times of his sojourn in Ethiopia. It was all too brief. On 17 May the advance guards of the army reached the town, and Bruce was given a guard to protect him from plunder. Emfras was plundered but not fired, and Bruce, after brushes with soldiers, escaped injury.

Next day he visited the emperor's headquarters, where he was received by Esther, who was making the campaign with her husband, Ras Michael, and the emperor. It was decided that Bruce should follow the army, not marching as a soldier but independently, under its protection, to carry out his own plans for observation and discovery. Part of that plan was to visit the Tissisat Falls on the Blue Nile—what Bruce called the Third Cataract—within easy riding distance from the point where it flows out of Lake Tana, on the southeast, an impressive feature described by Jesuit writers. At a place called Lamgue, according to Balugani, the wide plain by the lake was covered with a species of tree called, locally, Leham *(Syzygium guineense)*, on which the fruit was visible but not yet ripe. Even today the tree is plentiful along the shores of the lake. His drawing, entitled Leham, shows the climber, *Saba comorensis* (p. 72), which must have completely covered the tree pointed out to him as Leham,

The intention was to make for Dara, a village beyond the river Gomars, but Bruce had much still to observe by the lake—hippopotami as well as plants—and sent his Greek servant, Strates, ahead with the bearers and baggage. This group was set upon by horsemen under two young relatives of the empress, who were with the advance guard of Powussen. The servants were stripped of their clothing and belongings, including some of Bruce's firearms. This was bad enough but Bruce now received a report that Powussen, the governor of Begemder, and Gusho, the governor of Amhara, previous allies of Ras Michael, had switched their allegiance to Fasil. This was the more alarming as they all three planned to attack Michael and prevent his retreat to Gondar and the province of Tigré. Not only would Bruce's way to the source of the Blue Nile be commanded by enemies, but the support of the emperor and Michael might be cut off. He now took the precaution of sending his astronomical instruments back to Gondar.

The cataract was "one of the most magnificent, stupendous sights in the creation, though degraded and vilified by the lies of a grovelling fanatic priest." That priest was, of course, Jerónimo Lobo.[49] Bruce believed that he had exaggerated the height of the falls. They fell forty rather than fifty feet. Moreover, Lobo claimed to have got behind the falls, resting on a rock of the riverbed. Bruce said such a feat was impossible. He was wrong on both counts. The falls are much higher than either man estimated, and a later traveler, R.E. Cheesman, proved Lobo's claim when he succeeded in getting behind the falls but that was in the dry season, when less water was coming down.[50] In such a way Bruce sought to demolish Lobo's account and thus make his own claim of being the first European to reach the Blue Nile's source the more credible.

The army was moving back to Gondar, and Bruce had no alternative but to return with it. He was in command of his black horsemen, who acted as part of the emperor's bodyguard. What was in effect a temporary truce was arranged between Michael and Fasil, who pretended to disown his new supporters, Gusho and Powussen, while the emperor actually proclaimed Fasil governor of the Agows and of Maitsha, Gojam, and Damot. The proclamation was to be repeated in Gondar. The emperor was in a generous mood, and a feast was held. It was the right moment for Bruce to win fresh support in high places. He begged to be left behind in Gondar in order to be free to complete his mission. The emperor complied and advised him to stay for the time being in Koscam in the security of the palace. The empress and Esther remained there while the emperor and Ras Michael marched out of Gondar toward Tigré, leaving the city undefended. The two rebel governors came and went, and Bruce found them reasonably friendly.

Once the emperor and his most powerful support, Michael, had moved away with their forces, the power of Fasil was in the ascendant. The Empress Mentwab, who had never forgiven Michael for the murder of her grandson, the Emperor Joas, felt ready to make the supreme gesture in support of Fasil, who had been loyal to Joas. She proclaimed Susenyos emperor. The pretender was a reputed son of Yasous II and so, perhaps, a grandson of the empress. He was generally held to be dissolute and stupid but was apparently the only available candidate. One of his first actions was to sit in judgment on one of the men who had confessed to be in the group which, on Michael's orders, killed Joas. The man, a Galla, was hanged. He had revealed where Joas's body had been buried. It was exhumed, but no one, fearing Michael, was willing to prepare it for reburial. Bruce tells how he got a prayer rug and had the body laid on it, wrapped in a shroud. There was a secret funeral and his action, if it is to be believed, pleased all factions as showing, from a stranger, a remarkable respect for the office of emperor.[51]

Bruce had news at this time from Michael that he had formed an alliance with Fasil and would soon return to Gondar. He decided that now, after the rains, was the time to make another attempt to reach Geesh and the source of the Nile. He realized the dangers—from the pagan Gallas and no less from those, usually priests, who thought of him as a Frank and might rouse the local people against him.

With Balugani and the usual small party Bruce set out on 28 October 1770. The distance was comparatively short, not much more than a hundred miles. But he knew that unless he could obtain the protection of Fasil he had no hope whatever of getting there. They found Fasil in camp at Bamba, near the northwest shore of Lake Tana, which they reached on the evening of the third day. Bruce's reception was at first discouraging. Fasil pretended surprise that he should be willing to penetrate the territory of the wild Galla. How could he succeed? Bruce answered that he could not without Fasil's help. Fasil then taunted Bruce with the effeminacy of white men as against the virility of born warriors and made out, no doubt with truth, that Abba Salama, the powerful priest at Gondar, had asked him to prevent Bruce's progress, as it was against the law for Franks to travel through the country. To have his courage so dismissed and to be called a Frank was too much for Bruce. He declared he had never been so misjudged and challenged Fasil to match his two best horsemen against him alone. As if to underline his challenge his "nose burst out in a stream of blood." By one means or another Fasil seems to have glimpsed his true worth. It would be wise, he said, to start before the Galla were disbanded, and he offered him a horse. A groom paraded some poor specimens and suggested a bay pony, a "favourite" of Fasil's, "but too dull and quiet for him." Bruce was urged to mount. He did so. The horse kicked, reared, and "leaped like a deer, all four [feet] off the ground." He just managed to gain control, discovering the pony was unbroken. He rode him to exhaustion and realized how treacherously he had been treated. But his horsemanship had impressed the Galla, and he had moreover shot down two kites from his horse with his double-barreled shotgun.

Fasil now quite changed his attitude. To press home his moral victory Bruce told him that Ras Michael was on the march toward Gondar. This had an unnerving effect. He followed this up with presents of silk sashes and Egyptian glass bowls which he knew Fasil coveted. When finally he came to take his leave he was ceremoniously invested as lord of Geesh and given a bodyguard of seven Galla chiefs—"I never saw more thief-like fellows in my life"—as well as a fine horse of Fasil's, known to everyone, to be driven before him, so that all knew he traveled under Fasil's protection. In addition, Fasil had chosen a guide, Woldo, an Agow by birth, a sardonic character, who preferred to travel on foot. Bruce was to learn how well he understood men.

Woldo guided them first through the province of Maitsha and thought nothing, in making fires, of using the new wood, with which the local people were rebuilding their houses, burned down by Michael in the campaign against Fasil. He led them to the camp of a local chief, the Jumper, for safety. Since there was no greater thief or murderer he would be a reliable protector—"he resembled very much a lean keen greyhound." Here Strates, the Greek servant, who had been robbed on the first journey to reach the Nile, rejoined Bruce and gave him the latest news from Gondar. A message from Esther claimed she was dying and begged Bruce to return to her. It is more likely that, isolated from the empress and the usurper Susenyos, she was terror-stricken at the prospect of Fasil taking control. Bruce was torn between the prime object of his Ethiopian travels and his regard for her, for her influence had meant his very survival. In an agony of mind he decided he could not now turn back. He sent her a message that he would return as soon as he was able after reaching his objective, and stipulated a course of treatment to be administered to her by a Greek priest at Koscam.

On their journey to the south, the Jumper told them, they would meet a party of two hundred men near a place called Roo, the most westerly point on their route, commanded by his brother the Lamb. They had been sent by Fasil to protect Bruce's company after they had been parted from their Galla bodyguard. They reached the marketplace of Roo on 2 November, where hides, honey, butter, and cattle were sold, and made contact with the Lamb, who seemed to take more notice of Fasil's horse than of Bruce or Woldo. Soon after this their Galla guards left them. Later the same day, as they traveled south, Strates shot a bird with strikingly colorful plumage which Bruce (or Balugani?) began to sketch, intending to finish the drawing when time allowed. They then heard cries and the sound of galloping horsemen. These were the Lamb's men, who had heard the shot and feared that Bruce's party had been attacked by Agows. Bruce, much impressed by their sense of duty and their alacrity, spread a meal for them. Presents offered to the Lamb seemed to leave him indifferent, yet he did ask one thing, the tablecloth they had been using for their meal, to protect his head from the sun. Bruce was astonished that he should even consider his complexion but promptly agreed. Then the chief, putting it round his head and half his face, rode off with his men. Before he left, as Fasil had done, he sent a detachment of fifteen men ahead for Bruce's protection.

Bruce continued his journey south and shortly reached the province of Goutto, inhabited, he tells us, not by Galla, who were recent invaders, but by the original people of the land, who were more civilized and better governed. They spoke chiefly Agow and Amharic but a few, in more distant areas, the language of the Falasha, the Ethiopian Jews. The people were richer and had finer cattle and produced excellent honey, though most of it was derived from the nectar of "lupins" (not, in fact, *Lupinus* but *Crotalaria* species) growing everywhere, which gave it a bitter taste. Bruce, perhaps scenting his quarry at last, is unusually enthusiastic and descriptive of the natural life: "All this little territory of Aroossi is much the most pleasant that we had seen in Abyssinia; perhaps it is equal to anything the east can produce. The whole is finely shaded with acacia trees, I mean the acacia vera [in fact, *Acacia senegal*], or the Egyptian thorn, the tree which, in the sultry parts of Africa, produces the gum-arabic." And he writes of the Assar valley:

> The strength of vegetation which the moisture of this river produces, supported by the action of the very warm sun, is such as one might naturally expect from theory, though we cannot help being surprised at the effects when we see them before us, trees and shrubs covered with flowers of every colour, all new and extraordinary in their shapes, crowded with birds of many uncouth forms, all of them richly adorned with variety of plumage... But as there is nothing, though ever so beautiful, that has not some defect or imperfection, among all these feathered beauties there is not one songster; and, unless of the rose or jessamin kind, none of their flowers have any smell; we hear indeed many squalling noisy birds of the jay kind, and we find two varieties of wild roses, white and yellow, to which I may add jessamin (called *Leham*), which becomes a large tree.

This characteristic effusion shows Bruce's intense awareness of colors and forms but is lacking in specific observation. There are no yellow roses in Ethiopia and the tree Leham *(Syzygium guineense)* has flowers which, though fragrant, have no resemblance to those of jasmin.

They reached the Blue Nile on 2 November, still a large river though they are within two days' journey of the source. Here it was 260 feet broad and swiftly flowing. The western bank— and here Bruce is more specific—"is chiefly ornamented with high trees of the salix or willow tribe, growing straight without joints or knots, and bearing long pointed pods full of a kind of cotton. This tree is called, in their language, Ha [*Salix subserrata*, p. 98]." The local people, who worshipped the river, stopped them from riding across it or fording it on foot with shoes. "I sat by exceedingly happy at having so unexpectedly found the remnants of veneration for that ancient deity, still subsisting in such full vigour." Woldo got them across by a theatrical combination of deceipt and browbeating. The night was spent at Goutto, but before resting Bruce mounted Fasil's horse and with a local guide went to see the so-called First Cataract of the Nile, half an hour's easy gallop to the east. It was insignificant compared with the falls at Tissisat. He returned to find that Woldo was on the point of killing a cow that had come out of the place where the villagers had hidden it. Bruce felt that further impositions on the local inhabitants must now be prevented. He told them that henceforth all provisions obtained from the Agows would be paid for and that as lord of Geesh he intended to remit all taxes in his control, which would normally have gone to Fasil or the emperor.

On leaving Goutto the company passed level country full of pollarded acacia trees, the small, shooting branches of which were made into large baskets by the local people to be hung up near their houses as beehives, for this was honey country. They were heading for Sacala, lying in a half circle of mountains now appearing to the south, but, before they started climbing, they passed through flat marshland bounded on the west by the Blue Nile. "In this plain, the Nile winds more in the space of four miles than, I believe, any river in the world; it makes above a hundred turns in that distance." Struck too by this feature, Balugani drew a map of the serpentine river in his journal, normally confined to laconic verbal entries. From various vantage points, as they ascended, they could see the mountain ranges to the south and southeast. "This triple ridge of mountains, disposed one range behind the other, nearly in form of portions of three concentric circles, seems to suggest an idea, that they are the Mountains of the Moon, or the *Montes Lunae* of antiquity, at the foot of which the Nile was said to rise."

They crossed many small rivers, all tributaries of the Davola, which runs into the Blue Nile to the north. Around the villages they passed they found plantations of Ensete, the African wild banana *(Ensete ventricosum)*, "one of the most beautiful productions of nature, as well as most agreeable and wholesome food of man." Of this plant Balugani left no fewer than ten drawings (p. 112). Before they descended into the plain of Sacala they were struck by the freshness of the climate and the beauty of the flowers and trees. This and the nearness of their destination gave them new heart. Bruce and Strates shot birds and animals probably with the intention of recording them rather than for food. But Woldo seemed to grow more and more despondent, lagging behind before the last heavily wooded mountain which, in Bruce's words, the party "ascended... with great alacrity, as we conceived we were surmounting the last difficulty after the many thousands we had already overcome." At a quarter to two on 4 November 1770 they reached the top near the church of St. Michael Sacala, and Bruce had his first sight in the valley below of the Blue Nile, or the Little Abay as the Ethiopians call the river before it reaches Lake Tana. He writes, "I could not satiate myself with the sight, revolving in my mind all those classical prophesies that had given the Nile up to perpetual obscurity and concealment." He recited to himself lines of the Latin poet Lucan to the same effect.

When he came out of his "delightful reverie," he found that Woldo was missing. None of the servants knew where he was. There were even suggestions that he was plotting to ambush them, while Strates was convinced that he had been eaten by the large apes or baboons they had seen not an hour before. He was found, barely crawling, near the church. On examining him, Bruce realized that there was nothing wrong. After various unconvincing excuses he finally came out with the real reason for his behavior. Bruce had promised him a reward for his services, and he had coveted above all else a crimson silk sash that Bruce wore. He seems to have thought that when Bruce saw the modest springs of the Nile he would be so disappointed that he would not be willing to part with his sash. Bruce handed it over with the promise of severe action if he should play any more tricks. Woldo was then told to show him the exact place where the springs emerged, and he pointed out below a "hillock of green sod" in the middle of a marsh. He warned him to take off his shoes for the waters were held to be sacred by the pagans. Bruce ran the two hundred yards downhill, half undressed without his sash, without his shoes, through the flowers, falling twice before he reached the island of green turf, which was "in the form of an altar," and stood in rapture over the principal fountain. His moment of triumph was quickly followed by feelings of despondency at the dangers still ahead.

To create a diversion and give himself time to settle his thoughts he devised a theatrical show. He called Strates, who was on the hill nearby, to join him. "Strates, faithful squire! come and triumph with your Don Quixote, at that island of Baratavia... come and triumph over all the kings of the earth, all their armies, all their philosophers, and all their heroes." Strates didn't understand a word. But he was willing to drink a health to George III from a coconut cupful of Nile water, which Bruce held up and added, "Confusion to his enemies." Strates was also eager to drink to Catherine, empress of all the Russias, and to the defeat of the Turks but strangely refused a toast to the Virgin Mary. Reid believes Bruce really asked him to drink to Margaret Murray, the girl he had left behind in Scotland and still hoped to marry. But he could hardly have used her name in his book and so invented Strates's unlikely reluctance to honor the Virgin Mary.[52]

Meanwhile Woldo had met the headman of the village, the high priest of the Nile, Kefla Abay, Servant of the Father of the Waters. He was a venerable bearded figure of whom Balugani, presumably, drew a portrait (Murray (1808), pl. 20; for the original drawing see Ill. 6). Bruce was intensely interested in his religious beliefs and rites and questioned him at length. Kefla Abay was host to the party and Fasil sent a milk-white cow, sheep and goats, and other provisions and there was much feasting. Bruce stayed there five days, and never was a single small area so carefully measured, positioned, and recorded by him and Balugani as that which cradled the three springs of the Blue Nile.

They left Geesh on 10 November and reached the gates of the palace at Koscam on 19 November.[53] Gondar was apprehensively awaiting the return of Ras Michael. Fasil had at last arrived in the city and was promptly made ras by the usurping Emperor Susenyos. Esther had been assured by Fasil of his loyalty to the Emperor Takla Haimonot, which meant, of course, that when the right moment came he would abandon Susenyos. But the latter fled with the empress at the news of Michael's advance across Tacazze. Fasil himself withdrew to regroup, and the city was left without a government. Bruce went north to meet the emperor with Yasine and had brought the steel-gray horse of Fasil's as a present for the emperor, who asked him to ride ahead to show off its paces. They passed a stream over which a thorny kantuffa bush hung. Its thorns are so shaped that once they catch in woven material they tear it to pieces. Before an emperor rode through the countryside there was a traditional proclamation that these bushes should be cut down wherever he might go. Bruce successfully passed the bush, wearing his goatskin cape. The emperor was less fortunate. The cotton cloak wrapped about his face was caught and had to be pulled off so that his features became visible, shameful exposure in a ruler. The emperor remained calm but called out, "Who is the shum of the district?" The local governor and his son appeared smiling, without realizing what had happened. The emperor made a sign, the two men were seized and hanged on the spot by the executioner, who rode with him. Bruce was dismayed at the change in the emperor, who not long before had been revolted by Michael's savage executions.II The thorny kantuffa *(Pterolobium stellatum)* was drawn by Balugani and is illustrated in the *Travels* (see Ill. 7).[55]

The emperor's army reached Gondar on Christmas Eve. Then and on subsequent days Michael wreaked vengeance on all those that opposed him. They were hanged, blinded, or beheaded. Twenty entertainers who had come to welcome him but who had, he heard, lampooned him in his absence, were cut to pieces by his cavalry. The most important of those that were put on trial was the priest Abba Salama, accused of murder and high treason. He put up a spirited defense but was hanged, a fate which caused Bruce few misgivings, as the priest had plotted against him and always referred to him as a Frank.[56]

The wanton cruelty and general bloodshed sickened Bruce, and the news that Fasil, with Gusho and Powussen, was regrouping his forces was depressing. He saw his plans to leave Ethiopia by way of Sennar being frustrated. The emperor refused to allow Bruce and Yasine to go to Ras el Feel to buy horses and coats of mail for the cavalry since he feared the former might never return. Though Bruce managed to persuade the emperor to let him send a letter to the Sudan he was required to take an oath not to leave the country until the present crisis was over.

Some time after the middle of February 1771, Luigi Balugani died.[57] However much, out of vanity, Bruce may have tried to negate his assistant's part in the expedition—sharing his triumph at reaching the source of the Nile, and as artist—he must have felt the loss keenly. These were the gloomiest times he had yet experienced in Ethiopia. But he was kept busy training his three hundred black horsemen, in whom he took great pride.

At this time, and no doubt at previous periods of enforced stay in Gondar, Bruce was patiently gathering historical information on Ethiopia and collecting Ethiopian manuscripts either by direct acquisition or by commissioning copies. When Amha Yasous, prince of Shoa, then a southern outpost of Ethiopia and not involved in the struggle for power, now paid a formal visit to the emperor he had already heard about Bruce. The two were soon to meet frequently, and Bruce obtained through him much of the information he was seeking.[58] From the monastery of Debra Libanos in Shoa the prince obtained a copy of the imperial chronicle for Bruce's use, from which the explorer compiled much of his earlier history of Ethiopia. As for his collection of manuscripts, Ullendorff states that he returned with at least twenty-seven, "all of which are exquisite examples of Ethiopian manuscript art,"[59] including a complete and unique copy of the *Book of Enoch*, which he was to present to Louis XV in Paris. Among the manuscripts, most of which are now in the Bodleian Library, Oxford, is an extremely rare version of the *Song of Songs* in Gafat, a Semitic Ethiopian language now extinct.

But Bruce's stay in Gondar was occasionally enlivened by moments of sheer farce. There was the extraordinary visit of the Galla chief Guangoul, riding on a small saddleless cow with enormous horns. The appearance of this ludicrous figure, the body hung about with ox guts and running with butter, caused such uncontrollable mirth, even on the emperor's part, that, after he left, Esther insisted on the scene being reenacted in a charade, with a court dwarf playing Guangoul. It was an enormous success. Bruce, who helped with the costume and makeup, relates the story superbly, doubtless improving on the event.[60] But even this figure of fun soon forced the emperor to face the real and threatening situation, for Guangoul quickly joined the rebel forces.

The following days saw indecisive and curiously ritualistic battles in which there were few casualties. Bruce was involved as commander of his black cavalry, who, according to his account, fought to considerable effect. He was also used as an intermediary between the emperor and the rebels on the pretext of treating the enemy sick. It was clear that the enemy, by defections, was growing stronger all the time, and Ras Michael was finally forced to retreat to Gondar, where his men were disarmed and marched out of the city. Gusho was made ras in Michael's place, and the other rebel chiefs came to pay homage to the emperor. The empress returned to Koscam, and Bruce was able to reassure her that the emperor, against whom she had plotted, did not intend to harm her. Bruce himself stayed on in Koscam, where, too, Esther was still living. But now that the situation had changed into an uneasy calm, with new faces at court, Bruce saw his chance of disengaging himself, and surprisingly the Empress Mentwab and the Emperor Takla Haimonot consented to let him go. He even obtained the blessing of a senior and respected priest, Tensa Christos, with whom he had discussed the state of the Ethiopian church, praising its faith but frankly criticizing the morals of its clergy and people. Bruce paid homage to the empress and to the emperor for the last time (Ozoro Esther was away from Koscam) and left Gondar on 26 December 1771.[61]

He traveled with his usual mixed company: three Greeks, a Copt, a few Ethiopians in charge of the mules, and an old Turk, Hadji Ismael, a descendant of the Prophet Mohammed, on his way back to Egypt. He was Bruce's chief assistant and was never afraid of speaking his mind. The plan was to travel westward to Ras el Feel, where Yasine was organizing his departure through Sennar, but their initial objective was the house of Esther's son Confu, located at Cherkin, sixty miles northwest of Gondar. Upon reaching the house, they found it was built on a hilltop, surrounded by enormous trees, and fed by a clear stream. To Bruce the place seemed idyllic. To his intense surprise and pleasure he found not only Confu there, but Ozoro Esther and Takla Mariam, the daughter of the emperor's secretary. Two somewhat conventionalized miniature portraits of these two beautiful women, made presumably at Koscam by Balugani, illustrate the *Travels*.[62]

Bruce recounts in detail a big game hunt they took part in with elephant, rhinoceros, and buffalo. He describes how, when a cow elephant was wounded in the back legs ready for the kill, its small calf charged several times in support of its mother and how he had not the heart to lift his hand against it. He was surprised how easily this story was believed in England as against the steak on the hoof story, which was everywhere disbelieved.[63]

Messages had now arrived from the emperor for Esther and her son Confu to return to Gondar. They tried to persuade Bruce to go back with them but this time he did not hesitate to refuse. He was not to see them or hear from them or his friends at court again. After staying another week in this delightful place they made their way through difficult country to the village of Hor Cacamoot, which Bruce translates as the Valley of the Shadow of Death, an apt name, for here he fell dangerously ill with dysentery. He was nursed by Yasine, who had awaited him there, and was eventually cured by taking the powdered root of Wooginoos (*Brucea antidysenterica*, p. 100). He recounts how he drew the plant illustrated in the *Travels* "on the spot," "Heaven having put the antidote in the same place where grows the poison."[64] But as mentioned below (p. 58), the engraving is closely connected with a group of drawings made by Balugani at Sacala near the source of the Blue Nile.

On 18 March 1772 Bruce took his leave of Yasine, his loyal traveling companion on the journey into Ethiopia. They were now to travel to Teawa rather than direct to Sennar. Yasine had written to Fidele, the governor of the province of Atbara, whose home was at Teawa, to provide protection. The country they were entering was indeed a savage and desolate region, the Fung kingdom of Sennar, a country then virtually unknown to Europeans and where at this season all the water holes were dry. They found their way with difficulty through dead villages destroyed by famine or nomad raiders. The sixty-five miles between Yasine's village and Teawa took them a week. Though the basic weather recordings went on, it is extremely unlikely that, without Balugani, Bruce would have the opportunity of drawing plants or animals. It was enough to be traveling home with as much speed as possible and with what security they could devise.

Bruce was warned before they reached Teawa that Fidele was dangerous and untrustworthy. So it turned out. He was of the same treacherous, sly, but in the event cowardly, nature as the naib of Massawa. They came out of Teawa alive thanks to help from the governor's harem, which Bruce penetrated as a medical man, to the watchful intelligence of Moslem messengers, and to the arrival of two "holy men" sent by a friendly sheikh. A message had been sent back to Yasine and only on the appearance of his horsemen were Bruce and his men allowed to depart. They left on 17 April.

The country changed toward Sennar, where the land became flat and comparatively fertile. Before reaching the city proper, they came on the outskirts to a semicircle of separate villages which housed the Negro slaves who formed the king of Sennar's infantry. Bruce was impressed, as ever, with the signs of order and military efficiency he found there, for by means of this disciplined army of pagans, living apparently contentedly with their families, the power of the Moslem government in Sennar was maintained. He also had good reason to test the friendliness of the Nubian soldiery. Near to the ferry on the Blue Nile a whirlwind struck them. One camel was lifted up, then thrown down, and badly injured. Bruce and two others were thrown down and plastered with mud. They were cleaned up and given shelter by these people and rested in a village. Bruce was given his own hut and spent one of the pleasantest nights of his travels. "Some of the Nuba watched for us all night and took care of our beasts and baggage. They sang and replied to one another alternately, in notes of full and pleasant melody till I fell asleep, involuntarily and with regret."

The city of Sennar pleased him less. It consisted of a large number of flat-roofed mud houses, some with two storys and spacious enough, but the people were suspicious of all foreigners and particularly of Europeans. There was an atmosphere of foreboding which depressed him intensely. Had not a French envoy and his suite been murdered here? "War and treason," Bruce writes, "seem to be the only employment of this horrid people whom Heaven has separated by almost impassable deserts from the rest of mankind." Furthermore, far from being helpful King Ismain was obstructive and unwilling to give Bruce the help he needed. But he found two allies: Sheikh Edelan, the vice-vizier; and Ahmed, a very strange figure in Bruce's gallery of portraits, one whose official duty was to kill the king when the council of ministers judged him unfit to rule or out of touch with the people he incarnated. But surprisingly, in view of this, Bruce yet found him to be one of the gentlest people he had ever met. His advice to Bruce was that he should get out of the city as quickly as possible and until then, so far as he could, keep under the protection of Edelan.

At Sennar Bruce suffered two major setbacks and an alarming experience. The merchant who was supposed to have provided him with money to buy camels and provisions for the journey refused, and the guide with whom he was meant to travel through the desert left earlier with another party, apparently at the king's persuasion. As an influential eunuch and holy man, the guide's status was held to be some guarantee of security, and moreover he was beholden to Bruce, who had cured his malaria with quinine. Though Bruce later had good reason to be glad that he had not gone with him, he was now without a guide, but rather than wait for one it was more important that he escape from Sennar. Ironically, the opportunity was provided as the result of a night attack on Edelan's town house, which he had provided for Bruce's accommodation, by a group of ruffians led by Mohammed, a servant of the king. It was broken up when Hadji Ismael fired his gun.[65] Edelan had Mohammed and his accomplices arrested and since they had attacked his property they faced execution. The king summoned Bruce and asked him to intercede for Mohammed's life. If he was successful Bruce could have him as a guide. Thinking how much better off he would be without him, he pretended to agree. Setting out with his servants, ostensibly to visit Edelan, who was staying well outside the city, Bruce escaped by turning north toward their ultimate destination. As for the question of money, that had been settled by Ahmed, the executioner of kings, or, to give him his proper title, the Sid el Kum. He had argued with the merchant, who eventually agreed to give Bruce the money he needed in exchange for most of a large gold chain which the emperor of Ethiopia had given him.

They left Sennar on 6 September, a tiny group to undertake by far the most dangerous journey yet. Bruce had with him three Greeks, one half blind, and the elderly Turk Hadji Ismael. They had five camels, four well loaded with provisions and with Bruce's precious records, manuscripts, and drawings. The other was for riding. The route lay north northeast along the west bank of the Blue Nile toward its confluence with the White Nile, near what today is Khartoum. At first the going was good, the plain green with springing corn after the rains. At Herbagi they visited Sheikh Wed Ageeb, who accounted to Sennar for the Arab chiefs of the northern part of Atbara. They were well received, and with great interest, as Bruce was the first European the ruler had seen. He was in fact approaching territory which no European is recorded to have reached before him. The sheikh told him that Yasine had burned Teawa, expelled Fidele, and had punished Mohammed and the others who had made the night attack, news which must have given him a certain satisfaction. He also gave him a letter for his sister, the Sittina, or ruler of Chendi, a place north of the meeting of the two rivers.

They had to recross the Blue Nile at Halifoon, and soon afterward, traveling along its eastern bank, they came on 23 September to the meeting of the waters. Though Bruce realized that the White Nile had a more constant flow of deep water, he still thought the Blue Nile was the main stream. They were now approaching the line where the regular rains ceased and only the river produced a strip of vegetation. Here Bruce found a magnificent breed of horses and an Arabic more pure than that of Arabia itself.

Chendi had once been a great market where many trade routes met—from Egypt, the Red Sea, Ethiopia and Sennar, and countries near the Niger—but was now less important. The Sittina was anxious to help Bruce and provided all the supplies he needed. But she was genuinely alarmed that he had no guide for the desert journey. She inquired why he had not gone with Mahomet Towash, the eunuch, who in fact had taken from Chendi all the available guides. She begged him to wait until one should appear. Eventually one did, Idris, an Egyptian Arab, whom Bruce bought out of debt. At their final interview, she actually allowed Bruce to kiss her hand "in the most gracious manner." She was after all a handsome woman with "the finest teeth and eyes I have ever seen."

Bruce's small company had now been enlarged by Idris, the guide, two Nubians to look after the camels, and six Negro pilgrims (whom be permitted to join the party, against his better judgment). The camels were loaded with as much Nile water as they could carry and with powdered durra bread, their only food. They set out on 11 November, keeping always to the east bank of the Nile, going easily, past the ruins of the ancient capital of Nubia, Meroe, until they reached Berber. They then took a ferry over the river Tacazze, near Goos, the same river which marked the boundary of Tigré, and which some hundreds of miles further along its upper reaches, they had crossed on their journey to Gondar. Not far north of here the Nile begins its great loop to the west and here they left it and took their northerly course toward Aswan (Bruce's Syene). Bruce asked Idris for the direction of their destination, fixing the line with his compass and discovering later that it was almost exactly correct.

The Nubian Desert opened up around them with its burning rocks and sands. Besides enduring the heat, which inflamed their feet unbearably, and the harsh cold at night, they met with terrifying whirlwinds, "prodigious pillars of sand..., at times moving with great celerity, at others stalking on with majestic slowness; at intervals we thought they were coming in a very few minutes to overwhelm us... Again they would retreat so as to be almost out of sight, their tops reaching to the very clouds." When at dawn the sun's rays hit the sands it "gave them an appearance of moving pillars of fire... The Greeks shrieked out and said it was the day of judgement." Even in these conditions Bruce did not cease to make his weather records and enter them in the log which Balugani had previously kept. Even more terrifying than the pillars of sand was the simoom, the burning, poisonous, suffocating wind of the desert. Bruce was not able to rid himself of the effect of it on his lungs for two years, like a kind of asthma. His face became so swollen he could scarcely see. Six days out from Goos they were in despair, but Bruce, with an eloquent speech and the promise of an extra ration of water, managed to lift their spirits for they were in sight of Chigré (Shikrib), where there were wells. Here the camels drank enough to last them to Egypt. They were even able to bathe, but two of the West Africans died after drinking from the wells. Idris told them that they were about halfway to their destination, which Bruce was able to confirm from his observations.

They continued their journey to Terfowey, where they had their first encounter with Bishareen nomadic Arabs. On the night of 19 November, Bruce discovered a man and two women, one with a child, attempting to steal their camels. They were seized and questioned and would have been put to death had not Bruce intervened. The three Arabs confessed to having murdered the eunuch Mahomet Towash and his party. They decided to take the man along with them. Again they were attacked by the simoom; one of the camels died of fatigue. They dried some of its flesh for food but, in their famished state, found it repugnant. Their bread was almost finished, and Bruce's feet were so badly swollen that he could walk only with great pain. One of the West Africans went mad, refused to get up, and had to be left. They then stumbled on the body of the murdered eunuch and three of his servants and soon of others of his party. But a wild duck appeared at the next pool and they knew they could not be far from the Nile. Yet several of them could go no further. As they were resting, a group of friendly nomads came by and told them that Aswan was two days' riding away. When, finally, the camels would not stand up after a night of bitter cold Bruce had no alternative but to abandon their baggage—instruments, records, and drawings—retaining only their firearms.[66] The camels were killed and the water drained from their stomachs. Bruce felt that pride of achievement, which had buoyed him up during every ordeal, had now been extinguished. Yet, that very day, as they continued their journey they saw kites, which as scavengers were never far from human habitation, and a few trees. Next day, as they finished their last bread and water, Bruce began to recognize the rocky landscape near Aswan. That night he thought he heard water, the sound of the cataract above Aswan, and a flock of water birds flying low seemed to confirm it. They were almost there.

They reached Aswan on 29 November.[67] The journey from Berber had taken twenty days. The sheikh received Bruce well—though not at first recognizing him—no doubt in gratitude for medicines that Bruce had given him four years previously. When Bruce had recovered somewhat he persuaded the sheikh, rather against his will, to lend him camels to attempt to retrieve the abandoned baggage. They found it untouched, an outcome beyond all expectation. To the Bishareen guide whom he had saved from death and who had been loyal and served him well, he gave a camel, clothing for his family and a quantity of millet—to his inexpressible gratitude.

On the river passage to Cairo Bruce speaks of "a kind of stupor or palsy of the mind... I seemed to be as if waked from a dream, when the senses are yet half asleep, and we only begin to doubt whether what has before passed in thoughts is real or not." He was ill with what probably was malaria and the guinea-worm in his leg, and his feet were still in poor condition. In Cairo he was painfully prodded along on a donkey by soldiers, as a ragged infidel vagrant, to the bey's palace. Ali Bey's successor, whom he had met before, recognized him though not at once and thanked him for his kindness to Hadji Ismael. Bruce refused an offer of gold and when asked what favor he wished requested that British merchants be allowed to bring their ships to Suez and trade directly with the Egyptians rather than being stopped at Jidda, where they were obliged to pay heavy duties to the sherif of Mecca. This request eventually resulted in a license to trade at beneficial rates of duty in the Gulf of Suez.[68] Though this privilege was later rescinded it shows to what extent Bruce had British interests at heart.

Bruce set sail for Alexandria, where he found a French ship bound for Marseilles. He landed after a stormy voyage on 25 March 1773, almost exactly ten years after he had sailed from Leghorn to take up his post at Algiers.[69] He was still a very sick man but wisely avoided amputation of his leg, which eventually responded to treatment. News of his arrival had preceded him. The greatest of contemporary naturalists, the Comte de Buffon, who had helped to replace some of the instruments lost on Bruce's travels, welcomed his return. Bruce told him all he could about his observations in Africa and showed him the drawings. In volume 3 of his *Histoire des oiseaux* (1775), part of his monumental *Histoire naturelle*, Buffon makes a generous reference to the value of Bruce's "immense collection of drawings he had made and coloured himself—animals, birds, fish, plants, buildings, clothing, arms etc.... nothing seems to have escaped his curiosity."[70]

Bruce spent some months in France visiting Paris and other centers. He presented his unique Ethiopian copy of the apocryphal *Book of Enoch* to the king's library and seeds of African plants to the royal gardens (for Louis XV was now seriously ill and could not receive these gifts in person). He stayed long enough to see some paintings of his plants—exactly what these were, if they were not just copies, is not clear—and was highly critical of the artists' work.[71] It is also not clear if any of these artists executed drawings in the "Paris folder" which is catalogued below, though these are well drawn. Buffon agreed that paintings of plants raised from Bruce's seeds should be sent on to him in Scotland. Unfortunately, because of the inadvertant switching of labels and envelopes by Bruce's Greek servant, some of the plants were wrongly named. This resulted in adverse remarks about Bruce's identifications by the botanist Antoine de Jussieu, who in turn was lampooned by Bruce.[72]

Later in 1773 Bruce decided to go to Italy to try the baths at Poretta—he had not yet fully recovered his health—and to find Italian artists to "finish" the architectural drawings, that is, embellish them with landscapes, skyscapes, and figures of men and animals. Both at Bologna and Poretta he was a guest of Marchese Ranuzzi, Luigi Balugani's old patron and owner of the baths at Poretta. His good health gradually returned, and he seems to have remained on good terms with Ranuzzi in spite of the considerable misgivings held by others in Bologna (p. 50) concerning his treatment of Balugani and his work. There were various acrimonious exchanges of letters by Bruce with artists about their work and payment.[73] From biographical notes in the records of the Accademia Clementina we know who the artists were whom Bruce employed: Davide Zanotti (d. 1808) and Vincenzo Martinelli (1737–1807), who drew the landscape backgrounds; and Emilio Manfredi (d. 1801) and Giacomo Zampa of Forli (1731–1808), who drew the figures. (As an example of a "finished" drawing see Ill. 3). Oretti in his large, unpublished manuscript about artists of Bologna has also recorded the assistance these artists gave to Bruce.[74] In the biography of Manfredi he writes of the many figures he placed in the drawings of ancient buildings by Luigi Balugani for "Sir" James Bruce. In that of Zanotti he mentions that Martinelli did some of the landscapes and Zampa some of the figures, though in the biography of Martinelli there is no mention of his work for Bruce, and Zampa himself has not been given a biography. There are suggestions in Bruce's correspondence with Ranuzzi that these artists were paid from the supposed legacy of Luigi Balugani. However Bruce may have antagonized artists and academicians, Ranuzzi seems to have taken every trouble to make his stay at Poretta a pleasant one.

Bruce was also in Florence, where he presented seeds to the botanic gardens and where, the story goes, he actually met at the theater, by accident, Margaret Murray, his former fiancée. It seems that she had heard that Bruce had died in Ethiopia and had met and married an Italian nobleman, Marchese Filippo Accoromboni. Bruce believed that he had been betrayed and challenged Accoromboni to a duel in spite of Horace Mann's advice against such a rash venture. "I am your equal Marquis, in every way, and only God can give me justice for the wrong you have done me," wrote Bruce. The marquis, however, managed to deflect Bruce's anger, saying that when his marriage took place there was no mention of a previous promise to Bruce, whose name was entirely unknown to him.[75] To Horace Walpole and others Bruce's quarrel was ludicrous, an exaggerated piece of behavior which did not make it easier for those at home to accept his account of his Ethiopian experiences.

Otherwise in Italy he was lionized. He was received by Pope Clement XIV through the good offices of the French ambassador at the Vatican, with which Great Britain had no diplomatic dealings. The pope presented him with a number of gold medals, and Bruce became an honorary member of various learned societies throughout the country.[76] His fame had reached all parts of Europe, and now that he had fully recovered he was ready to make his long-postponed return home. However, finding France always congenial he stayed there on the way back until June 1774, when he at last reached London.

He arrived to great acclaim and was graciously received by George III, to whom he presented his drawings of classical remains in North Africa and Syria in two large folio volumes, with a number of unbound views of Palmyra and Baalbek. Though Bruce received thanks and some payment for his work, he was not given the baronetcy he felt he had been led to expect. Possibly the monarch remembered that his consulship was not thought by the ministry to be an unqualified success. Bruce must have felt that he received something of a rebuff particularly when Lord North doubted whether Bruce's efforts to open Suez to British trade would be found particularly useful by the East India Company.

The fact is that Bruce was given adverse publicity. Boswell wrote several articles on him for the *London Magazine* which played down his achievements since Bruce had not deigned to publish his own account of his travels. He made light of his claims with a rather juvenile wit, poking obvious fun at the raw meat story. Dr. Johnson, after first meeting Bruce on 1 April 1775, told Boswell, accurately enough, that he "was not a distinct relater."[77] He later unworthily came to doubt whether Bruce had ever been to Ethiopia. This may seem strange from the author of *Rasselas Prince of Abissinia*, but in this work he gives a picture of a wise ruler and a civilized kingdom so entirely different from Bruce's revelations of cruelty and barbarism that many of the latter's stories must have seemed wildly improbable. Bruce's temperament did not make it easy for others to believe him, as he was unwilling to communicate his views freely, as with equals, unless with friends. To others he was arrogant and aloof. In France and Italy he had been accepted as a remarkable explorer whose discoveries would benefit learning, agriculture, and trade. In his own country, though there were honors—he was elected a member of the Royal Society and had his name, commemorated in the genus *Brucea*, added to botanical nomenclature by its president, Sir Joseph Banks—he was conscious of an attitude of disbelief not experienced elsewhere. He pitied Omai, the Polynesian whom Banks had brought to London, because when he returned to Tahiti he would pass "for a consummate liar," for "how can he make them believe half the things he will tell them?" His personal idiosyncracies afforded a butt for sophisticated wit, satire, and caricature, of which Horace Walpole's was the most adroit and the most destructive.[78] And as Beckingham remarks, "Bruce was not at home in the world of Lady Sneerwell."

Perhaps Fanny Burney gives us if not one of the liveliest pictures of Bruce about this time, yet a keen and judicious one. Though not without a touch of caricature it is by no means hostile.

> He is the tallest man you ever saw in your life—at least *gratis*. However, he has a very good figure, and is rather handsome... He is warmly attached to Mrs Strange and her family... where, when they are alone, he is not only chatty and easy, but full of comic and dry humour; though, if any company enters, he sternly or gloatingly, Miss Strange says, shuts up his mouth, and utters not a word—except, perhaps, to her parrot, which, I believe, is a present from himself. Certainly he does not appear more elevated above the common race in his size, than in his ideas of his own consequence.

Mrs. Strange was the wife of the engraver, and sister of Bruce's Jacobite friend in Rome, Andrew Lumisden. She called Bruce affectionately, "His Abyssinian Majesty." He was often at her home in the spring of 1775 when he was hoping to get some government support for the expensive business of publishing his drawings. Though he showed them to influential and interested people, he evidently failed in this object, and in fact only a few were published before his death, in the first edition of the *Travels*.[79] Boswell summed up the position, probably correctly, "I conjectured," he wrote in his journal, "that he had come to London with high expectations from Government but had been disappointed. This had soured his temper, not very sweet originally."[80] In the meantime little of his achievements was reaching the public, his researches on Ethiopian history and customs, African animals, plants, and music, his map-making in the Red Sea, his knowledge of classical and Egyptian buildings, and of Ethiopian languages. Perhaps no other living person knew so much about so many of these things, yet he did almost nothing to satisfy public curiosity. In the face of public scepticism he retired to Scotland.

Bruce was a marrying man. He wooed Lady Anne Lindsay, the poetess, but she refused him. He succeeded with Mary Dundas, who belonged to a rich and influential Stirlingshire family, and they were married in May 1776. Though the families had had bitter quarrels, they shared a common loathing of the Carron Company, which smelted iron close to Kinnaird. The marriage was very successful, and Bruce was now encouraged to embark on a long and complicated lawsuit with the company, who had built the works illegally, he contested, within view of his house. He now began to abandon his correspondence, particularly with his learned French friends, who were always pressing him to publish. He locked away his manuscripts and spent his time on domestic pursuits, the rebuilding of part of his house, and his litigation, but before that was completed his wife died, in 1785. He was alone again and his friends, chiefly Daines Barrington, urged him once more to think of publishing. He engaged a secretary, William Logan, and set to work. With this his hopes of royal recognition and reward revived. He prepared a memorandum, aiming still at a baronetcy, but the king by now had begun his first long period of insanity, and William Pitt did not respond.

It seems that Bruce chose to dictate much of his book, relying on an excellent memory and on his and Balugani's journals. The Rev. Benjamin Latrobe, a Moravian pastor to whom Bruce had given subscriptions for mission work, claimed that Daines Barrington engaged him to transcribe Bruce's dictation in May 1788. He describes how Bruce would sit dictating, without interruption, from 8 a.m. to 9 p.m., and this went on, at intervals, for more than a year. There had been no formal agreement and when it was finished and Latrobe had done what he could to bring coherence and order into the work, he inquired about his fee. In reply he received five guineas and an insulting letter.[81] It is impossible to verify Latrobe's story but it is remarkable how much it resembles Bruce's treatment of Balugani and the question of the money believed owing to the family.

Travels to discover the source of the Nile, published in 1790, was an immediate success and was soon translated into French and German. It was dedicated to the king. Bruce took great care with the choice of format, paper, and ink and in preparing some of the drawings for engraving. It was a handsome production in five volumes (see Ill. 8). Within a few years a second edition was in prospect. Bruce had already begun to correct and revise but he was not to see it through. On the evening of 26 April 1794, while handing an elderly lady to her carriage at Kinnaird, he slipped and fell headlong down a flight of steps. He never recovered consciousness and died next morning. He was buried nearby in Lambert churchyard.[82]

His son Robert inherited the estate, and the family entrusted the editing of the second edition to a remarkable scholar and orientalist, Alexander Murray. The new work, which included Bruce's corrections, excerpts from his correspondence and journals, as well as the fundamental life of Bruce by Murray, was published, with further engravings in eight volumes, in 1804–05. Murray went on to edit a third edition, that of 1813. In the preface he writes, "At the close of life, after twenty years' repose and much domestic affliction, the Author of these Travels seems to have viewed his former life as in a dream. Each interesting event found a glowing place in his description, though indolence often prevented him from fixing, by his journals, the true time and place." Privately, Murray was more critical of his carelessness, his "vein of romance," his "too little regard to fact."[83]

It is not the place here to enter into any sort of discussion on the merits or demerits of Bruce's work. It is enough to add that Henry Salt, the next serious traveler to Ethiopia, and one of Bruce's severest critics, in conversation with Sir Walter Scott, "corroborated... Bruce in all his material facts," although he thought that he "considerably exaggerated his personal consequence and exploits." And of his achievements, perhaps the two greatest, beyond his feats as an explorer, were the collection of Ethiopian manuscripts which he acquired or had copied and brought to the West; and the great quantity of drawings which he and Balugani created and which he succeeded in bringing out intact under the severest possible conditions.

Notes to Chapter I

1 Details of his family, life, and career, apart from his travels in Egypt, Ethiopia, Sennar, and the Nubian Desert, are taken largely from Murray in Bruce (1805) or Murray (1808).
2 Bruce was to claim that his prowess as a horseman and marksman won the day in a number of difficult situations.
3 Printed in Murray (1808), pp. 141–42.
4 See Grave (1905), p. 237.
5 Murray, p. 16.
6 These were given expression later in Ethiopia, where he was given command, if we may believe his own account, and fought in several battles. See p. 30 below.
7 See Reid (1968), p. 32.
8 Murray, pp. 26–28; Reid, pp. 32–34.
9 Murray, p. 29.
10 Bruce (1790), vol. 1, p. vi.
11 We may presume this from his interviews with the king. See Reid, p. 37.
12 Murray, p. 33.
13 *Remarks on the antiquities of Rome and its environs.* London, 1797.
14 Murray, p. 34.
15 Ibid., p. 36*n*.
16 Robert Strange (1721–92), later knighted, was one of the finest reproductive engravers of his time. He wrote to Bruce in July 1766 (the letter printed in Murray, pp. 182–84) that he had completed the Paestum engravings, but I have found no record of them.
17 Murray prints a series of letters mostly from Bruce to Lord Halifax, secretary of state (pp. 143–78), showing how the former gradually lost the support of the government.
18 See the letter printed in Murray, p. 178.
19 See Playfair (1877), pp. 23 ff.
20 See Murray, p. 181.
21 Bruce (1790), vol. 1, pp. viii–x.
22 Murray, pp. 43 ff.
23 Playfair, p. 2 and passim.
24 Murray, pp. 184–90.
25 Bruce (1805), vol. 1, pp. 47–50.
26 Murray, p. 52.
27 See Bruce's draft letter, unaddressed, the only record of his journeys to Baalbek and Palmyra, printed in ibid., pp. 190–95.
28 Ibid., pp. 55–56.
29 Ibid., pp. 197–98.
30 Ibid., p. 59.
31 Ibid., p. 60. As a result of these Bruce received much help from Greeks and Moslems in Ethiopia and nearby countries. See the letters addressed to him at Massawa and Gondar printed in Murray, pp. 214–29.
32 Bruce (1790), vol. 1, p. 126.
33 Balugani's drawings are now in the Bruce archives at Yale and are engraved in Bruce (1790), at pp. 128 and 130.
34 Letter Bruce–Wood in Murray, p. 207. The engravings of the boat, after Balugani, are in Bruce (1790), at pp. 43 and 44.
35 The fish drawings are preserved in the Bruce archives.
36 Murray, p. 63.
37 Letter Bruce–Wood, Murray, p. 209.
38 Ibid., pp. 292–93.
39 See Ullendorff (1953), p. 129.
40 See Beckingham (1964), p. 12.
41 Beckingham (1984), p. xxxi.
42 See Beckingham (1964), p. 8.
43 These and subsequent quotations from Bruce's narrative are taken from the second edition of the *Travels*, Bruce (1805), from books 5, 6, 7, 8,

vols. 4, 5, 6, and are seldom referenced individually.

44 Bruce (1790), vol. 3, at p. 130.

45 Beckingham (1984), p. xxxii.

46 Balugani was to draw a miniature portrait of him (for a medallion) now in the Bruce archives. See Reid pl. 6a at p. 129.

47 This incident is described in Bruce (1805), vol. 4, pp. 420–21.

48 His independence of local interests, his understanding of Amharic, and something of other Ethiopian languages, his medical knowledge, and his military virtues made his place at court all too secure.

49 For a comparison of Lobo's description with Bruce's, see Beckingham (1984), p. xxxi, and for Lobo's own account, see ibid., pp. 232–33.

50 Cheesman (1936), p. 228.

51 Bruce (1805), vol. 5, pp. 166–67.

52 Reid, p. 170.

53 Bruce (1805), vol. 5, p. 488.

54 Ibid., vol. 6, pp. 12–13.

55 Bruce (1790), vol. 5, pl. at p. 49.

56 Bruce (1805), vol. 6, pp. 20–23.

57 The question of the date of his death is discussed fully below, chapter II, pp. 48–50.

58 Bruce (1805), vol. 6, pp. 41–42.

59 Ullendorff, p. 133.

60 Bruce (1805), vol. 6, pp. 46–47.

61 Ibid., vol. 6, p. 204.

62 Ibid., vol. 8, pl. 2.

63 Ibid., vol. 6, pp. 233–34.

64 Bruce (1790), vol. 5, pp. 70–72, pl. at p. 69.

65 Bruce (1805), vol. 6, pp. 401–03.

66 Ibid., vol 6, pp. 500–01.

67 Ibid., vol. 6, p. 504.

68 Ibid., vol. 6, pp. 532–33.

69 Ibid., vol. 6, p. 549.

70 Buffon (1775), "Avertissement," pp. iii–iv; translated in Reid, pp. 276–77.

71 Bruce (1790), vol. 5, pp. 64–65.

72 Ibid., vol. 5, p. 60.

73 Chiovenda (1940), documents nos. 43 and 46, pp. 490–91, 492–93.

74 Oretti MS, vol. 10, p. 258.

75 Bruce's letter to the marchese and his reply are printed in Murray, pp. 247–49.

76 Ibid., p. 115, and Lewis (1961), p. 234.

77 Boswell (1934–1950), vol. 2, pp. 333–34.

78 Burney (1907), vol. 2, pp. 14–26.

79 Bruce (1790) contains, besides a very few miscellaneous plates in the text, the following numbers of natural history engravings in vol. 5: twenty-four plants, six mammals, eight birds, three reptiles, a fish, three shells, and a fly.

80 Boswell (1960), pp. 271–72.

81 Moorehead (1962), pp. 39–40.

82 Murray, p. 129.

83 Reid, p. 314.

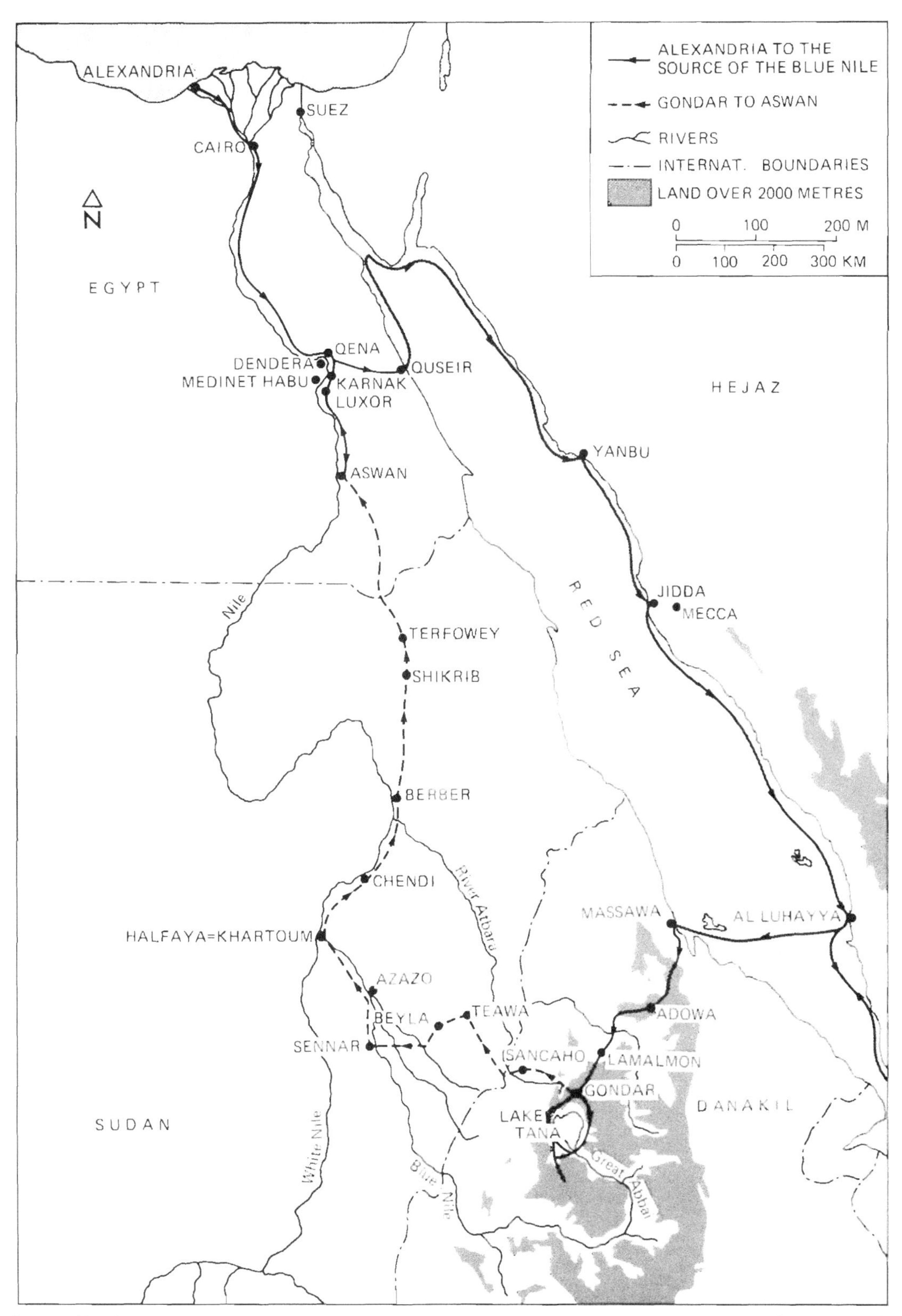

Map 1. Sketch-map showing the route taken by James Bruce and Luigi Balugani to the source of the Blue Nile, and James Bruce's return route to Cairo. See also Index iii: Names of places.

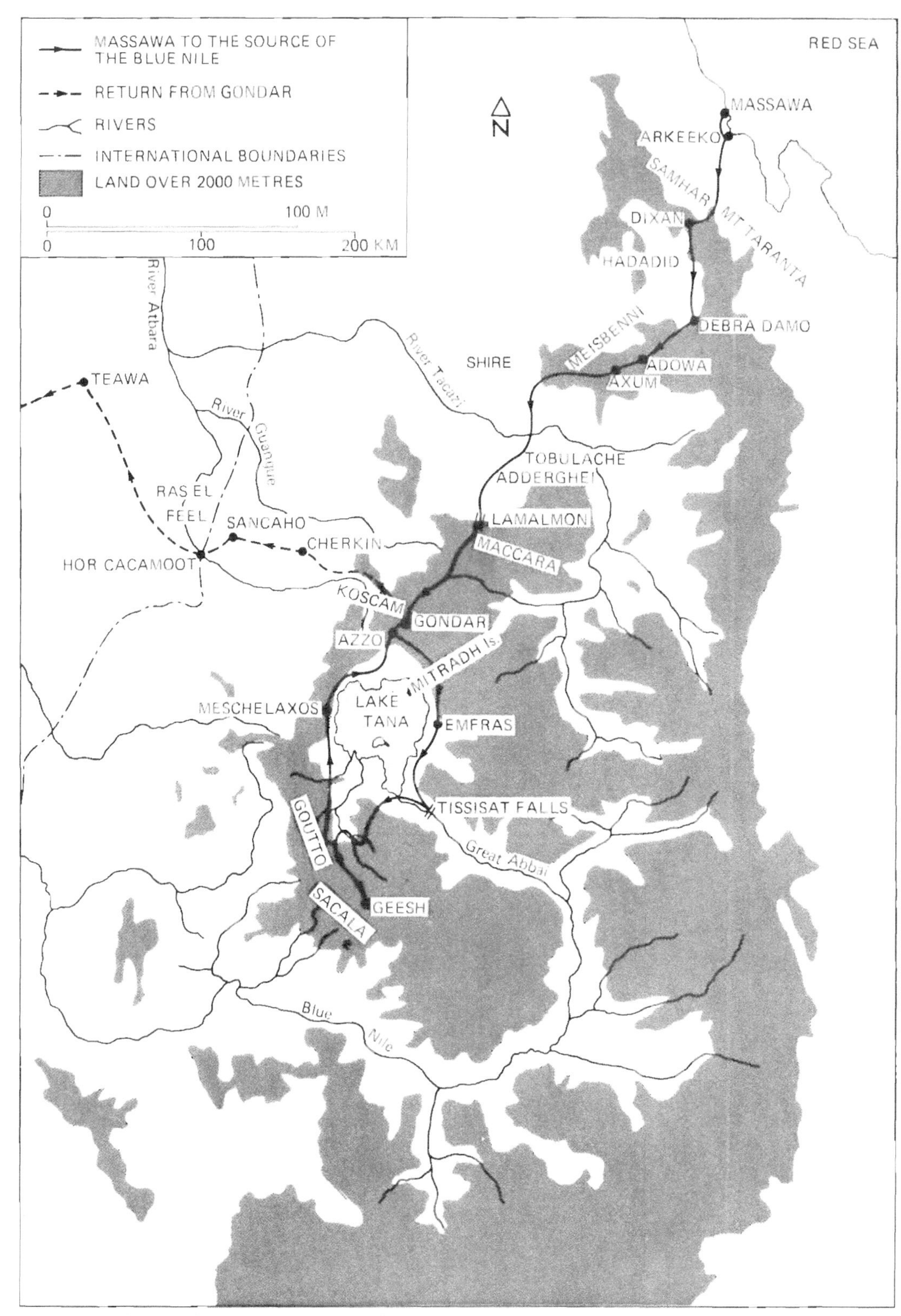

Map 2. Sketch-map of Ethiopia with names of places mentioned in the text. See also Index iii: Names of places.

Ill. 1. James Bruce and Giuseppe Zocchi (?). *Temple at Paestum.* Grey wash over pencil; 256 × 225. It is possible that Zocchi painted the landscape in this drawing (see p. 7). Yale Center for British Art. B1977.14.8496.

Ill. 2. Luigi Balugani (?). *Triumphal Arch at Maktar*. Grey wash over pencil; 397 × 550 (see p. 10).
Yale Center for British Art, B1977.14.8863.

Ill. 3. Luigi Balugani (?) and others. *Triumphal Arch at Maktar.* Grey wash over pencil; 484 × 645 (see p. 10). The landscape was painted in Bologna by an anonymous Italian artist. Yale Center for British Art, B1977.14.8862.

Ill. 4. James Heath after Luigi Balugani. *Canja under sail.* Line engraving; 275 × 230 (plate) (see p. 13). Yale Center for British Art, B1977.14.8614.

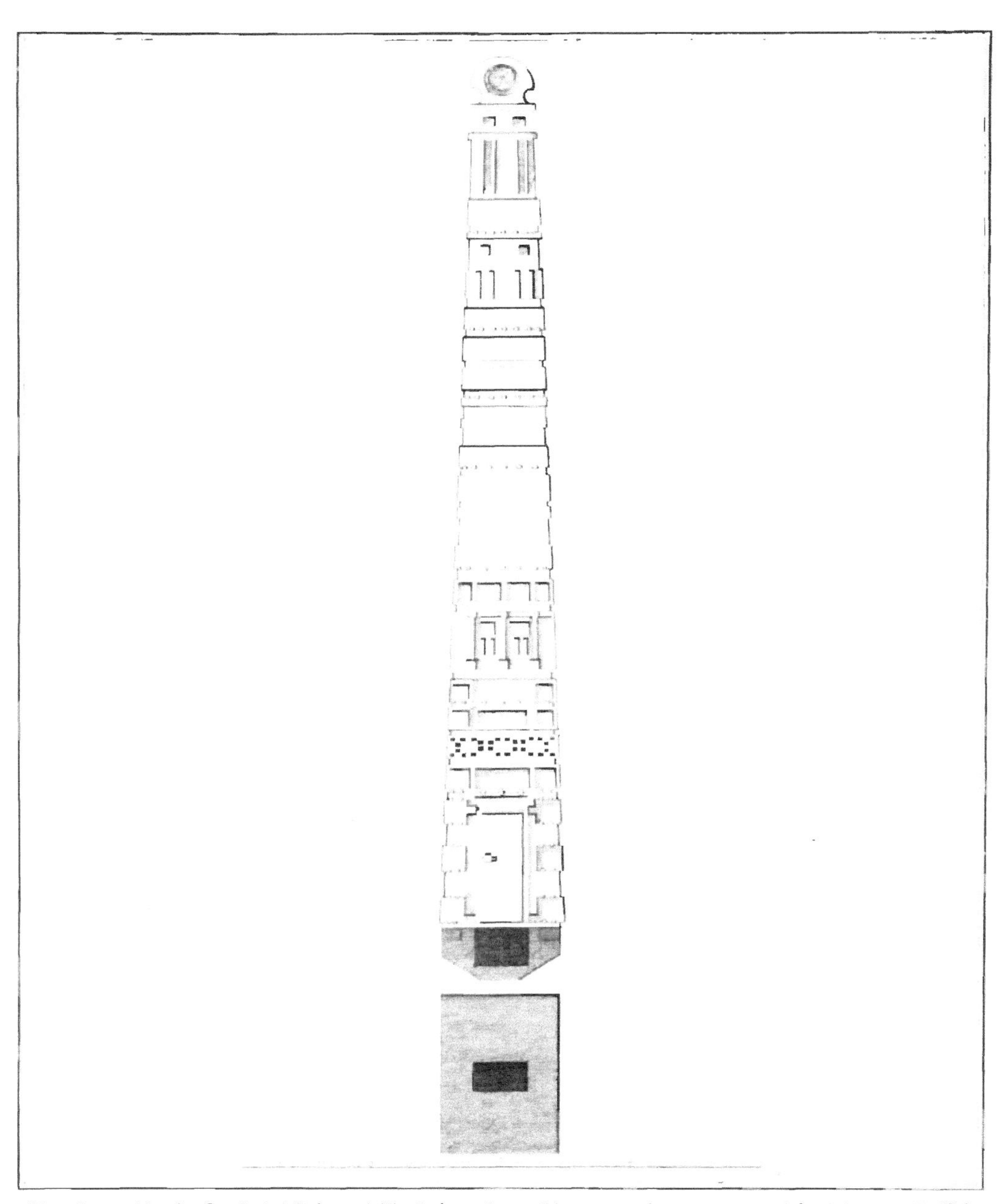

Ill. 5. James Heath after Luigi Balugani (?). *Stele at Axum.* Line engraving; 276 × 234 (plate) (see p. 18). Yale Center for British Art, B1977.14.8643.

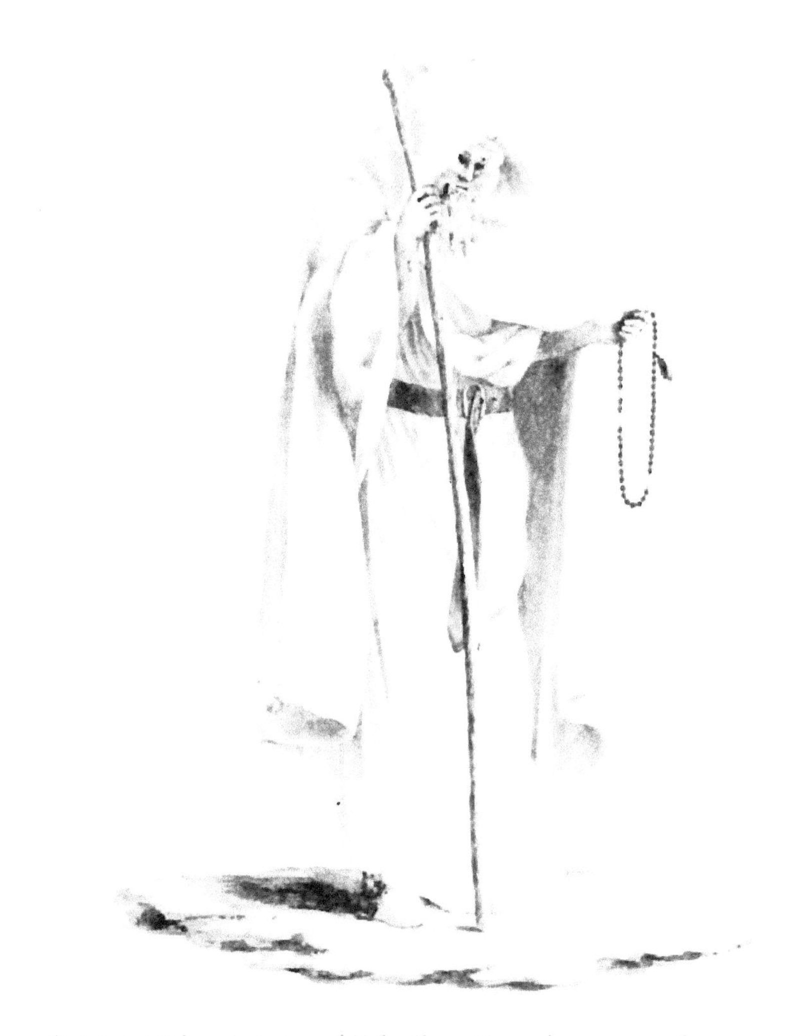

Ill. 6. Luigi Balugani. *Portrait of Kefla Abay*. Watercolor over pencil; 300 × 229 (see p. 28). Yale Center for British Art, B1977.14.8717.

Ill. 7. James Heath after Luigi Balugani. *A prickly shrub called Pterolobium stellatum or Kantuffa* (in Amharic). Line engraving and etching; 315 × 209 (plate) (see p. 29). Yale Center for British Art, B1977.14.11543.

T R A V E L S

TO DISCOVER THE

SOURCE OF THE NILE,

In the Years 1768, 1769, 1770, 1771, 1772, and 1773.

IN FIVE VOLUMES.

BY JAMES BRUCE OF KINNAIRD, ESQ. F.R.S.

VOL. II.

Vixere fortes ante Agamemnona
Multi, sed omnes illachrymabiles
Urgentur ignotique longâ
Nocte, carent quia vate sacro. HORAT.

E D I N B U R G H:
PRINTED BY J. RUTHVEN,
FOR G. G. J. AND J. ROBINSON, PATERNOSTER-ROW,
L O N D O N.

M.DCC.XC.

Ill. 8. James Heath after Luigi Balugani. Title-page to Volume 2 of the first edition of Bruce's published *Travels* (1790), with a medallion portrait of Ras Michael; 62 × 131 (plate) (see p. 38). Yale Center for British Art, DT 377 +B88 1790.

· II ·

Luigi Balugani *and his Relationship with* James Bruce

Luigi Balugani has received scant attention from Bruce's biographers. This is not surprising since the explorer's published story rarely mentions him and only twice by name.[1] When he does speak of him it is usually to stress his artistic shortcomings so that the reader of the first edition of the *Travels* (1790), if he remembers him at all, is left with the impression that he contributed very little to Bruce's achievements, in particular to the drawings he brought back with him from the travels. Since, in our opinion, Balugani's contribution was considerable —he was the author of all the preliminary botanical drawings and all the finished watercolors from those drawings, to mention no others—it is necessary, now that these works are published in full for the first time, to examine the evidence for a fuller account of Balugani's brief career.

It is true that a number of writers have published material on Balugani. Playfair explained what he believed to be Balugani's part in the making of the drawings,[2] but he was concerned only with the architectural drawings of classical remains in Algeria and Tunisia. Of the other writers, nearly all Italian, only Chiovenda made any real attempt to publish the relevant documents, but he had access to archives only in Bologna and Parma. Moreover, his article was addressed to the specialist reader rather than to the general public.[3] The result is that writers on Bruce have since made little use of his researches.

Luigi Antonio Melchiorre Balugani was born in Bologna on 14 January 1737.[4] He was therefore about seven years younger than Bruce, a significant difference in age if we bear in mind that Bruce was still a young man of thirty-five when he set out with Balugani on his travels. He was the second son of Luca Balugani and his wife Eleanora, née Ceccani. Luca was described as "bravo in ogni corpo di lavori meccanici"[5] (highly skilled in every sort of mechanical craft), a talent which was inherited in a more artistic sense by his two sons. Filippo, the elder by two years, became an established sculptor and medallist and official die-cutter at the mint in Bologna. He was closely concerned with his brother's career, exchanged letters with him on his travels, and after his death was intent on seeing that Bruce dealt justly with the family.

Luigi began drawing at an early age and apparently with much promise, producing ornamental designs and developing a strong desire to study architecture. He begged his father to find him a good master and was apprenticed to Giuseppe Civoli, the Bolognese painter and professor of architecture at the Accademia Clementina, the Bolognese academy of art. This distinguished institution had been reestablished by Pope Clement XI in 1709 and was named after him, assuming increasing importance in a city with a long artistic and university tradition. Balugani made rapid progress and won prizes for architectural drawing. Hardly any work from his early period is known to have survived, but a drawing of 1753 in pen and watercolor, of the facade of a temple with its plan, is preserved in the Accademia di Belle Arti of Bologna and was exhibited in 1979.[6] It had won the prize in the highest class, the *premio Marsili.*

In 1759 the R. Accademia di Belle Arti of Parma held an open competition in architectural design on the theme of "una magnifica rotunda" (a large circular building). Balugani competed, submitting his design with the motto *Ars longa vita brevis*, which was to prove sadly apt in view of his own short life. Much against the odds, for he was in competition with some of the best young architects, native and foreign, studying at the Italian schools (the favorite, a Fleming, Louis Feneulle, who later became the royal architect in Parma, was supported by a powerful minister of state), he won the gold medal. His plans were particularly praised for excellence of detail, style of ornamentation, precision and method which clearly revealed his capability as an architect. Other virtues singled out were ingenuity and simplicity.[7]

Balugani's success so impressed the Accademia Clementina that it elected him an academician at twenty-two, waiving the regulation which stipulated twenty-five as the minimum age. At the same time he was nominated "director" of the architectural school of the academy,[8] though it is probable that this was never more than an honorary appointment. These brilliant achievements do not seem to have diverted him from carrying on normally with his architectural studies and his practice of drawing. To judge from his accomplished drawings of plants of a few years later, when he was working with Bruce, he must already have been developing his talent as a botanical draughtsman, acquiring some understanding of botany. There is, however, no evidence of how this came about.

It was now essential for him to establish himself in his profession. Since his family was poor (his father died when he was young and his mother seems to have been deprived of a legacy she believed due to her), they could not support him. The protection of a patron was essential, and Balugani was fortunate in finding an unusually sympathetic and considerate patron in Marchese Girolamo Ranuzzi, count of Poretta and senator of Bologna. The latter's family had for centuries owned the hot mineral springs of Poretta, situated outside Bologna, and Ranuzzi was at this time busy improving and expanding the facilities there, which James Bruce was to make use of in 1773, recuperating after his travels.

The Palazzo Ranuzzi, now the Palazzo di Giustizia, had been the home of the Ranuzzi family since 1679.[9] According to tradition Andrea Palladio had designed the main building, though this is not now universally accepted. On the other hand, the work of later artists such as Ferdinando Bibiena, who constructed the grand salon in 1720, has been documented. Nearer to the time of the second Girolamo Ranuzzi, Balugani's patron, Vittorio Bigari, the master of Filippo Balugani, executed some of the ceiling decorations and Filippo himself contributed sculptures symbolizing the council of the gods for the grand staircase. Of considerable interest to us is the remarkable series of etchings after Luigi Balugani's drawings, many of them executed by Luigi himself,[10] of the various parts of the palace—elevations, plans, cross sections inscribed with the names of the architects involved and signed "Aloysius Balugani" (Ill. 9). These provide a most valuable record of the building as it was about 1760.

Two of the prints were sent by Girolamo Ranuzzi to Bruce in 1765 when Balugani was engaged as his assistant, as evidence of the architect's capability. Oretti mentions other works done by Balugani in Bologna, no doubt also for Ranuzzi, including a drawing in pen and watercolor of the chapel of St. Dominic in the church of St. Dominic and the altar of St. Francis of Paola in the church of St. Benedict.[11]

On 17 November 1761 Balugani traveled to Rome to further his architectural studies and stayed there for four years at the house of Girolamo Ranuzzi's brother, Monsignor Vincenzo Ranuzzi. Here he seems to have become acquainted, to judge from letters written on his travels, with some of the wealthiest and most influential families. He maintained a lively interest in the Accademia Clementina and corresponded with Count Gregorio Casali, its secretary, about the accession of newly qualified members.[12] He appears to have begun to receive important commissions, for while working on an etching of the interior of St. Peter's for the pope he came to the notice of Andrew Lumisden, the authority on Roman antiquities and secretary to the Jacobite court in Rome. Lumisden recruited him to assist Bruce as architectural draughtsman with the aim of recording classical remains in North Africa and Asia Minor.[13]

It is doubtful whether Bruce at this stage envisaged making any natural history drawings. We may guess that, finding Balugani adept and enthusiastic in this kind of work, Bruce saw it as a kind of bonus that would enhance his achievement when he came to publish his travels. Balugani had been engaged purely on his qualifications as an architectural draughtsman and these were exceptional. He would certainly have sought the advice of his patron and perhaps of Casali and his fellow academicians before accepting such an unusual and difficult commission. Lumisden was familiar with the problems of recording classical antiquities and was well informed about artistic life in Rome. He was therefore exceptionally well placed to recognize the right man for Bruce's requirements. Having received excellent testimonials from Monsignor Ranuzzi and his brother, Marchese Ranuzzi, he contracted with Luigi Balugani to work for Bruce, in the first instance for a year, at thirty-five Roman "crowns" a month, with board and other expenses defrayed.[14] Girolamo Ranuzzi makes it clear that this is the kind of money Balugani earned in Italy, and he thought it little enough for a hazardous journey to distant lands. Chiovenda translates "crowns" as *scudi*, Roman silver coins of medium value. Balugani asked for and received an advance of three months' pay to cover petty debts and initial expenses. Lumisden explains to Bruce that though the terms might seem high, it was impossible to find anyone willing to leave his affairs and native land for less. Balugani himself seemed eager to take up the challenge, believing that Bruce's plans to publish their records gave him the chance of gaining a European reputation. He was honest enough to admit his own weakness as a figure draughtsman, and Lumisden suggests to Bruce that here "you must assist him yourself." On the credit side were Balugani's special skills in drawing architecture and his ability to etch. So sure was Lumisden that he had discovered the right man that arrangements were made for Balugani to begin his journey to Algiers before Bruce had time to question the terms of his contract. Girolamo Ranuzzi was more concerned that Bruce should adopt his (Ranuzzi's) own role with regard to Balugani, not only as patron but also as protector and "father."[15]

Balugani left Rome on 9 February 1765, sailing from Leghorn to Algiers, where he arrived on 20 March.[16] After joining Bruce he seems to have kept up a regular correspondence with his old patron, Marchese Ranuzzi, with Count Casali as secretary of the Accademia Clementina, and with a wide circle of friends and acquaintances. Surviving letters provide the main evidence for his life with Bruce.

Those to Casali relate mainly to academy matters and are full of polite phrases revealing almost nothing of Balugani's inner thoughts. However, one letter is of particular interest. Writing from Algiers on 17 July 1765, he proposes that Bruce should be made an honorary member of the Accademia Clementina, arguing that Bruce was not only an excellent draughtsman but was well versed in science.[17] The proposal was without doubt sincere, though his own humility and deference toward his new patron could have caused him to exaggerate his master's abilities. He gave in evidence Bruce's work on the ruins of Paestum and his proposed history of Africa, both of which the latter intended to publish in the not distant future. Casali answered in support of the proposal and mentioned the opportunities for Balugani in his travels not only to observe and document antiquities but to record natural history.[18] From this it seems clear that Balugani's gifts as a natural history draughtsman were known in Bologna. In another letter to Casali he offers to collect medals from Carthage for the academy as he will not be able to collect other (presumably larger) items.[19] (It was Bruce who returned with a collection of medals from Carthage, but he does not seem to have given any to the academy.)

The letters to Ranuzzi reveal a more open and friendly relationship and indicate that his old patron was genuinely concerned with Balugani's fate. Writing from Algiers after about a month with Bruce, Balugani cheerfully reports in what a friendly and courteous manner Bruce had received him, how influential Bruce was with the dey of Algiers, what a pleasant country house he had, and in what comfort they would travel to record ancient remains.[20] Balugani's optimism extends to the chance of his returning to Italy with some zecchini (valuable Venetian coins) in his purse, of great concern no doubt to his impoverished family. Rather in contrast to this optimistic outlook is the next surviving letter to Ranuzzi, written from Tunisia after about one and a half years with Bruce.[21] It expresses both the energy and enthusiasm of Balugani but now openly reveals the pressures under which he worked and something of Bruce's attitude toward him. He excuses himself for not having written before because of pressure of work. His master has not allowed him a day off to attend to his own affairs for several months, not even Sundays. But he hastens to add that this was quite to his own taste and that he might soon return to Italy when his duties were finished.

Another letter to Ranuzzi, this time from Sidon, on the coast of modern Lebanon, dated 12 July 1767, says how much he would like to repay him for his favors with work of his own. Such work would be incomparably better than any he had done before his travels. For he had profited more from his experience since leaving Rome than from all the study he had done before then.[22] He concludes by saying that he is about to undertake with his master a journey to Baalbek and Palmyra but that Ranuzzi should tell no one, as Bruce had forbidden him to mention it. A later letter from Sidon of 13 November speaks of some of the dangers they were exposed to in the Syrian desert during these journeys, and again Balugani asks Ranuzzi not to mention their whereabouts to anybody.

When Balugani wrote this last letter Bruce had already left for Aleppo but would soon return and "honor" him with information about future traveling plans. A last letter from Sidon to Ranuzzi of 25 April 1768 reveals the arrangements for a journey into Egypt.[23] There is no mention yet of Ethiopia, though by now Bruce had probably decided to travel to that country and attempt to reach the source of the Nile. They were now waiting for a boat to take them to Alexandria so that they could pass into Egypt, a journey Balugani expected to last a few months after which, he believed, they would return to Asia Minor. The last letter Balugani addressed to Ranuzzi was written from Cairo and dated 16 August.[24] He thanks him for his help in sending a bank draft to his brother Filippo to reimburse money lent him before he left Rome. He mentions that he had not received a letter from his brother for three years until finally one reached him at Sidon. His conscientiousness as a correspondent can be gauged when he says that he had written more than thirty letters home but that nearly all seem to have been lost (which was only partly true). He had written the same number of letters to Ranuzzi and hoped that he did not sound ungrateful as he still had not received his letters, which so easily went astray. He also mentions the plague in Egypt, which now seemed to be over so that the French community in Cairo could once again open their doors (to such interesting visitors as themselves). Finally, he gives the latest details of their traveling plans, so far as he knows them. Again, extraordinarily, there is no word of Ethiopia. They were to travel for some months in Egypt and continue to Syria, Caramania (Asia Minor), Morea (Peloponnese), and back to Italy. But we can be sure that by now Bruce had other ideas.

In addition to Casali and Ranuzzi, Balugani corresponded with a circle of family friends and acquaintances. The sudden journey into Egypt had so far made it impossible for him to reply to his brother Filippo's letter received at Sidon. The copies of letters of Filippo's that survive give news of relatives and friends such as Giuseppe Civoli, Balugani's former teacher at the academy; Mauro Tesi, the painter, whose death is announced and whom Balugani had recommended to Bruce as one artist who might supply the figures his master felt it necessary to add to the architectural drawings; and Giacomo Zampa, the painter from Forli. A letter of 23 February 1768 speaks of their mother, "much aged" and longing before she dies to enjoy "the fruits of your efforts."[25] Another family correspondent, a cousin, F. or P. Sebastiano, is enlightening about the family's financial circumstances. He supplies news of the death of an uncle of Balugani whose estate was inherited by a brother who made no provision for his sisters. One of them, Balugani's mother, had been turned out of the house without redress. She and Filippo were now lodging at San Felice with a Dr. Catani.[26]

Among Balugani's correspondents one, Anna Ferrari, reveals quite a different side of his activities. She makes it clear that he was acquainted with the Doria Pamfilii family in Rome, one of the wealthiest and most influential of the time in Italy, and with the Cremanesi family. She confirms what can be guessed from scraps of information elsewhere in the correspondence, that he was given commissions to buy antiques and other rare objects for some of his rich acquaintances.[27] Other letters give us glimpses of Balugani's trading, or attempts to trade, in medals, sculpture (a "head"), antique stones (cameos?), coins, and shells.[28] Yet others speak of consignments of clothing probably for Balugani's personal use though, if so, they sound as if he had expensive tastes. The list of correspondents compiled by Balugani reveals the extent of his contacts and suggests certain business instincts and surprising social graces given his rather humble origin.[29]

From his own letters one thing particularly impresses, considering the pressures and the constantly changing travel plans: the complete absence of complaint about the hardship of the journey, or the load of work, or of any sign of a quarrel with his employer. Nowhere is there even a hint of bitterness.

It is clear from the letters and other documents that when not making drawings of classical ruins, which he was employed to do, he was always busy with some other artistic or scientific work. Murray has analyzed Balugani's part in the making of journals, commonplace books, and other writings from which Bruce composed his *Travels*:

> Luigi kept the weather journals in Barbary, Egypt, and Habbesh [Ethiopia], and assisted his master in making and calculating the astronomical observations, after which he entered them in the books. He designed much of the architecture and many of the articles of natural history, as is evident from the first sketches still preserved [he probably made all the botanical drawings, as argued below]. The journals relating to Egypt and Abyssinia were kept in the following manner:
>
> Luigi marked the state of the thermometer, winds, weather, &c. at particular times of the day, on separate pieces of paper, and afterwards transcribed them into a large folio book kept for that purpose. The celestial observations were also recorded in this book; but no remarks on the countries...
>
> As soon as they left Cosseir on the Red Sea, Luigi began a separate set of journals in the Italian language, by his master's order, who wished to have their routes exactly measured by the time; number of computed miles; occasional observations; and other circumstances;—reserving for himself the historical part, freed from minute detail, and interspersed with general reflections. These journals of Luigi are written on cream-coloured paper on which the Arabs write their books.[30]

Murray then goes on to describe the large folio volume in which the weather recordings were entered and the observations of latitude and longitude made from the time of leaving Negade and Badjoura in Egypt until the return to Gondar from the source of the Blue Nile in November 1770. The weather journals were kept from "about the time Mr. Bruce landed at Masuah [Massawa], (September 1769), and continued, with very little interruption, till December 1771, the month in which he left Gondar. All these are in Italian, written by Luigi, till the 14th day of February 1771, when Mr. Bruce's hand appears in them."[31] The rest of the volume contains Bruce's notes on the history of Ethiopia.

Murray continues with details of the journal kept by Balugani, which he describes as "diaries of the road and rates of travelling." They consisted of six closely written parts, beginning at Quseir on the Red Sea and finishing on 28 October 1770 with the return to Gondar after the journey to the source of the Blue Nile. "The last contains a complete detail of the hours and the days in which they travelled, of the villages, rivers, mountains, and, in short, every remarkable object they met with."[32]

It is also known that Balugani recorded descriptions of things of general interest on loose sheets of paper, involving such diverse subjects as the design of the boat they used on the Nile and local marriage customs in Ethiopia. Examples of these records and descriptions are published by Murray in Bruce's *Travels* (1805) at relevant points in the travel account. In addition, of course, Balugani was making records and drawings of birds (see Ill. 10), fishes, snakes (see Ill. 11), mammals, and plants. One letter, the last of Balugani's to reach Europe, provides a glimpse of his scientific activities. Written to Filippo Balugani when they were traveling on the Red Sea and were occupied in making astronomical observations for the mapping of the ports, it contains a request for observations of the moons of Jupiter to be made by the professor of astronomy at Bologna, in order to make comparisons with observations made at the Red Sea ports.[33] This would ultimately allow calculation of the longitude of these places. In 1725 an astronomical observatory had been added to the main university building in Bologna, opposite the Accademia Clementina, and it is certain that Balugani would have been familiar with it. Apparently no use was made of the Bolognese astronomical observations, if indeed the request was followed up, and Murray's edition of the *Travels* (1805) contains in fact comparisons between observations made in the field and at Greenwich, rather than at Bologna, as the basis of the calculations of longitude. Nevertheless, the letter strongly confirms that Balugani was not only acting as draughtsman, but was active in all the different kinds of scientific observations conducted by Bruce. It is difficult to estimate how much initiative the latter allowed him, but it seems likely that the request to Bologna was Balugani's idea. Bruce would almost certainly have preferred astronomers in London or Paris to carry out the necessary observations.

But the greatest single piece of evidence of Balugani's diligence and capacity, and the one which is of particular interest to the purpose of this book, is the range of his natural history records, the drawings of about three hundred animals and plants and, in numerous instances, the extensive notes which are found with them, especially with the botanical drawings. They reveal how thoroughly Balugani had immersed himself in the task of describing and illustrating what were often new species, to an extent far beyond what might have been expected of him by Bruce. They show him working with an almost feverish haste yet with the method and order characteristic of the trained architect, even though the structures he was concerned with were organic, not man-made. Though he was clearly a fast worker he skimped nothing of importance. It is difficult to understand how, with all his other duties, he could have found time for all this work, considering the amount of detailed information he was able to convey.

The leisure to write letters must have all but disappeared as they traveled in Ethiopia.

But in addition to the entirely factual records of Balugani's journal there is one document which seems to be part of a draft letter, probably intended for Ranuzzi, which Bruce brought back to Scotland with the drawings and the journals. It is a detailed description of the journey to the source of the Blue Nile, on a loose sheet, giving not just factual details but something of Balugani's reactions to what was for Bruce the climactic experience of the whole adventure. After describing how they were robbed three days after setting out from Gondar on the first attempt (May 1770) to reach the source, he goes on,

> But patience! The journey to the fountains of the Nile, our principal object, is accomplished, and we can say, in the face of many sovereigns of antiquity, that we have seen what for so long they have desired to see, but always, for want of information, took those roads which led them from their purpose. Now that this is done, if it please God, we shall not delay long to return home, and the world shall have a true account of Ethiopia, with a map of these places which we have visited, and their positions ascertained by most accurate observation with large instruments, shewing what errors have been committed by those who have given maps of Ethiopia, and what nonsense and false assertions have been uttered concerning the manners, religion, government, and, in short, all that relates to the history of the country—to the most part of which, I, who am on the spot, can bear witness, that it has been absolutely falsified, or stated very far from the truth, whether through ignorance, I know not; but the fact is so clear as to be indisputable.[34]

This perfectly echoes Bruce's own opinions, which pour scorn on earlier accounts of Ethiopia written by Jesuits, and is another pointer to Balugani's subservience to the dominant personality of his master. Apart from the weather journals this writing is the last record we have from Balugani's hand.

A sad end came to all this activity when Balugani died at Gondar, sometime after 14 February 1771. From this date on, Bruce took over the weather journals because his assistant, through illness, had been forced to discontinue the daily entries. The precise date of his death is not known. Only Bruce could have known it but forgot it or for some reason chose to leave it vague, making a number of conflicting statements. The first announcement came in a letter written by Bruce to Count Casali from Marseilles on 12 July 1773. Written in rather muddled French, without punctuation, it contains the information that Balugani died of an incurable dysentery after they had arrived at Gondar (first reached in February 1770) and that Bruce had supervised his burial in the cemetery of St. Raphael, Gondar (probably the church Cuddus Rafael, founded by the Emperor Bakaffa [reigned 1721–30] but now demolished), where Roman Catholics had traditionally been buried. After this brief announcement Bruce immediately and tactlessly goes on to emphasize Balugani's shortcomings as an artist. Though he had paid him a good salary, believing him to be a capable draughtsman in every way, he found him so weak in perspective, though good at architecture and ornament, that he was obliged to add the landscapes himself. But the drawings remained "unfinished," without figures and skies.[35]

On 31 October 1773, Domenico Pio, who had succeeded Casali as secretary of the Accademia Clementina, inserted a notice of Balugani's death in the proceedings of the academy, the news of it having in the meantime been passed by Casali to Filippo Balugani and to the academy.[36] Pio had found the information about his death in the letter to Casali "confused," but with the arrival of Bruce in Bologna, probably sometime in August of that year, he had learned that Balugani, having taken part in the journey to the source of the Blue Nile (something which Bruce never otherwise seems to have admitted), had died on 15 May 1770. The year was clearly wrong since the famous journey had taken place toward the end of 1770.

The earliest modern writer on Balugani, E. Panzacchi, states that Bruce gave the date of death as 3 May 1770.[37] He must therefore have relied on a different local report of Bruce's information. And as he had access only to the first edition of the *Travels*, in translation, he was not aware of the error in the year which Murray comments on. Chiovenda accepts the day and the month as given by Panzacchi, believing that Bruce was more likely to have been right about these than about the year so that, in his opinion, Balugani died on 3 May 1771, in spite of Pio's notice.

Bruce's accounts of his assistant's death, as published in the first edition of the *Travels*, are vague and muddled. On page xii of volume I he writes a few lines about Balugani's ending: "He contracted an incurable distemper in Palestine and died after a long sickness, soon after I entered Ethiopia, after having suffered constant ill-health from the time he left Sidon." In volume 2, page 241, he writes,

> For my own part, I never was so dejected in my life. The troublesome prospect before me presented itself day and night. I more than twenty times resolved to return [home] by Tigré, to which I was more inclined by the loss of a young man who accompanied me through Barbary, and assisted me in the drawing of architecture which I made for the King there, part of which he was still advancing here, when a dysentery, which had attacked him in Arabia Felix, put an end to his life at Gondar. A considerable disturbance was apprehended upon burying him in a church-yard. Abba Salama [the high priest] used his utmost endeavours to raise the populace and take him out of his grave; but some exertions of the Ras [Michael] quieted both Abba Salama and the tumults.

Neither of these accounts gives a date of death and each gives a different version of where Balugani contracted his illness. The second adds more confusion, for Bruce says that the priest, Abba Salama, had tried to disinter Balugani's body when, according to him, that priest had already been executed in late December 1770, during the period of terror which made Bruce long to get away from Gondar and return home.[38] At that time Balugani was still alive and making entries in the weather journal and was to continue doing so for more than another six weeks.

Murray, as previously mentioned, noted Bruce's error about the date of Balugani's death, explaining it, not very convincingly, as "an anachronism in the account of his death...

into which Mr Bruce seems to have been led by inattention or forgetfulness." Murray himself gives the date as about the middle of February 1771, that is, almost immediately after the last entry in the weather journal. There is little likelihood now that any new evidence will emerge about the time and circumstances of Balugani's death.[39] However, there is no reason to give any credence to the extreme views current in some circles in Bologna that Bruce had actually caused his death.[40] To put Bruce's attitude toward his assistant at its lowest valuation, his help on the journey ahead was far too valuable for him to wish that any harm should befall him. On the other hand, there is every reason to accept Bruce's stated cause of Balugani's death, an incurable dysentery, an illness which was to bring Bruce himself very low within a short time.

All the drawings made by Balugani and Bruce were brought back to Europe by Bruce. They were seen by a number of people in Bologna. The biographical note on Balugani in the records of the Accademia Clementina noted that his architectural drawings showed buildings dating from the time of Hadrian to that of Septimus Severus, that there was also a book with miniatures of royalty (probably the drawings of Ethiopian royalty used for the miniature portraits in Bruce's *Travels*; see Ill. 12), two books of birds, others of wild animals and quadrupeds, another of fruits, and two of medicinal plants, all superbly colored with "Chinese" colors.[41] The note concludes with a statement about the loss that the academy had suffered from Balugani's death. The feeling of loss was sharpened and embittered by another passage in Bruce's letter to Casali from Marseilles, already mentioned. To guarantee the completion of the drawings, that is, to add figures, skies, and in some instances landscape backgrounds, he wrote that he had kept back a considerable sum of money (from Balugani's salary, presumably) for engaging and paying the artists who would be involved. The letter was received with astonishment and dismay by the academy. Here were drawings whose quality was plain to the academicians yet Bruce, at the very moment of announcing Balugani's death, had belittled his ability as an artist and intended to deprive him, and now his family, of the money that he owed.

We can assume that at the beginning of Bruce's stay in Bologna the feelings between him and Balugani's family were friendly. Filippo Balugani expressed a wish to meet Bruce in order to thank him for what he had done for his brother and to help him obtain artists to finish the drawings. But soon the situation became embittered, and the point about a possible legacy from Luigi, who had served Bruce for about six years at a salary of 35 Roman *scudi* a month (and thus had earned about 2,500 *scudi*, if Bruce had not withheld part of the sum), was raised by the family. We know from a Filippo letter to Bruce, which the former thought polite but the latter found most offensive, that Bruce maintained that he never had a written contract with Luigi and did not owe Filippo anything.[42] Bruce's attitude certainly caused bad feeling among Balugani's friends in Bologna, so that some even suspected that he had murdered him, not that the more responsible of them ever went along with this theory. But it is difficult to avoid noticing a certain coolness, even sarcasm, in the note, probably by the secretary of the Accademia Clementina, Domenico Pio, about academician Bruce: "Bruce, James, who from 1765 to 1772 went to Africa with the Bolognese Luigi Balugani, member of the Accademia Clementina, in the department of architecture, in order to draw antique monuments, and so to Abyssinia right to the springs of the Blue River (Blue Nile), which he took for the source of the Nile."[43]

From the evidence Bruce's treatment of Balugani plainly was unfair, not to say irrational and indefensible. Modern writers on Bruce have sought to explain his behavior in different ways. Beckingham gives one main reason:

> His desire to be acclaimed as the only European to have been at the source of the Nile is presumably responsible for his denigration of Balugani. There is no reason to credit the sinister gossip of Bologna, but the treatment of his talented companion in the *Travels* is unpardonable. Bruce nowhere mentions him by name [he does so in fact twice], he rarely mentions him at all, and when he praises him it is in the most patronizing terms. It is difficult to believe that the mistake about the date of his death was not deliberate.[44]

Reid has speculated on the relationship between Bruce and Balugani:

> Part of the truth may be that Balugani was an unassertive being, for whom Bruce always remained "il mio principale," while to Bruce the young Italian was never more than a useful hand: so that throughout all their wanderings they never became friends. Warmth of feelings, strength of character, originality (for good or bad) were the qualities which Bruce appreciated most easily: they were indeed his own.[45]

This judgment seems rather unfair to Balugani, who certainly had shown originality as an artist (his prize in the Parma contest) and some initiative during their travels (his suggestion of obtaining astronomical observations from Bologna). Moorehead, attempting to be fair to Bruce, fails to find much that is attractive in his personality: "It is strange that with all his obvious merits one does not like Bruce very much... Some vital ingredient was missing in his nature, perhaps it was humanity."[46]

There is truth in all these observations. Certainly the two men were complete opposites: Bruce was hot-tempered, proud, prejudiced, egocentric, romantic; he could embroider a fact until it verged on fantasy; but he was also brave, daring, strong-willed, and decisive, traits of character which, together with his considerable physical assets, helped him in many a dangerous situation. Balugani, on the other hand, seems to have been of a gentle, quiet, unobtrusive, even deferential disposition, but also conscientious, helpful, honest, and not without charm—as he was apparently much liked by his friends and fellow academicians. These differences meant that there was no firm basis for friendship between the two, though both possessed qualities which were essential for the successful outcome of the travels. The journey would never have been completed without Bruce's strong, almost ruthless leadership; on the other hand, the results of the travels would have been a good deal less impressive without Balugani's industry, attention to detail, and ability as a draughtsman.

Perhaps one other element in Bruce's makeup played a considerable part in his attempt to obscure Balugani's role in the production of the drawings: his admiration, never straightforwardly admitted, of Balugani's skill, which made him as patron highly possessive of the drawings and at the same time jealous of Balugani and envious of his ability. A man rarely praises the object of his jealousy. Rather he attempts to distract from his achievement or ignore him. Bruce could not totally ignore Balugani, though he tried to, and the few times he does mention him he attempts to belittle him. When disparagement is not altogether credible, the little praise he allows him is given only in the most patronizing terms. In the passage in the introduction to the first edition of the *Travels* where he mentions Balugani and actually gives his name, he continues, "He knew very little when first sent to me. In the twenty months which he stayed with me at Algiers, by assiduous application to proper subjects under my direction, he became a very considerable help to me." Balugani's ability has become almost a product of Bruce's critical acumen and tutelage.

Then there is the strange story of the fennec, the small, big-eared, foxlike animal which is illustrated in volume 5 of the first edition of the *Travels* (for the original drawing see Ill. 13). Bruce writes,

> I made several drawings of it [the fennec], particularly one in watercolours of its natural size... A young man, Balugani, of whom I have already spoken, then in my service, in which, indeed, he died, allowed himself so far to be surprised, as, unknown to me, to trace upon oiled paper a copy of this drawing in watercolours, just now mentioned. This he did so servilely, that it could not be mistaken, and was therefore, as often as it appeared known to be a copy by people the least qualified to judge in these matters... The creature itself passed, by very fair means, from my possession into Mr. Brander's [the Swedish consul in Algiers]... The drawing was not justly acquired, as it was obtained by a breach of faith, and seduction of a servant, which might have cost him his bread. It was conducted with a privacy seldom thought necessary to fair dealing; nor was it ever known to me, till the young man began to be dangerously sick at Tunis, when he declared it voluntarily to me, with a contrition that might have atoned for a much greater breach of duty.[47]

What are we to make of this story? At face value it reveals Balugani as someone not above petty deceit and disloyalty. If, on the other hand, he was the originator of the drawing from which the above-mentioned plate was made, as seems certain from its quality and the evidence of Andreas Sparrmann, mentioned below, the fault appears less serious, though an element of disloyalty remains; and Bruce himself becomes the deceiver. Bruce goes out of his way in his account of the fennec to heap insults on the Swedish academician Sparrmann and scorns his description of the animal (though it is a good deal more accurate than Bruce's). The reason for this is not only that Sparrmann had different opinions about the fennec's habitat and feeding habits—and Bruce did not take even implied criticism lightly—but that he had dared to let the world know that a "painter," who was not Bruce, had portrayed the animal in Algiers.[48] Brander, the Swedish consul, who came into possession of the fennec formerly owned by Bruce, is also accused of the "seduction of a servant," though Bruce here restrains his language, believing Brander had not published the drawing—which in fact he had, very accurately—with his account of 1777.[49]

Bruce's opinion of Balugani here is completely at variance with the picture we have formed of him as an honest, loyal, and conscientious assistant—and a better artist than Bruce. It is a by-product of his accusations against Sparrmann and Brander. But whether his story is true or not we may be puzzled why Bruce felt it necessary to go the considerable length of publishing it. Though pretending to describe a peccadillo he is clearly concerned to discredit Balugani. Perhaps the tale is most interesting for what it reveals about Bruce himself. It seems that just as he wished to be the only European to have stood at the source of the Nile, so he wished to be the sole originator of the drawings which came out of his travels. It is as if he saw them as an extension of his own image of the distant lands, which he explored against great odds and succeeded in penetrating solely as the result of his determined leadership. Balugani, as Reid says, was considered by Bruce as "never more than a useful hand", yet he helped to form that image which Bruce came to persuade himself, more and more after Balugani's death, was one which he alone had created.

Notes to Chapter II

1 Bruce (1790), vol. 1, p. xii; vol. 5, p. 129.
2 Playfair (1877), pp. 6–8.
3 The article appeared in a publication concerned with the natural sciences. See Chiovenda (1940).
4 Oretti MS, as quoted in Chiovenda, document no. 1, p. 454.
5 Ibid., p. 454.
6 See Catalogue (1979), pp. 270–71.
7 Chiovenda, document no. 8, p. 459.
8 Degli Azzi (1908), p. 427.
9 Cuppini (1974), p. 92.
10 Bertalà-Ferrara (1974), p. 1.
11 Chiovenda, document no. 2, p. 455.
12 Ibid., document no. 15, pp. 464–65.
13 See letter Lumisden to Bruce in Murray (1808), p. 179.
14 Ibid., p. 180.
15 Letter Ranuzzi to Bruce, Feb. 1763, Bruce archives no. 326.
16 Letter Luigi Balugani to Ranuzzi, 24 Apr. 1765, in Chiovenda, document no. 30, p. 477.
17 Letter Luigi Balugani to Casali, ibid., document no. 24, p. 471.
18 Letter Casali to Luigi Balugani, 3 Dec. 1765, Bruce archives no. 1236.
19 Chiovenda, document no. 25, p. 472.
20 Ibid., document no. 30, p. 477.
21 Ibid., document no. 31, p. 479.
22 Ibid., document no. 32, p. 480.
23 Ibid., document no. 34, pp. 481–82.
24 Ibid., document no. 35, pp. 482–83.
25 Letter 23 Feb. 1768, Bruce archives no. 1248.
26 Letter 18 Feb. 1768, Bruce archives no. 1244.
27 Letter 1 March 1766, Bruce archives no. 1239.
28 Letter Chiesa to Luigi Balugani, 2 Apr. 1765, Bruce archives no. 1235; letter Filippo Balugani to Luigi, 23 Feb. 1768, no. 1268.
29 Memo, Bruce archives no. 1261.
30 Murray (1808), pp. 291–93.
31 Ibid., p. 293.
32 Ibid., p. 293.
33 Chiovenda, document no. 27, pp. 273–74.
34 Murray, pp. 230–32.
35 Chiovenda, document no. 28, pp. 474–75.
36 Ibid., document no. 13, pp. 462–63.
37 Panzacchi (1897), p. 297.
38 See Murray, p. 95.
39 Since this was written evidence has come to light from Bruce's daybook in which he notes, on 3 March 1771, that when he died, Balugani was "in custody" of a purse and its contents belonging to Bruce. This makes it clear beyond doubt that Balugani was dead by 3 March. We are grateful to Elizabeth Fairman of the Yale Center for

British Art for discovering this and for communicating it to us.

40 Panzacchi, p. 298; Chiovenda, document no. 2, p. 455.

41 Chiovenda, document no. 2, p. 456.

42 Ibid., document no. 43, p. 490.

43 Ibid., document no. 12, p. 462.

44 Beckingham (1964), p. 17.

45 Reid (1968), p. 43.

46 Moorehead (1962), p. 22.

47 Bruce (1790), vol. 5, pp. 128–38, pl. at p. 128.

48 Sparrmann (1783), chap. 14.

49 Brander, writing after he was ennobled; see Skjoldebrand (1777).

Ill. 9. Luigi Balugani. *East front with sections of the Palazzo Ranuzzi.* Etching and engraving by Luigi Balugani after his own design, *c.* 1760 (see p. 42). Bologna: Gabinetto Disegni e Stampe (by courtesy of Soprintendenza per i Beni Artistici e Storici).

Ill. 10. Luigi Balugani. *Ground Hornbill, in Amharic Abba Gumba* (*Bucorvus leadbeateri*). Watercolors over pencil: 313 × 246 (see p. 47). Yale Center for British Art, B1977.14.8922.

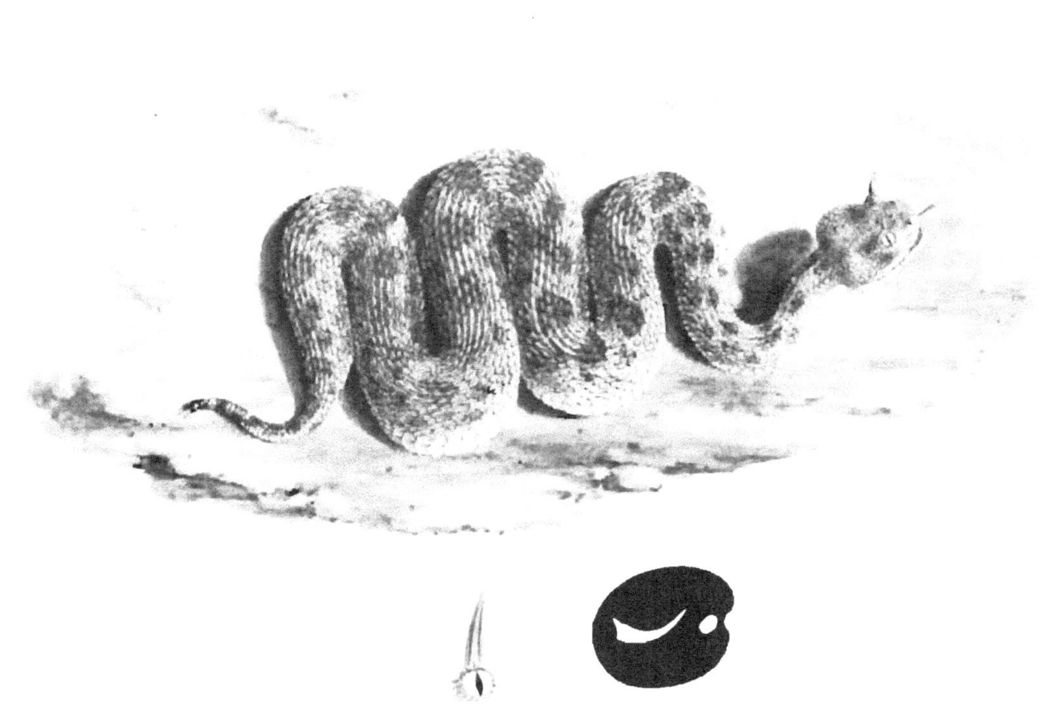

Ill. 11. Luigi Balugani. *Horned Viper* (*Cerastes cerastes*). Watercolors over pencil; 301 × 248 (see p. 47). Yale Center for British Art, B1977.14.8926.

Ill. 12. James Heath after Luigi Balugani. *Portraits of Ethiopian royalty and nobility*. Stipple engraving; sheet 279 × 230 (see p. 50). Yale Center for British Art, Ef bb/768 Bb.

Ill. 13. Luigi Balugani. *The Fennec (Fennecus verdus)*. Watercolors over pencil; 247 · 315 (see p. 52). Yale Center for British Art, B1977.14.8701.

· III ·

The Authorship *and* Quality *of the Plant Drawings*

It is clear from numerous statements in the first edition of the *Travels* (1790) that Bruce claimed the drawings of plants as his own work. Early in the introductory section of volume 5, where a small selection of this material is described and engraved, he makes a point of stressing the value of drawings of plants in comparison with collected specimens: "One drawing of this kind, painfully and attentively made, has more merit and promotes true knowledge more certainly than a hundred horti sicci."[1] He later emphasizes the care he took to see that nothing should be allowed to divert him from presenting the most accurate possible representation of the subject. In his description of the Ergett Dimmo (*Dichrostachys cinerea*, p. 86) he writes, "none of the parts, however trifling or small, being neglected in the representation, and none of them supposed or placed there out of order, for ornament, or any other cause what ever; a rule which I would have the reader be persuaded is invariably observed in this collection, whether tree or plant, beast, bird, or fish."[2] All this could of course have referred to his supervision of Balugani's work, but there is no mention of his assistant having any part in the drawings. On the contrary, in the descriptions of plants Bruce frequently goes out of his way to emphasize the trouble he personally took over the accuracy of the illustrations. In one instance, in writing of the Farek (*Bauhinia farek*, p. 85), he says, "I do confess it [botany] never was my study, and I believe from this the science has reaped so much the more benefit. I have represented to the eye, with the utmost attention, by the best drawings in natural history ever yet published, and to the understanding in plain English, what I have seen as it appeared to me on the spot, without tacking to it imaginary parts of my own."[3] His own favorite illustration is the Walkuffa (*Dombeya torrida*, p. 103): "the inimitable beauty of the subject itself has induced me to bestow much more pains upon it than any other subject I have published, and according to my judgement, is the best executed in this collection."[4] One other example is sufficient to demonstrate beyond doubt Bruce's claim to authorship, and it had a particular importance for him. This is the Wooginoos (*Brucea antidysenterica*, p. 100), the plant which cured him of his dysentery; for discovering and recording it Sir Joseph Banks honored him by naming it after him. Bruce writes, "The present figure is from a drawing of my own on the spot at Ras el Feel."[5]

Contemporaries as well as later commentators have questioned Bruce's claim. Daines Barrington (better known as the correspondent of Gilbert White of Selborne), in a letter to Bruce dated 9 July 1792, wrote, "The European Magazine of last month contains more letters from that worthless fellow Montague; from one of which is clear that he first propagated that all your drawings are those of Luigi."[6] Now Edward Wortley Montague was notorious for reporting anything but the plain truth. Yet in this instance, residing in Venice in 1775, the date of the letter in question, he was sufficiently close in time and place to catch reports coming out of Bologna about the drawings brought back there by Bruce on his way home in 1773. He wrote, "There is a report prevails in Italy that Mr.——'s drawings were not done by himself but by the young man he took from Bologna with him, and who died there; and it is universally believed as all the Connoisseurs (who are well acquainted with him) assert they know his hand."[7] We know that the drawings aroused the admiration of Oretti and other academicians in Bologna who saw them and who considered them to be largely Balugani's work.

Bruce could certainly draw, but how far did his ability as a draughtsman take him? R.L. Playfair, his nineteenth-century successor as British consul in Algiers and a great admirer, as the title of his book *Travels in the footsteps of Bruce in Algeria and Tunis* (1877) suggests, produces evidence from the Bruce archives which throws some light on the question. He sets out the evidence in answer to the "grave doubts" so often expressed over Bruce's share in the making of the drawings. Here he confines himself to the architectural drawings made in Algeria and Tunis. His extracts make clear that while Bruce could draw he recognized his own limitations: "I had all my life applied [myself] unweariedly," Bruce writes, "perhaps with more love than talent, to drawing, the practice of mathematics, and especially the part necessary to astronomy."[8] Bruce explains the difficulties he found in drawing architectural remains without the help of an assistant, which his experience at Paestum made plain. Looking ahead to his travels in North Africa and his intention of recording its classical remains, he began the search for an assistant. "A.M. Chalgrin, a Frenchman [a connection of the architect Jean-Francois-Thérèse Chalgrin (1739–1816)], engaged himself, was terrified and then drew back. All the assistance I could get was a young man, a Bolognese, called Luigi, surnamed Balugani, which signifies short-sighted. This was very feeble help; but being of good disposition, in twenty-two months which he stayed with me in Algiers, by close application and direction, he had greatly improved himself in what I chiefly wished him to apply to, foliage and ornaments in sculpture."[9] This faint praise is the best he ever had to say of Balugani in his extremely rare references to him. And "Mr. Robert Strange, now Sir Robert Strange, knows well I have been an indifferent draughtsman in ruined architecture near these forty years, for about that time he himself recommended me my second drawing master, poor Bonneau [Jacob Bonneau, English landscapist (died 1786)]... Till then I had only been used to drawing military architecture; and with ruler and compass I have ever since mostly drawn."[10]

Such were Bruce's self-confessed limitations as a draughtsman. He took pains to improve himself, according to Murray, taking drawing lessons in Florence before proceeding to his appointment in Algiers. And he evidently mastered the use of the camera obscura. From his admitted shortcomings in "foliage and ornaments" and his use of the ruler and compass it is clear that while he was capable of producing adequate drawings of classical ruins he was hardly qualified to make the skillful drawings of plants we are considering here.

The drawings themselves provide plain evidence that the many preliminary studies are the work of one man. The manuscript notes in Italian with which they are inscribed, sometimes copiously, are, almost without exception, in Balugani's hand. The interconnection of notes and drawings, in design as well as subject matter, makes it certain beyond doubt that Balugani was responsible for all of this work, which is consistent in style and of high quality. But what of the finished watercolors? Here the same characteristics apply: consistency of style and draughtsmanship of the same high order. (We are not concerned here with the watercolors from the "Paris folder," which are by a number of different, presumably French, hands.)

It will be seen from the catalogue that nearly all the watercolors are taken directly from one or other of the preparatory drawings, or are very closely connected with them. They would have had to have been colored on the spot to assure accuracy of coloring. Detailed color notes made at the time with the brush might have been a reasonably accurate guide, but they do not exist. Though comments on the colors frequently appear in the written notes these are not precise enough to serve for the illustrations. The watercolors must have been painted *in situ*. Bruce speaks of taking with him "a prodigious quantity of pencils [that is, brushes], India ink, and colours."[11] An interesting passage in the *Travels* relates the ordeal by heat and thirst of his journey through the Nubian Desert on his return to Egypt, how he and his companions only just succeeded in coming out alive, having had in their extremity to abandon his drawings, journals, and all his precious equipment in the sand. These were afterwards retrieved intact, to his "unspeakable satisfaction." He describes the astonishment of the aga, the Turkish governor at Aswan, when shown the drawings, and how he held them to his nose as if to smell them.[12] They must have been colored to be so lifelike.

It might perhaps be argued that Bruce was capable of making finished watercolors from Balugani's drawings. Many of them have their local names, occasionally with the Latin, inscribed by Bruce, on the front or back of the sheet. But these inscriptions do not integrate with the design as do Balugani's inscriptions and merely indicate Bruce's concern to label them. And no less ability is required to paint a finished watercolor from a preliminary study than to make the original study, even if the technique is of a somewhat different order. A quick eye and free hand are necessary for the first sketches, careful craftsmanship involving the difficult medium of watercolor for the finished drawing. Yet the feeling for line, the essential ability to draw, are a common factor, the lack of which would immediately be apparent. Bruce was hardly equipped to make either a really convincing first study of a plant or a finished watercolor. In any event it would not have been in character for Bruce to have made carefully finished watercolors from his assistant's drawings, reproducing as they do every mannerism of his line, the artist then adding the necessary detail and color with the brush. That is not to say that Bruce did not on occasion make sketches of plants or paint a finished watercolor. Some examples, a drawing of *Brucea antidysenterica* (p. 100) among them, show uncertain pencil outlines of leaves, lacking in precision, which have been corrected in pen by a bolder, surer hand that we can recognize as Balugani's. The feebler outlines may be by Bruce. As for watercolors, one of a fish signed with Bruce's monogram, is in the collection of Lord Elgin at Broomhall and must be accepted as a certain watercolor by Bruce (see Ill. 14). While the drawing is competent, the way the tailfin and body scales are treated is somewhat stylized and too precisely linear to make the fish appear altogether lifelike. Were such a technique to be applied to the drawing of plants the result would be obviously much less convincing. A further probable pointer to Balugani's authorship of the watercolors is the appearance sometimes of a faint scale in pencil at the bottom of the sheet, not divided into inches but presumably *pollici*, the thumb breadth measurement then used in Italy and so often referred to in the written notes of the preliminary drawings.

What then is one to make of Bruce's claims? We must distinguish between his published comments and his private communications. The former assert apparently his sole authorship of the plant drawings, at least. The latter, with friends and superiors, acknowledge his own artistic deficiencies and, for the architectural drawings, the help of other artists. He wrote that when he showed the king and queen the drawings of classical ruins that he presented to George III, "The Queen remained with me at the drawings, and I was a good deal surprised at her asking if I had not had help? I answered, Undoubtedly, every help I could get to make them worthy of the King."[13] The remark is substantiated, as we know, by his taking so much trouble to find artists in Italy to "finish" them. He was never pressed on the botanical drawings. Balugani had died long before they were shown to discerning Europeans who, apart from a few Italians in Bologna, were in no position to dispute his authorship.

Bruce in his own edition of the *Travels* (1790) says specifically that the figure of *Brucea antidysenterica* (see Ill. 15) was "from a drawing of my own on the spot, at Ras el Feel," where he stayed on his return journey to Egypt early in 1772. This cannot have been so. The engraving was made from a watercolor itself taken from a drawing belonging to a group made by Balugani at Sacala, where the plant was discovered on "5 November" (1770), as noted by him on one of them (fig. 247). A drawing (fig. 248) similar to but sketchier than the model for the watercolor has inscribed on it by Balugani, "Woginus arbusto Velenoso" (Woginus a poisonous shrub) and by Bruce, "Roots dried & reduced to powder good against the bloody flux" (that is, as a medicine for dysentery). The engraved figure is therefore certainly derived from Balugani, not from any drawing which Bruce may have made at Ras el Feel. In fact, none of the engravings in the *Travels* can be traced to drawings made after Balugani's death, which occurred sometime after mid-February 1771, for none such are known to exist. Each, on the other hand, can be traced to a known watercolor derived from a pencil drawing by Balugani. We are therefore forced to conclude that Bruce deliberately misled his readers to gain the reputation of having created the watercolors which he clearly so much admired. Perhaps the same deep-seated motive was at work which caused him to present himself as the discoverer of the source of the Blue Nile, which he must have known, whatever his protestations to the contrary, was found by Portuguese Jesuits long before his time. The plant drawings needed no argument in support of his authorship, for he had virtually eliminated Balugani, long dead, from any part in their production so far as the public was concerned.

The drawings show that Balugani was not only skilled and experienced in drawing plants but was clearly preoccupied in recording as many new species as time would allow. He probably spent more time in making drawings and watercolors of plants than on all the other records he was required to make. He must have done so willingly, judging from the care he lavished on them and the detail into which he felt it necessary to go both in drawing and in his manuscript notes. They are not the work of a gifted architectural draughtsman turned botanical recorder, as required by Bruce, but of someone already practiced in the art. More than just a keen interest in the superficial appearance of plants, they show a grasp of their structure and form. His intention was to make a faithful record of the habit of the plant, the scale and dimensions of its parts, the form of its leaves and that of its flowers and fruit where possible, supported in numerous instances by detailed drawings of dissected parts. This was accomplished by quick and often rudimentary sketches, from which some idea of the character of the plant could be gained, followed by a detailed study of the whole plant or of a characteristic shoot, again usually in pencil. The latter was often indented in order to transfer the outlines to another sheet as the basis for the finished watercolor. Where possible, the various parts of the plant were drawn life-size. Very often written notes were added giving details of where and when the plant was found, its season of flowering, the size and nature of its parts, its coloring and sometimes the uses made of it by the local inhabitants. Balugani was nothing if not a most conscientious and intelligent recorder, taking trouble not to neglect any detail of significance.

We do not know how Balugani acquired this interest and knowledge of plants and his skill in drawing them. It is unlikely that plant drawing was part of his curriculum at the academy in Bologna, but the city had one of the earliest botanic gardens in Renaissance Italy, founded in 1567, as well as a museum of natural history founded by Aldrovandi. Some of the watercolors of one of the greatest of sixteenth-century plant artists, Jacopo Ligozzi (1547–1626), are in the university library, and Balugani may have been familiar with them. The scientific study of plants and animals was at this early period further advanced in northern Italy then elsewhere in Europe. Balugani's drawings of plants were very much in this scientific tradition; form, structure, and habit receiving at least as much attention as the more superficial aspects of texture and colour. He does not set out simply to emphasize the attractive qualities of the plant or to place it on the sheet in the most elegant way as a portrait of the flower beautiful in the manner of his great contemporary Georg Ehret. That was not required of him by Bruce. On the other hand, he attempts to give the fullest idea of the character of the plant as he found it, precisely in accordance with Bruce's principles, and succeeds remarkably well.

Looking more closely into his methods, we see his adeptness at suggesting by the simplest pencil strokes the often subtle shapes of the rounded fruits. Equally he can give a convincing idea of the irregular surfaces of leaves and the way a branch, leaf, or bract grows from the stem. Characteristic examples of these features are *Minusops kummel* (p. 99) and *Ensete ventricosum* (p. 112). His ability rests on a quick, discerning eye, a grasp of the problem of illustrating a complex organic form, and, perhaps most important, a rare capacity to draw with pencil and brush. Add to these the ability to apply opaque and clear washes of color sensitively and accurately, a trait not always possessed by some of the more outstanding watercolorists of his time when they painted plants.

Balugani was a botanical draughtsman in the strict meaning of the term rather than a painter of flowers. Like Sydney Parkinson, who as a young man sailed with Cook and Banks in the *Endeavour* on the first voyage to the South Pacific (1768–71), he was working against time. But while Parkinson made few, if any, finished drawings, Balugani somehow found the time to paint the finished watercolors. Both young artists died during their missions, Parkinson in 1771 in the Indian Ocean, Balugani in 1771 in Gondar. F.P. Nodder, who finished many of Parkinson's plant drawings, curiously, painted a watercolor of *Brucea antidysenterica* in our collection (fig. 30)—independently of course of Balugani—which can be compared with the latter's watercolor of the same plant (p. 101).

The scientific quality of Balugani's work has been emphasized (pp. 58–68) and the high quality of his draughtsmanship and faithfulness of his coloring. The reproductions of his drawings and watercolors in this volume give the fullest indication of the range of his artistic ability. He was one of those artists of exploration whose talent for drawing developed and flowered in the most hostile conditions. His artistic personality is clearly visible in his drawings, perhaps less clearly in his watercolors. His reputation as an architectural draughtsman is as yet confined to a few specialists. As a plant artist he deserves at least the kind of reputation which some of Captain Cook's artists have earned, not indeed so great as that of the outstanding plant illustrator Ferdinand Bauer (1760–1826), but at least the equal of Parkinson's, rather greater than John Webber's and well above that of Johann Georg Forster, all artists who sailed with Cook and drew plants. Though Bruce succeeded in negating Balugani's talent as a natural history artist, ironically, his careful preservation of the drawings makes it possible to judge Balugani as a botanical illustrator. They show him as an artist of considerable talent who merits a place among the more outstanding botanical artists of the later eighteenth century.

Notes to Chapter III

1 Bruce (1790), vol. 5, p. vi.
2 Ibid., p. 35.
3 Ibid., p. 60.
4 Ibid., p. 68.
5 Ibid., p. 72.
6 Murray (1808), p. 286.
7 *European Magazine*, June 1792, p. 420.
8 Playfair (1877), p. 3
9 Ibid.
10 Ibid., p. 5.
11 Ibid., p. 4.
12 Bruce (1805), vol. 6, p. 518.
13 Playfair, p. 5.

Ill. 14. James Bruce. *Wrasse* (*Cheilinus lunulatus*). Watercolors and bodycolors; inscribed by the artist "J B pint."; 407 × 305 (see p. 57). Lord Elgin's collection.

Ill. 15. James Heath after Luigi Balugani. *Wooginoos in Amharic* (*Bruce antidysenterica*). Line engraving and etching; sheet 357 × 270 (see p. 58). Yale Center for British Art, B1977.14.11579.

· IV ·

The Scientific Value *of* Balugani's *Plant Illustrations and Notes*

The middle of the eighteenth century was a period of great activity in the study of plants and animals. The Swedish naturalist Carl Linnaeus was publishing a vast number of new names for living organisms, while his students at Uppsala University became so enthused by his lectures that they set out for distant lands in order to collect more specimens. In fact, at that time remarkably few organisms were known from tropical regions, and especially neglected was tropical Africa. Nothing was known about those in Abyssinia (present-day Ethiopia) so the return to Europe of James Bruce in the 1770s with living seeds and roots as well as numerous drawings and descriptions of other plants was the cause of great interest among naturalists, especially in those countries that had benefitted from his expedition, namely, France, Italy, and Great Britain. In each of these countries some plants were raised and subsequently illustrated by artists on the spot, but these were few in comparison with the mass of artwork produced in Abyssinia and taken back to Kinnaird House in Scotland. It is a pity that only a representative sample of twenty-four engravings eventually appeared in Bruce's *Travels* of 1790, to which were added some more in Murray's second and third editions of 1805 and 1813. Altogether these amount to 28 species, whereas we know now that some 160 plant species were recorded by Bruce and Balugani, most of them being unknown to botanists of the day. Several species were named and described from living specimens grown at Paris, Florence, Kew, Chelsea, and possibly Edinburgh, and the published plates were subsequently given acceptable Latin names, as we shall see below. Bruce also published illustrations of various mammals, birds, fish, and other animals—many remain unpublished—but it is for zoologists to unravel their scientific identities.

Plants new to science grown from Bruce's seeds and roots

There is a letter at the Musée national d'histoire naturelle, Paris, listing the bulbs, seeds, and roots that were presented by Bruce to André Thouin, one of the gardeners of Louis XV at the Petit Trianon, Versailles. The king himself was in terminal illness at the time, otherwise they might have gone directly to him. Bruce insisted that the drawings of plants raised in Paris should be sent to him in Britain, which indeed they were and they are still kept together in a "Paris folder," now at Yale (B 1977.14.8882–8902). Unfortunately, the packets and contents of seeds had been muddled by Bruce's Greek servant Michael during the voyage from Alexandria to Marseilles, hence the resulting plants did not always correspond to the notes. Bruce was chided for this muddle by Professor de Jussieu, who thought Bruce had made these mistakes, such as calling a *Verbascum* an "aromatic herb." In return Bruce lampooned de Jussieu in his *Travels* (1790), vol. 5, p. 60: "A large species of Mullein likewise, or as he pleases to term it Bouillon Blanc, he has named Verbascum abyssinicum; and this the unfortunate Mr Bruce, it seems, has called an aromatic herb growing upon the high mountains. I do really believe that Mr. de Jussieu is more conversant with Bouillon Blancs than I am; my Bouillons are of another colour; it must be the love of French cookery, not English taste, that would send a man to the high mountains for aromatic herbs to put in his Bouillon, if the Verbascum can have been really one of these".

Several of these plants grown in France were figured and described as species new to science by Maria Theresa's botanist, Professor N.J. Jacquin, who was publishing in Vienna sumptuously illustrated volumes with descriptions of rare and interesting plants. In the first volume of *Collecteana* (1787) he figured the bulbous plant *Albuca abyssinica*, which had recently been described by both Murray and Dryander under the same name. In the third volume of *Hortus Vindobensis* (1776) Jacquin illustrated a sage he named *Salvia nilotica*, and his name has priority over that of Linnaeus the Younger, who five years later (1781) described the same plant under the name of *Salvia abyssinica*.

After leaving Paris, Bruce went to Italy, where he gave some seeds to the botanical garden in Florence, (the Banks letters (1958), p. 177). Plants of the famous cereal that was known to the Ethiopians as teff were raised there, and a dissertation on the subject was published by A. Zuccagni in 1775. Although this was in Florence, it is strange that Jacquin in Vienna should have had a water-colour made of this grass and preserved a specimen, too, unless he also grew it in Austria or had collected a specimen during his visit to France. Seeds of it must also have been grown in Scotland, since there is a watercolor signed by A. Fyfe dated 1775 at the Edinburgh "Botanick Garden." The naming of this grass has been confusing to say the least. Zuccagni first called it *Poa tef*, but Bruce unfortunately published an engraving of an entirely different grass (a *Panicum*), while in fact teff really belongs to the genus *Eragrostis*.

In Florence Bruce also gave seeds to Mr. Giovanni Mariti, a common acquaintance of Bruce and Balugani with whom Balugani had corresponded during the travels. Mariti passed the seeds to the Panciatichi garden, a well kept garden at Fiesole, just outside Florence. The gardener there, G. Piccivoli, reports in an account of the garden (Hortus Panciaticus, 1783) that he had grown the following plants from Bruce's seeds (although the mixing of names on the seed packets is also in evidence): *Bauhinia variegata*, a legume of unknown identity, marked "Ambovai", which is the name of a species of Solanum. *Rhamnus spina-christi*, probably *Ziziphus spina-christi*, from a parcel marked "Guggiola", of unknown significance. *Cordia myxa*, probably *Cordia abyssinica*. A plant referred to as "Govania", *Rumex abyssinicus*, which was identified with the illustration published by Jacquin. *Solanum niveum*, probably the *Solanum marginatum* mentioned below; it is interesting here to note that Vitman, who described the same species under the name *Solanum abyssinicum*, is mentioned among the botanists who studied in the Panciatichi garden. But perhaps the most interesting plant from this garden is the *Panciatica purpurea* which is described as new and illustrated on a full size plate in the publication. This plant is now known as *Cadia purpurea* (Picc.) Ait. Aiton recognized the plant as being identical with Forsskål's *Cadia* when he raised the same species from Bruce's seed at Kew. The species is not represented among the drawings in the Bruce collection.

Linnaeus the Younger described two more of Bruce's Ethiopian plants grown at Versailles. Although he does not actually mention Bruce's name, there is no doubt that they were grown from his seeds as nobody else had brought any from "Abyssinia" at that period. One was an economically important daisy (*Guizotia abyssinica*) that yields edible oil from its seeds, known in Amharic as *nuk*. The other was the nightshade *Solanum marginatum*, a handsome shrub with gray-felted leaves which soon came into cultivation as an ornamental and has found its way around the world.

An extraordinary story is attached to a leguminous tree called Kuara (and later named *Erythrina brucei* by Schweinfurth), whose leaves, flowers, and fruit were supposedly illustrated in Bruce's *Travels*. Seeds were grown at the Petit Trianon but the herbarium specimens prepared from these plants were sent by Thouin to Geneva, where the famous French botanist Alphonse De Candolle was writing a monumental botanical work, *Prodromus systema*. Noticing that the leaves were different from those illustrated, De Candolle named the species *E. abyssinica*. It was not for 137 years that it was realized by the Kew botanist J.B. Gillett that Balugani's illustration reproduced by Bruce consisted of a mixture of two species of *Erythrina*: only the fruits belonged to *E. abyssinica* while the flowers and leaves were those of *E. brucei*.

Another leguminous woody plant known by the Amharic name of *Farek* and later described as *Bauhinia farec* is still a scientific mystery. Although there is an excellent description and an illustration in the Paris folder, presumably prepared from a living plant grown in France, it has never been re-found in the wild state. In 1982 the Swedish botanist Mats Thulin made an expedition to Geesh, where Bruce found Farek, but there was no sign of it. The woody vegetation of that area is so depleted that this shrub must have been exterminated too, if it really is an Ethiopian species.

Bruce's seeds of *Phytolacca dodecandra* were also sown in Versailles and the plants were described and so named by L'Héritier in 1791, but it is surprising that no herbarium specimen can be traced in the Paris collections. The illustration must therefore be regarded as the "type"; that is a technical term for the standard reference (usually a specimen) on which the name is based. However, an excellent dried specimen of the gourd *Coccinia abyssinica* exists in Lamarck's Paris herbarium. Lamarck named this for plants grown from Bruce's seeds and painted by an unknown Parisian artist: the painting was later sent to Bruce in accordance with his conditions.

In England most of the plants were grown at the King's garden at Kew, which at that time was the personal property of George III with Sir Joseph Banks directing policy and William Aiton in charge of the cultivation. It was to Aiton that Bruce had sent seeds in October 1773 through the British consul at Leghorn in Italy. Some of them were probably grown nearby at Chelsea Physic Garden, where Phillip Miller was in charge.

The new genus named by Banks in honour of James Bruce—*Brucea antidysenterica*—was actually first published by John Miller in 1779, although F.P. Nodder painted it in 1777, and the Kew-grown specimen is in Banks's herbarium at the British Museum (Nat. Hist.). The epithet recalls the time when bark of this plant, which is called *Wooginoos* by the Ethiopians, cured Bruce of dysentery while he was at Hor-Cacamoot: ominously meaning the Valley of the Shadow of Death. Today there is international research taking place in the medicinal properties of this small tree and the bark is in great demand for analysis of the active chemical principle.

The only other seeds known to have been grown at Kew was a new species of a leguminous herb which was later named *Crotalaria pallida* by Aiton. A specimen is housed in the British Museum (Nat. Hist.).

As far as can be ascertained the above were the only plant species to be described from living specimens raised in Europe, all the others being represented as drawings—Table I summarises the nomenclature. (For a fuller botanical account see F. Nigel Hepper, "Taxonomic analysis of James Bruce's and Luigi Balugani's Ethiopian travels" *Monographs in Systematic Botany from the Missouri Botanical Garden* (1988), vol. 25: pp. 577-80.

Table I

Vol. 5 (1790) Opposite	*Bruce's name published in Travels*	*Former scientific name based on Bruce's illustration*	*Correct name today* (* = based on Bruce's material)
p. 16, 17	Balessan	—	*Commiphora gileadensis* (Linnaeus) C. Christensen
p. 28, 29	Sassa	*Sassa gummifera* J.F. Gmelin	* *Albizia gummifera* (J.F. Gmelin) C.A. Smith
p. 34	Ergett dimmo	*Mimosa sanguinea* J.F. Gmelin	*Dichrostachys cinerea* (Linnaeus) Wight & Arnott
p. 35	Ergett el krone	*Mimosa cornuta* J.F. Gmelin	*Mimosa pigra* Linnaeus
p. 36, 37	Ensete	*Musa ensete* J.F. Gmelin	*Ensete ventricosum* (Welw.) Cheesman
p. 42, 43	Kol-quall	*Euphorbia abyssinica* J.F. Gmelin	* same
p. 44	Rack	*Racka torrida* J.F. Gmelin	*Avicennia marina* (Forsskal) Vierhapper
p. 47	Geshe el Aube	*Andropogon afer* J.F. Gmelin	* *Ischaemum afrum* (J.F. Gmelin) Dandy
p. 49	Kantuffa	*Cantuffa exosa* J.F. Gmelin *Acacia cantuffa* Poiret	*Pterolobium stellatum* (Forsskal) Brenan
p. 52, 53	Gaguedi	*Protea gaguedi* J.F. Gmelin	* same
p. 54	Wanzey	—	* *Cordia abyssinica* R. Brown
p. 57	Farek, Bauhinia acuminata	—	* *Bauhinia farec* Desveaux
p. 65	Kuara	—	* *Erythrina abyssinica* De Candolle (fruits) * *Erythrina brucei* Schweinfurth (flowers)
p. 67	Walkuffa	*Walcuffa torrida* J.F. Gmelin	* *Dombeya torrida* (J.F. Gmelin) Bamps
p. 69	Wooginoos, Brucea antidysenterica	*Brucea antidysenterica* J. Miller	* same
p. 74, 75	Cusso, Banksia abyssinica	*Hagenia abyssinica* (Bruce) J.F. Gmelin	* same

Vol. 5 (1790) Opposite	*Bruce's name published in Travels*	*Former scientific name based on Bruce's illustration*	*Correct name today* (* = based on Bruce's material)
p. 76	Teff	*Poa tef* Zuccagni	* *Eragrostis tef* (Zuccagni) Trotter
Vol. 7	(1805, 1813)		
No. 44	Cassia	—	*Cassia fistula* Linnaeus
No. 45	Toberne montana	—	*Saba comorensis* (Bojer) Pichon
No. 46, 47	Krihaha No. 1, No. 2	—	*Arundinaria alpina* K. Schumann
No. 48	Anguah No. 1	—	*Boswellia papyrifera* (Delile) Hochstetter
No. 49	Anguah No. 2	—	*Sterculia cf. africana* (Louriero) Fiori
No. 50	Geshe	—	*Rhamnus prinoides* L'Héritier
No. 51	Merjombey	—	*Solanum adoense* Hochstetter ex A. Richard
No. 52	Nuk, Polymnia frondosa	*Polymnia abyssinica* Linnaeus fil.	* *Guizotia abyssinica* (Linnaeus fil.) Cassini
No. 53	Umfar or Lentana	—	*Buddleia polystachya* Fresenius
No. 54	Kummel	—	*Mimusops kummel* De Candolle

The quality of the botanical observations in Balugani's plant drawings and notes

The scientific description of the appearance of a plant is part of the branch of botany called plant morphology. Not a very advanced science at the middle of the eighteenth century, morphology was brought considerably forward by the Swedish botanist Carl Linnaeus and his pupils. The structure and function of many plant organs were only incompletely understood, but the understanding of the flower was making particularly rapid progress after Linnaeus had centered the botanists' interest on this part of the plant by using the number and arrangement of the various floral parts, especially the stamens and styles, in his classification.

How do Balugani's careful drawings and notes compare with the observations of contemporary botanists? Balugani's drawings in pencil or pen and ink are mostly full of detail and supplemented by dissections and enlargements of particular parts, such as flowers, fruits, stamens, ovaries, and seeds. From the notes and the designations of the dissections and enlargements one gets some idea of Balugani's botanical terminology and his ability as an observer of plant morphology.

The drawings themselves show a very competent morphological observer. Only rarely can one establish that Balugani has made wrong observations, and when he did make mistakes it is mostly due to the fact that he has combined leaves, flowers, and fruits from different plant individuals on the same drawing. In one case this has led to the confusing combination of the flowers of *Erythrina brucei* with the fruits of *Erythrina abyssinica* (p. 87), but that is a unique mistake in the entire collection. In other similar cases, for example, when he combined flowers, fruits, and leaves of various individuals of *Boswellia papyrifera* in one drawing (p. 75), no such confusion of species has occurred, and the result is a drawing which shows only structures that were not visible at the same time on the same individual.

The morphological terminology used by Balugani in his notes is a mixture of Italian words and scientific terms, as is indeed the morphological terminology in most languages today. Trees, even small ones, or what we would call shrubs, Balugani generally termed *albero* (tree). Small shrubs, such as *Myrsine africana* (p. 92) and *Osyris lanceolata*, are called *arbusto* (shrub). Herbaceous plants are generally called just *pianta* (plant), apart from grasses and sedges, which are called *erba* (herb, grass). The Mountain Bamboo, *Arundinaria alpina*, is called *canna* (reed). A woody parasite, *Loranthus globiferus*, is called *aborto* (malformation). Balugani also uses a reasonably detailed terminology for the various parts and organs of the plants. Roots are referred to as *radice* (root), but the same term is used to refer to specialized organs, such as bulbs or corms. Trunks of trees are called *tronco* (trunk), bark *scorza* (bark, skin); stems and branches are referred to as *ramo*, and a shoot is termed *tiglia* (fiber, thin branch). A leaf is called *foglia* (leaf), but sometimes the term *fronda* (leaf, usually large ones, now mostly used about the leaves of ferns) is used instead. When the leaves are provided with a petiole, this is again most often called *tiglia* or sometimes *fibra* (fiber). Balugani clearly distinguishes between *tiglie primarie*, the petiole of the leaf as a whole, and *tiglie secondarie*, petiolules of the leaflets, but also here the word *fibra* might be used instead of *tiglia*. The midrib of the leaf is termed *costa meridia* (midrib). The leaf margin may be *dentato* (toothed), but a term for the opposite, an entire margin, is not used.

For the flower Balugani uses the common Italian word *fiore* (flower), and a fairly consistent terminological system is used for the various floral parts. The pedicel, or stalk of the flower, is again called *tiglia*. The calyx is consistently called *perianthio*, occasionally written *perianthion*, a term which in modern Italian is written *perianzio*, and is a collective term for both calyx and corolla. The *perianthio*, as conceived by Balugani, consists of *foglie* (literally meaning leaves, but here obviously referring to the calyx segments). The corolla is referred to in two different ways, either as *foglie del fiore* (leaves of the flower) or *calice* (literally meaning calyx, a term now exclusively used for the outer floral parts, the calyx). If the flower is sympetalous, that is, when the corolla is fused into one structure, it may be described as having a *tubo* (floral tube), but also in a sympetalous flower the perianth segments are referred to as *foglie del fiore* (leaves of the flower).

Stamens are always termed *stamina* (stamen), anthers are called *testa di stamine* (head of stamens), and occasionally the filament of the stamen is called *cordetta* (small rope). The style, or probably sometimes the entire pistil, is termed *pistillo*. If the style is branched, each of the branches can be called *foglietto* (small leaf). The ovary is simply *ovario*. Fruits are always called *frutto*, without distinction between various fruit types. Seeds are called *semo*. There is no consistent terminology for endosperm, embryo, etc., although Balugani occasionally has described the interior of a seed.

Some specialized or complex structures that were well understood by scientists at the time of Linnaeus, such as the head (capitulum) of the Compositae, the daisy family, were not correctly interpreted by Balugani. In his analysis of the head of a *Guizotia abyssinica* (p. 77) he has made an absolutely correct drawing, but used a terminology at variance with the Linnaean interpretation. The outer bracts, or phyllaries, of the head are called *perianthio* (calyx), as if the head were a single flower. On the other hand, he does seem to have been aware of the fact that the florets are individual flowers, as his description continues like this: *Il cuore del Fiore e formato da un ammasso di picoli fiori...* (The heart of the flower consists of a large amount of small flowers...). More exotic plants with complicated flowers, such as *Mimusops kummel* (p. 99) and *Dombeya torrida* (p. 103), are absolutely correctly drawn, but the terminology used in the descriptive notes is hardly adequate for the complex structures. However, in these cases even the professional botanists of Europe had not made better analyses.

In some cases Balugani was certainly ahead of, or very closely following, the scientific discoveries made in Europe. His analysis of the inflorescence of *Rhamnus prinoides* (p. 96) shows, together with the notes, that he had, apparently quite independently, conceived the idea of the particular inflorescence type called a monocasium. His little diagram shows the basic principle. In no previous botanical publication had there been such a clear diagrammatical representation. Even more surprising perhaps is his analysis of the leaf arrangement in *Protea gaguedi* (p. 94). Here he describes and gives numerical and diagrammatical representations of what is now in plant morphology termed a 2/5 leaf spiral, that is, a leaf arrangement where the leaves are placed in a spiral running twice round the stem and passing five leaves, before the same pattern is repeated. The arrangement of leaves on plant stems was first studied scientifically by the French physiologist Charles Bonnet, who, in his work *Recherches sur l'usage des feuilles dans les plantes* (1754), had published an analysis of common patterns of leaf arrangement. No other publication on this subject appeared before 1812. It is almost impossible to imagine that Balugani should have known of Bonnet's work, and we must therefore credit him with the independent discovery of the 2/5 leaf spiral, which was later to become an important tool in the interpretation of the flower as basically a shortened shoot with spirally arranged leaves.

Altogether, Balugani's morphological terminology is simple but adequate, and his power of observation is sometimes surprising. His notes are, in combination with his drawing, always intelligible. The most confusing term in his morphological vocabulary is the word *tiglia*, but here he may have set his own standards of usage. The word *cordetta* for filament may be a coinage. Most of the other words of his morphological vocabulary are common Italian words used also outside botanical circles, such as the terms for tree, shrub, herb, and leaf. A few descriptive words, and the most interesting ones at that, are specialized botanical terms, such as *ovario*, *stamina*, *pistillo*, and *perianthio*, but the latter two are used by Balugani in a more restricted sense than in botany in general.

Balugani's floristic knowledge is more difficult to assess. All the plants he saw in Ethiopia were new to him, but we do get the impression that he was able to see a relationship between some of the Ethiopian plants and some of his native Italy. Occasionally he has identified the plants in his drawings with genera he knew from home, as is the case with *Aleandro* (oleander, *Nerium*), *gelsomino* (jasmine, *Jasminum*), *salice* (willow, *Salix*), *ginepro* (juniper, *Juniperus*), *trifoglio* (clover, *Trifolium*), *Aloe* (*Aloe*), and *oliva selvaggia* (wild olive, *Olea europaea* in the broadest sense). In a few cases he has used garden plants for comparison, in size and shape, for example *carciofolo* (artichoke), *mandola* (almond), and *ceraso* (cherry). He never seems to use the scientific names, only the Italian ones.

What do these observations tell us about Balugani, his training in botany, and about the scientific quality of his work? The morphological terminology he employs, especially for floral detail, seems to indicate that he had some botanical training or was perhaps self-taught from books. On the other hand, the comparative simplicity of his terminology, and the fact that some of his terms are wrongly used, suggests that his botanical studies were not extensive. He does not appear to be aware of the Linnaean sexual system, in which the plants are classified by the number of stamens and styles, and he does not seem to be familiar with the Linnaean botanical terminology. But the simple terminology is used very consistently throughout the notes, and the detail of observation is sometimes astonishing. Balugani was certainly a very careful observer. Probably the architect turned botanical student was an ideal combination—the structural discoveries he made in the inflorescence of *Rhamnus prinoides* and the leaf arrangement of *Protea gaguedi* are perhaps due to that combination.

A comparison of Balugani's descriptions in Italian with those published by Bruce in the *Travels* shows the difference between the two men's ability to convey an impression of a plan. The most striking example is in the description of the *Albizia gummifera* (p. 84). Balugani's description is long and quite difficult to follow, but does basically make sense. Bruce's published description is clearly derived from that of Balugani, but is almost totally incomprehensible in places. Bruce seems to have felt his trouble because he states at the end of the text: "Without a very distinct drawing, it would be difficult to make a description that should be intelligible."

Bruce sometimes added information to Balugani's notes when he wrote up the entries about the plants in the natural history section of his *Travels*. Some of this information is confused or impossible to believe. In the description of *Albizia gummifera* he states that the tree was originally brought to the Lake Tana region from the hot lowland in which myrrh (*Commiphora*) grows. But *Albizia gummifera* is restricted to the more humid parts of the highlands in Ethiopia and nowhere grows together with the myrrh, which in northern Ethiopia is restricted to the dry lowlands along the Red Sea. Moreover, everything seems now to indicate that *Albizia gummifera* is a natural component of the vegetation at Lake Tana. In short, to discern real observation from later unconfirmed additions one must consult Balugani's original Italian notes and not only rely on the information as rendered in the *Travels*.

The *Catalogue of Plant Drawings* & Their Identifications

Explanation

The catalogue is arranged alphabetically by plant families in the classes Dicotyledons (p. oo), Monocotyledons (p. oo) and Gymnosperms (p. oo) and Algae (p. oo). Within the families the genera and species are also arranged alphabetically.

For each entry the scientific Latin name is used and, when possible, an English name is also provided, though sometimes this is a convenient adaptation from the scientific name. The location and date on which each plant was drawn is indicated where possible, based either on notes written on the drawings or on published comments in the *Travels*, but the two sources occasionally do not agree. Most of the locations are marked on map and are included in the index of place-names though some have not been traced. The spelling of these names is normally that used by Bruce or Balugani, as few have standard modern equivalents.

Vernacular names of plants recorded by Bruce or Balugani are cited as inscribed on the drawings or as published in one or more editions of the *Travels*. They are sometimes inscribed in Amharic script, but were mostly transcribed by Balugani and it should be borne in mind that they were noted by an Italian speaker. It appears that they were mostly derived from Amharic, but frequently too from Tigrinia. Where similar names are noted by modern botanists the fact is mentioned. Only rarely is a vernacular name of Arabic or Greek origin.

The description of each drawing is headed by its accession number. As both sides of the sheets of the preliminary drawings were frequently used (often for unconnected material), the recto (*r.*) or verso (*v.*) for these sheets is always indicated.

Inscriptions on the preliminary drawings are always in Balugani's hand except in very rare instances where the appearance of a note by Bruce is indicated. All the inscriptions on the finished drawings have been added by Bruce (by way of identification) except where indicated.

Dimensions are in millimeters and height preceeds width.

In addition to the drawings and engravings, a few dried herbarium specimens are also cited. Apart from the Yale Center for British Art, where most of the former are located, other material is found at the following places:

Broomhall – home of the Earl of Elgin, Dunfermline, Scotland.

British Museum (BM) – herbarium of the British Museum (Natural History), South Kensington, London.

Linnean Society (LINN) – Linnaeus's herbarium, Linnean Society, Piccadilly, London.

Paris (P Lamarck's herbarium: P-LA) – general herbarium, Laboratoire de Phanérogamie, Muséum National d'Histoire Naturelle.

Geneva (G) – herbarium of the Conservatoire et Jardin botanique.

Vienna (W) – Naturhistorisches Museum.

Botanical references have been selected to include only the original work of publication of the species, and a few other relevant publications, such as the enumeration of Ethiopian plants by Cufodontis, and a modern flora, such as the *Flora of Tropical East Africa*, where fuller descriptions and references are provided. There are also references to the microfiche set of *Travels* (1790), volume 5 (plates), prepared by the International Documentation Centre (IDC microedition 5501).

Bibliography

(References to scientific publications abbreviated in the Catalogue)

Cufodontis (1953-1972). Dr. Georg Cufodontis, "Enumeratio plantarum aethiopiae spermatophyta," in *Bulletin Jardin Botanique de l'État*, Bruxelles, Suppl. pp. 1-708.

Mooney (1963), H.F. Mooney, *A glossary of Ethiopian plant names*. Dublin, 1963.

Schweinfurth (1893). Prof. Dr. G. Schweinfurth, "Abyssinische Pflanzennamen," in *Abhandlungen der Königl. Preuss. Akademie der Wissenschaften zu Berlin 1893*, pp. 1-84. Berlin, 1893.

——— (1896). "Sammlung Arabisch-Aethiopischer Pflanzen," in *Bulletin de l'Herbier Boissier*, Appendix 2, pp. 1-266. Genève, 1896.

Strelcyn (1973). S. Strelcyn, *Traitis médicaux II. Les noms des plantes*. Naples, 1973.

Wilson and Mariam (1979). R.T. Wilson and W. Gebre Mariam, "Medicine and magic in central Tigré," in *Economic Botany*, vol. 33, pp. 29-34. New York, 1979.

DICOTYLEDONS

ACANTHACEAE Acanthus family

Acanthus sennii

A gregarious shrub 3-6 ft. (1-2 m) high with very prickly leaves and bracts; known only from Ethiopia in montane thickets. Its conspicuous scarlet flowers are probably bird pollinated. It was found at Sakalla and presumably at a place called Ajadar (see .8950 *r*).

VERNACULAR NAMES: Cashishillo, Cosciscilla (MS), Coshillilla (*Travels* [1805] 7, 325; [1813] 7, 341); Murray, (1808) 450. Similar Amharic names are recorded by Cufodontis (1964).

a. B 1977.14.8949 *r*., left. Outline sketches, one very rough, of inflorescence, habit and details of flowers.
Pencil, indented for transfer; 311 × 443; inscr., at the top, in pen and black ink: Cosciscilla; below, eight lines of notes beginning: Quest è un Cardone ritrovato a Sakalla li 5 Novbre 1770 (This is a thistle found at Sakalla the 5th November 1770). *Fig. 49*.

b. B 1977.14.8950 *v*. Outline drawing of flowering shoot and leaves.
Pencil, indented for transfer, with touches of pen and black ink; 222 × 194. *Fig. 50*.

c. Lord Elgin's collection, Broomhall. Finished drawing from b.
Watercolors and bodycolors over pencil outlines; 450 × 313. *Fig. 1*.

d. B 1977.14.8950 *r*.
Pencil, indented for transfer; inscr., top left, in pen and black ink: Ajadar li 26— (Ajadar the 26th—). *Fig. 51*.

e. B 1977.14.9099. Finished drawing from d with added details.
Watercolors and bodycolors over pencil outlines; 405 × 312; inscr. *v*. in pen and brown ink: Cashishillo. *Figs. 2, 52*.

BOTANICAL REFERENCES: *A. sennii* Chiovenda in *Atti Reale Accademia Italia, Memorie* (1940) 11, 50; Cufodontis (1964) 954.

Hygrophila auriculata

A stout, erect herb 3-6 ft. (1-2 m) high, growing in wet places throughout the Old World tropics. Its inflorescence is fiercely armed with six strong spines to each whorl of flowers; the flowers themselves are bluish-purple.

VERNACULAR NAME: None recorded.

B 1977.14.8949 *v*., left. Sketch of stem with inflorescences only.
Pencil; 311 × 443. *Fig. 53*.

BOTANICAL REFERENCES: *H. auriculata* (Schumacher) Heine in *Kew Bulletin* (1962) 16, 172 and in *Flora West Tropical Africa*, ed. 2, (1963) 2, 295; Cufodontis (1964) 930.

Ruttya speciosa

A straggly shrub inhabiting rather dense montane thickets and dry forests. The scarlet flowers have shining black patches around the mouth.

This species occurs in Ethiopia and possibly in the Yemen. It is very similar to *R. fruticosa* Lindau, which has a wider distribution in eastern Africa and southern Arabia, but *R. speciosa* appears to be shorter in habit and to have longer, narrower calyx lobes. It was found at El Kaha.

VERNACULAR NAMES: Semezza, Semezze (MS and *Travels* [1805] 7, 325; [1813] 7, 341). A similar Tigrinia name is recorded by Schweinfurth (1896), but Mooney (1963) applies the Amharic *Samizza* to another plant (*Adhatoda schimperiana*).

a. B 1977.14.9044 *r*. Outline sketch of flowering shoot, leaf, and details of flowers.
Pencil; 280 × 222; inscr., top left, in pen and black ink: Semezza; on either side with two columns of notes beginning: Questo/rappresenta un ramo/di una Pianta detta Semezza/ritrovata in Fiore li 10 Decembre 1770/erano al Kaha (This represents a shoot of a plant called Semezza found in flower on the 10th of December 1770. They were at El Kaha). *Fig. 54*.

b. B 1977.14.8955 *r*. Sketch of a flowering plant.
Pencil, indented for transfer; 221 × 210; inscr., top right, in pen and black ink: Semezza. *Fig. 55*.

c. B 1977.14.9098. Finished drawing from b.
Watercolors and bodycolors, within ruled pencil borders; inscr. *r*. below, in pencil: Semezza; and *v*. in pen and brown ink: Semezze (?). *Fig. 3*.

BOTANICAL REFERENCES: *R. speciosa* (Hochstetter) Engler, *Hochgebirgsflora Tropisch Afrika* (1892) 393; Cufodontis (1964) 965.

ANACARDIACEAE Cashew-nut family

Rhus glutinosa

A small tree restricted to the mountains of Ethiopia in the thickets and dry forests of the juniper zone. The young, trifoliolate leaves are reddish and the small, green flowers produce white fruits about 5 mm in diameter.

VERNACULAR NAMES: Menghe, Menghi (MS), Mengi (*Travels* [1805] 7, 334; [1813] 7, 348; Murray [1808] 450). Similar Tigrinia names are quoted by Schweinfurth (1896) and Mooney (1963).

a. B 1977.14.8937 *v.*, right. Three lines of MS notes on the tree's fruits. 221 × 313; inscr., in pen and black ink: Menghi: Albero che fa picoli frutti assai Copiosamente... (Menghi. A tree which bears abundant small fruits).

b. B 1977.14.8988 *r.* Outline sketch of a fruiting leafy shoot.
Pencil, indented for transfer; 305 × 203; inscr. above, in pencil: Menghi. *Fig. 56.*

c. B 1977.14.8954 *r.* Outline sketch of a leafy shoot with fruit.
Pencil, indented for transfer; 309 × 209. *Fig. 57.*

d. B 1977.14.6468f. Finished drawing from c.
Watercolors and bodycolors over pencil outlines; inscr. *v.*, in pen and brown ink: Menghi. *Fig. 59.*

BOTANICAL REFERENCES: *R. glutinosa* A. Richard, *Tentamen florae abyssinicae* (1847) 1, 144; Oliver in *Flora Tropical Africa* (1868) 1, 438; Cufodontis (1958) 473.

Rhus ?retinorrhoea

The drawings are only tentatively identified with this species, as the leaves do not match exactly: they are shown to be almost rounded at the apex, whereas the leaves of *R. retinorrhoea* are acutely attenuate. In *Travels*, it is noted as "a tree bearing small seeds like "Dora" [i.e., *Sorghum*] which are eaten."

VERNACULAR NAME: Bohah (MS and *Travels*, [1805] 7, 325; [1813] 7, 341). There is no recently recorded similar name in this family, but Schweinfurth cites "Bohha" for *Debregeasia bicolor* in Urticaceae, and that species does not have trifoliolate leaves like *Rhus*.

a. B 1977. 14.8937 *v.*, right. Three lines of MS notes; 221 × 313; inscr., sixth item down, in pen and black ink: Bohah. Albero che fa picoli frutti... (Bohah. A tree which bears small fruits).

b. B 1977.14.9054 *r.*, right. Outline drawing of a leafy spray with fruit.
Pencil; 302 × 397; inscr., above left, in pen and brown ink: Bohah. *Fig. 58.*

c. B 1977.14.6468a. Finished drawing similar to b. Watercolors and bodycolors over pencil outlines, within ruled pencil borders; 451 × 312; inscr. *v.*, in pen and brown ink: Bohah. *Fig. 62.*

BOTANICAL REFERENCES: *R. retinorrhoea* Oliver in *Flora Tropical Africa* (1868) 1, 438; Cufodontis (1958) 475.

ANNONACEAE Custard-apple family

Annona squamosa *Sweet-sop*

A small tree bearing large, green, scaly fruits known as sweet-sop or sugar-apple. Since it is native in the West Indies, presumably it was in cultivation in Ethiopia about 1770, before there were regular European contacts.

VERNACULAR NAMES: Chista-cachemantia (MS). A similar Amharic name is noted by Mooney (1963) for *Annona* species.

a. B 1977.14.8968 *r.*, left. Rough sketch of tree's habit.
Pencil; 180 × 295; inscr., top left, in pen and brown ink: Chista - Cachemantia; beneath, notes on proportions. *Fig. 60.*

b. B 1977.14.9089b. Finished drawing from a.
Watercolors and bodycolors over pencil outlines; 533 × 371; inscr., upper right, in pencil: 2. *Fig. 63.*

c. B 1977.14.8977 *v.*, right. Outline sketch of spray with fruit and leaves and details of dissections of flowers.
Pencil; 181 × 296; inscr. in pencil with letters against details. *Fig. 61.*

d. B 1977.14.9089c. Fisnished drawing from c, with added leaves and flowers on spray.
Watercolors and bodycolors, over pencil outlines; 533 × 371; inscr., upper right, in pencil: 3. *Fig. 64.*

BOTANICAL REFERENCES: *A. squamosa* Linnaeus, *Species Plantarum* (1753) 537; Cufodontis (1954) 117; Verdcourt in *Flora Tropical East Africa*, Annonaceae (1971) 113.

APOCYNACEAE Periwinkle family

Acokanthera schimperi

A small, rounded tree with thick, evergreen leaves and white and pink fragrant flowers. Although the black fruits mey be edible, the leaves are used for the preparation of arrow poison. It occurs in dry, evergreen bushland and is limited to northeastern Africa, but has closely related species throughout eastern Africa.

VERNACULAR NAMES: Mepta (MS). Similar Amharic names have been recorded by Schweinfurth (1896), Mooney (1963), and Cufodontis (1960).

B 1977.14.6467j. Finished drawing of flowering shoot. Watercolors and bodycolors over pencil outlines within pencil borders; 451 × 313; inscr. *r.*, top right, in pen and brown ink: 10; and *v.*: Mepta. *Fig. 4.*

BOTANICAL REFERENCES: *A. schimperi* (A. De Candolle) Schweinfurth in *Bollettino Societa Africana Italia* (1891) 10(11-12), 12.
Carissa schimperi A. De Candolle, *Prodromus* (1844) 8, 675; Cufodontis (1960) 685.

Carissa edulis

An evergreen bush or woody scrambler with sharp, stout thorns. Its flowers are white inside and red outside and sweetly scented. The red fruits are edible, and in Ethiopia are crushed to make a juice. It has a wide distribution in the drier parts of tropical Africa and Asia, where if often forms dense thickets on abandoned farmland. This is particularly the case in Ethiopia, where shifting cultivation is a long-established practice. It was found on the way to Gondar.

VERNACULAR NAME: Agam (MS and *Travels* [1805] 7, 324; [1813] 7, 340). The same Amharic and Tigrinia name is noted by Schweinfurth (1896) and Mooney (1963).

a. B 1977.14.9028 *v.* Outline sketch of a flowering shoot.
Pencil, indented for transfer; 233 × 171; inscr., at top, in pen and brown ink: Agam; with three lines of notes. *Fig. 69.*

b. B 1977.14.6467i, *r.* Finished drawing from a.
Watercolors and bodycolors, over pencil outlines; 451 × 314; inscr., top right, in blue and brown ink: 9; and *v.*: Agam. *Fig. 67.*

BOTANICAL REFERENCES: *C. edulis* (Forsskal) Vahl, *Symbolae Botanicae* (1790) 1, 22; Cufodontis (1960) 684; Huber in *Flora West Tropical Africa*, ed. 2, (1963) 2, 54.

Saba comorensis

A tall, woody climber yielding white latex. The flowers are white, yellow in the throat; the yellow fruits are edible. It is widespread in forest in tropical Africa and Madagascar. Bruce saw this plant at Lamgue near Lake Tana on 18 May 1770 and also "on the way to Gondar."

VERNACULAR NAME: Leham (MS and *Travels* [1805] 7, 324, 330; [1813] 7, 340, 346). Both Schweinfurth and Mooney cite *Laham* (Amharic, Tigrinia) and *Leham* (Tigrinia) as *Syzygium guineense*, not *Saba.*, so it is interesting to note that Bruce observed in the Assar valley large trees he called Leham which were presumably the tree *Syzygium* and not the climbing *Saba.*

a. B 1977.14.22703 *r.* Outline sketch of flower head.
Pencil, indented for transfer, 211 × 222, mutilated along top edge. *Fig. 66.*

b. B 1977.14.22703 *v.* Outline sketch of leaves and dissection of flowers.
Pencil; inscr. in black ink incomplete. *Fig. 65.*

c. B 1977. 14.9092 *r.* Fisnished drawing of flowering shoot.
Watercolors and bodycolors, over pencil outlines; 406 × 305; inscr., lower right, in blue and pen and brown ink: no 1; in pencil: D; and *v.*, center left, in pencil: Toberne montana; lower left: Leham. *Fig. 5.*
Engr., in reverse, by Heath *Travels* (1805) 8, pl. 45; (1813) 8, pl. 45.

BOTANICAL REFERENCES: *S. comorensis* (Bojer) Pichon in *Mémoires Institut Française Afrique Noire* (1953) No. 35, 303.
S. florida (Bentham) Bullock in *Kew Bulletin* (1959) 13, 391; Cufodontis (1960) 687; Huber in *Flora West Tropical Africa* ed. 2, (1963) 2, 61.
"Toberne montana" *Travels* (1805) 7, 330–32, & 8, pl. 45; (1813) 7, 346–48 & 8, pl. 45; Murray (1808) 454–56, pl. viii. This name is certainly a corruption of the Latin *Tabernaemontana*, another genus of the same family.
"Leham" *Travels* (1805), 7, 330; (1813) 7, 346; Murray (1808) 449, 454.

ASCLEPIADACEAE Milk-weed family

Calotropis procera *Apple of Sodom, Auricular Tree*

A rather grotesque shrub or small tree yielding latex, with large, leathery leaves covered with a white bloom. The waxy, white flowers produce large, applelike fruits that are inflated, very light, and, when ripe, full of downy seeds. It has a wide distribution in the semidesert parts of Africa, the Middle East and India, and this drawing was probably done in Egypt.

VERNACULAR NAME: None recorded.

B 1977.14.9123. Finished drawing of flowering and fruiting plant.
Watercolors and bodycolors, over pencil outlines; 543 × 383. *Fig. 68.*

BOTANICAL REFERENCES: *C. procera* (Aiton) Aiton f., *Hortus Kewensis*, ed. 2, (1811) 2, 78; Cufodontis (1960) 702; Bullock in *Flora West Tropical Africa*, ed. 2, (1963) 2, 91.

Cynanchum altiscandens

A slender herbaceous climber, with cut stems

yielding abundant white latex. The small white flower develops into two-lobed fruits with plumed seeds. Known only in Ethiopia.

VERNACULAR NAME: None recorded.

B 1977.14.8964 *r.*, left. Outline sketch of fruiting shoot, with detail of fruit.

Pencil; 409 × 308; inscr., upper left, in pen and black ink: Pianta rampante (climbing plant); and, upper centre, against detail of fruit: BB. *Fig. 71.*

BOTANICAL REFERENCES: *C. altiscandens* K. Schumann in *Abhandlungen Preussischen Akademie Wissenschaften* (1894), 64; Cufodontis (1960), 705.

Dregea abyssinica

A climbing shrub growing among thick, streamside vegetation in eastern Africa. Flowers white; fruits covered with corky wings and containing plumed seeds. It was found at Adderghei.

VERNACULAR NAMES: Scangho, Shanfo, Shango (MS). A similar Tigrinia name is recorded by Mooney (1963).

a. B 1977.14.8952 *v.* Outline sketch of fruiting shoot and separate leaf.

Pencil; 302 × 200; inscr. at the top, in pen and black ink: H. Scanghò a Adderghei Shango (Scanghò, at Adderghei [called] Shango). *Fig. 70.*

b. B1977.14.8953 *r.* Outline sketch of habit of plant climbing a stick. Pencil, indented for transfer; 302 × 199; inscr., top left, in pen and brown ink: Scanghò. *Fig. 72.*

c. B 1977.14.6469k. Finished drawing from b.

Watercolors and bodycolors over pencil outlines; 451 × 310; inscr. *r.*, upper left, in pen and brown ink: 11; and *v.*: Shanfo. *Fig. 75.*

d. B 1977.14.9056 *v.* Notes on plant without drawing.

302 × 199; inscr., lower page, in pen and brown ink, with eleven lines of MS headed: Scanghò H.

BOTANICAL REFERENCES: *D. abyssinica* (Hochstetter) K. Schumann in Engler, *Pflanzenwelt Ost-Afrika* (1895) C, 326; Cufodontis (1961) 724.

Gomphocarpus fruticosus

A perennial herb with several erect stems 3–6 ft. (1–2 m) high, occurring in upland grassland throughout Africa. The flowers are creamy-colored, each giving rise to two long, narrow follicles full of downy seeds. It was found at El Caka.

VERNACULAR NAME: None recorded.

B 1977.14.9041 *r.* Outline sketch of fruiting shoot with details of flowers and dissected seed case.

Pencil; 302 × 199; inscr., top right, in pen and black ink: Specie di Alleandro ritrovata al Caka (Species of oleander found at El Caka); and with note on size of leaf. *Fig. 73.*

BOTANICAL REFERENCES: *G. fruticosus* (Linnaeus) Aiton f., *Hortus Kewensis*, ed. 2, (1811), 2, 80; Cufodontis (1960) 700; Bullock in *Flora West Tropical Africa* ed. 2, (1963) 2, 92.

Kanahia laniflora

A small shrub with narrow leaves and white, woolly, hairy flowers, growing in rocky seasonal stream beds throughout much of Africa and southern Arabia. It was found in the Tacazi River.

VERNACULAR NAME: None recorded.

B 1977.14.8995 *r.*, left. Outline sketch of flowering plant with details of flowers.

Pencil; 302 × 397; inscr. below, in pen and brown ink: CC; and with note beginning: Questo disegno rappresenta un alleandro ritrovato al Tacazi... (This drawing represents an oleander... [i.e., *Nerium oleander*] found at Tacazi...). *Fig. 74.*

BOTANICAL REFERENCES: *K. laniflora* (Forsskal) R. Brown in *Memoirs Wernerian Natural History Society* (1810) 1, 40; Cufodontis (1960) 699; Bullock in *Flora West Tropical Africa*, ed. 2, (1963) 2, 91.

Tacazzea galactagoga

A woody climber with milky sap. The numerous yellowish flowers produce a few long, narrow fruits containing numerous plumed seeds. It is restricted to eastern Africa, where it occurs in montane streamside forest. It was found at Lamalmon.

VERNACULAR NAME: None recorded.

a. B 1977.14.9042 *v.* Outline sketch of flowering shoot with details of flowers.

Pencil; 302 × 199; inscr., at the top, in pen and brown ink: Pianta rampante ritrovata su la Malmon si chiama/e porta un frutto doppio come l'aleandro (Climbing plant found at Lamalmon called [no name given] and bears a double fruit like oleander [i.e., *Nerium oleander*]). *Fig. 77.*

b. B 1977.14.8964 *v.* left. outline sketch of fruiting shoot.

Pencil; 409 × 308; inset, lower left: AA. *Fig. 79.*

BOTANICAL REFERENCE: *T. galactogoga* Bullock in *Kew Bulletin* (1954) 9, 358; Cufodontis (1960) 693.

Tacazzea venosa

A rare Ethiopian species with rather woody stems and widely diverging bilobed fruits containing plumed seeds.

VERNACULAR NAME: None recorded.
B 1977.14.8964 *r.*, right. Outline sketch of fruiting shoot.
Pencil; 409 × 308; inscr., at the top, in pencil. DD. *Fig. 71.*

BOTANICAL REFERENCES: *T. venosa* Hochstetter ex Decaisne in De Candolle, *Prodromus* (1844) 8, 493; Cufodontis (1960) 693.

AVICENNIACEAE White mangrove family

Avicennia marina *A white mangrove*

A mangrove tree with silvery-gray foliage inhabiting the coasts bordering the Indian Ocean and the Red Sea. Breathing roots (pneumatophores) arise in quantity beneath the trees. The four-lobed flowers are yellowish and produce rounded fruits with a beak. Bruce records this tree from Raback on the Arabian coast and from Massawa on the African coast.

VERNACULAR NAME: Rack (MS and *Travels* [1790] 5, 44–46).
B 1977.14.8912. Finished drawing of flowering shoot with details of flowers.
Watercolors and bodycolors over pencil outlines; 310 × 244; inscr. *v.*, in pen and brown ink: Rack No. 17. *Fig. 76.*
Engr. Bruce, *Travels* (1790) 5, pl. at p. 44; (1805) 8, pl. 12; (1813) 8, pl. 12.

BOTANICAL REFERENCES: *A. marina* (Forsskal) Vierhapper in *Denkschriften Akademie Wissenschaften, Wien, Math.-Nat.* (1907) 71, 435; Cufodontis (1962) 803.
"Rack" Bruce, *Travels*, (1790), 5, 44-46 and plate opp. p. 44, microfiche IDC 5501–3: III.6; (1805), 7, 157–58 and 8, pl. 12; (1813), 7, 173–74 and 8, pl. 12.
Racka torrida J.F. Gmelin, *Systema Naturae*, ed. 13, (1791) 2, 245. Type: Bruce's plate.

BALSAMINACEAE Balsam family

Impatiens rothii *A balsam*

A herbaceous plant about 2 ft. (60 cm) high, with succulent stems, inhabiting damp, shady places in Ethiopia. Its flowers are red with an orange spur. Bruce found it at Sakalla.

VERNACULAR NAMES: Gerget, Gers seth, Ghersehetth, Ghersseth (MS). Similar Amharic and Tigrinia names are recorded by Schweinfurth (1896) and Mooney (1963).
a. B 1977.14.9034 *v.* Outline sketch of flowering shoot, with details of leaves and dissected flowers.
Pencil with touches of pen and black ink; 310 × 181; inscr., top left, in pen and black ink: Ghersehetth (deleted and amended); Ghe (deleted); and: Ghersseth; and, below, with eight lines of MS notes beginning: Quest e una pianta ritrovata a Sakalla (This is a plant found at Sakalla). *Fig. 81.*
b. B 1977.14.8963 *r.* Rough pencil sketch of flowering plant.
Pencil, over rough pencil strokes, indented for transfer; 222 × 190. *Fig. 78.*
c. B 1977.14.6469i. Finished drawing of flowering plant from b.
Watercolors and bodycolors over pencil outlines, with ruled pencil borders; 451 × 310; inscr. *r.*, upper right, in pen and brown ink: 9; *v.*: Gerget; and in pencil: Gers seth. *Fig. 80.*

BOTANICAL REFERENCES: *I. rothii* Hooker f. in *Flora Tropical Africa* (1868) 1, 302; Cufodontis (1958) 497; Grey-Wilson, *Impatiens of Africa* (1980) 161.

BORAGINACEAE Borage family

Cordia abyssinica

A moderate-sized tree up to 50 ft. (15 m), often growing in isolation in the mountains, and one of the characteristic trees of Ethiopia. It also occurs in the Yemen, East Africa, and a few places in West Africa. The flowers are very conspicuous since they are massed in compact heads and are pure white. The tree is not cut by the local people in the northern Ethiopian highlands, although other trees have been completely cleared for firewood. It is probably kept because it is a good bee tree, and beehives are often hung all over them. Bruce (*Travels* [1790] 5, 54–56) recorded it as common around Gondar, where it flowered in September, after the rainy season.

VERNACULAR NAMES: Wansey (MS), Wanzey (*Travels*). Similar Amharic names are recorded by Schweinfurth (1896) and Mooney (1963).
a. B 1977.14.9052 *r.* Outline sketch of a flowering shoot.
Pencil over rougher pencil strokes, indented for transfer; 302 × 204. *Fig. 82.*
b. B 1977.14.8916. Finished drawing from a.
Watercolors and bodycolors over pencil outlines; 317 × 246; inscr. *v.*, in pen and brown ink: Wansey No. 20. *Fig. 86.*
Engr. and lettered "Wanzey": *Travels* (1790) 5, pl. opp. 56; (1805) 8, pl. 17; (1813) 8, pl. 817.

BOTANICAL REFERENCES: *C. abyssinica* R. Brown in Salt, *Voyage Abyssinie*, Appendix (1814) 64—although the name appears without a description it is validated by reference to Bruce's "Wanzey" plate.
"Wanzey" Bruce, *Travels*, (1790) 5, 54–56 & pl. opp. 54, microfiche IDC 5501–4: II.2; (1805) 7, 164–66 & 8, pl. 17; (1813) 7, 180–82 & 8, pl. 17.

Cordia africana auctt. non Lamarck (1792)—Cufodontis (1961) 766.

Ehretia cymosa

A shrub or small tree with numerous white flowers and small, round, red fruits. It has a wide tropical African distribution, where it occurs mainly along the edges of upland forest.

VERNACULAR NAME: None recorded.

B 1977.14.6468b. Finished drawing of leafy, fruiting shoot.

Watercolors and bodycolors over pencil outlines, within pencil borders; 451 × 310; inscr. *R.*, upper left, in pencil: 2. *Fig. 87.*

BOTANICAL REFERENCES: *E. cymosa* Thonning in Schumacher, *Beskrivelse Guineiske Planter* (1827) 129; Cufodontis (1961) 770; Hepper, *West African Herbaria of Isert & Thonning* (1976) 32.

BURSERACEAE Frankincense family

Boswellia papyrifera *Ethiopian frankincense tree*

A medium-sized tree with spreading branches and thin, papery bark. The compound leaves occur briefly after rain as tufts toward the ends of the shoots together with inflorescences of red flowers. The fruits appear later. It occurs on dry hills in northeastern Africa. The people of the northern Ethiopian highlands practice a form of tapping of the resin (Ethiopian frankincense), which is used as a tranquilizer and to control fever and to ward off "evil spirits." Bruce recorded it (MS) near the Tacazze River, that the Ethiopians considered it as incense, and that it produced a similar gum.

VERNACULAR NAMES: Angurah, Anguà (MS and *Travels* [1805] 7, 325; [1813] 7, 341), Anguah (*Travels* [1805] 8, pl. 48; [1813] 8, pl. 48; Murray [1808] pl. xi). Similar Tigrinia names are recorded by Schweinfurth (1896) and Mooney (1963).

a. B 1977.14.9027 *v.* Outline sketch of leafy shoot.

Pencil; 233 × 181; inscr. above with nine lines of MS notes beginning: Questo rappresenta un Ramo d'Anguà con Frondi Giovani Verdi... (This represents a stem of Anguà with foliage of fresh green...). *Fig. 83.*

b. B 1977.14.9027 *r.* Sketch of leafy branches.

Pencil; inscr. above with thirteen lines of MS notes beginning: Quest è l'Idea di un ramo di un Albero che stesse al Tacazi. Si chiama Anguà. (This is the sketch of a branch of a tree which was at the Tacazi [river]. It is called Anguà.). *Fig. 84.*

c. B 1977.14.9029 *r.* Outline sketch of flowering leafy branches and detail of leaf.

Pencil, indented for transfer; 302 × 200; inscr. above: Angua. *Fig. 90.*

Engr. by Heath for *Travels* (1805) 8, pl. 48; (1813) 8, pl. 48; Murray (1808) pl. xi.

d. Lord Elgin's Collection, Broomhall. Finished drawing from c.

Watercolors and bodycolors; 450 × 304; inscr. *v.*, in pen and black ink: No 12 Angurah (?). *Fig. 85.*

BOTANICAL REFERENCES: *B. papyrifera* (Delile) Hochstetter in *Flora* (1843) 26, 81; Cufodontis (1956) 337; Wilson and Mariam (1979) 30.

"Anguah No. 1" Bruce, *Travels* (1805) 7, 334 and 8, pl. 48; (1813) 7, 350 and pl. 48; Murray (1808) 458, xi (note that the Anguah on pl. 49 of *Travels*, eds 2 and 3 is the fruit of *Sterculia africana*).

Commiphora gileadensis *Balm of Gilead, Opobalsam*

A small, gnarled tree, known from both sides of the Red Sea, with branches spreading horizontally and dark gray bark. The flowers, although red, are inconspicuous, either solitary or two or three together. Incisions in the trunk and branches of this species produce a pale resin known since ancient time as *opobalsam*, or balm of Gilead. According to the first edition of *Travels*, Bruce's plates were drawn from trees growing on the Red Sea coast of Arabia, but in later editions it was claimed that they were drawn at Massawa.

VERNACULAR NAMES: Balessan, balissan, balsam (*Travels*). Similar Arabic names are noted by Mooney (1963).

a. B 1977.14.8693. Finished drawing of fruits, dissected fruits, and leaves.

Watercolors and bodycolors over pencil outlines; 304 × 243; inscr. *r.*, in the center, in pencil, with instructions to the engraver: Leaves to be left and fruit to be transferred to; *v.*, in pen and brown ink: Balessan No. 3/The fruit to be transposed to the foot of No. 2/in the order it stands in this No. 3 but/no notice to be taken of the leaves [in red pencil] No. 3 so that/there are to be only two drawings of the/ Balissan No. 1 & No. 2. *Fig. 88.*

b. B 1977.14.8904. Finished drawing of tree's habit (with roots).

Watercolors and bodycolors over pencil outlines; 247 × 316; inscr. *v.*, in pen and brown ink: Balessan No. 1; and: B. *Fig. 6.*

Engr. Bruce *Travels* (1790) pl. after p. 16; (1805) pl. 2; (1813), pl. 2.

c. B 1977.14.8905. Finished drawing of fruiting branch.

Watercolors and bodycolors over pencil outlines; 308 × 248; inscr. *v.*, in pen and brown ink: Balessan No. 2; and: B (inverted). *Fig. 89.*
Engr. Bruce, *Travels* (1790) 5, pl. before p. 17; (1805) 8, pl. 3; (1813) 8, pl. 3. This plate shows fruits as in a.

BOTANICAL REFERENCES: *C. gileadensis* (Linnaeus) C. Christensen in *Dansk Botanisk Arkiv* (1922) 4(3), 18.
"Balessan" Bruce, *Travels*, (1790) 5, 16 & two plates opp. p. 16; IDC microfiche 5501–2: I.8, II.1, 2; (1805) 7, 131–41 and 8, pl. 2 and 3; (1813) 7, 147–57, and 8, pl. 2 and 3.
C. opobalsamum (Le Moine) Engler in De Candolle, *Monographiae Phanerogamarum* (1883) 4, 15; Cufodontis (1956) 389.

CAMPANULACEAE Bell-flower family

Michauxia campanuloides

A rough-hairy plant about 2 ft. (60 cm) high growing in rocky places in Asia Minor and Syria. The large, white flowers have reflexed corolla lobes and protruding anthers and styles.

The sketch must have been done in Syria, but no notes are included and a final drawing evidently was not prepared.

VERNACULAR NAME: None recorded.
B 1977.14.9032 *r.* Outline drawing of two flowers.
Pencil; 153 × 100. *Fig. 94.*

BOTANICAL REFERENCE: *M. campanuloides* L'Héritier ex Aiton, *Hortus Kewensis* (1789) 2, 8.

CAPPARACEAE Caper family

Capparis tomentosa *A caper*

A slender, climbing shrub with numerous sharp, stipular spines and small, white flowers, occurring in thickets throughout the drier parts of tropical Africa. In the MS it is noted that it grew in the province of Caruta at a place called Andatatus.

VERNACULAR NAME: Gemero (MS). Mooney (1963) records similar Amharic names.
a. B 1977.14.8995 *v.*, right. Sketch of flowering shoot with leaves.
Pencil; 302 × 397. *Fig. 93.*
b. B 1977.14.9040 *r.* Outline sketch of flowering shoot.
Pencil; 336 × 153; inscr., at top, in pen and black ink: Ghimaru (deleted); and: Gemero; below the drawing, nine lines of MS notes beginning: Quest è un Ramo di un albero picolo chiamato ritrovato nella provin/cia di Caruta in Luogo chiamato Andatatus (This is a shoot of a small tree called Gemero found in the province of Caruta in a place called Andatatus). *Fig. 91.*
c. B 1977.14.9064 *v.* Outline sketch of flowering and fruiting shoot.
Pencil, indented for transfer; 302 × 199. *Fig. 92.*
d. B 1977.14.6467f. Finished drawing of flowering and fruiting shoot.
Watercolors and bodycolors over pencil outlines, within pencil borders; 451 × 312; inscr. *r.*, upper right, in pen and brown ink: 6; *v.*: Cappaas (sideways). *Fig. 95.*

BOTANICAL REFERENCES: *C. tomentosa* Lamarck, *Encyclopédie Méthodique Botanique* (1783) 1, 606; Cufodontis (1954) 128; Wilson and Mariam (1979) 30.

Ritchiea albersii

Shrub or small tree with a short trunk and spreading crown. Leaves usually three-foliolate; a few large, greenish-white flowers in terminal inflorescences and green, plumlike fruits borne on a long stipe. It inhabits upland forest margins in wide areas of tropical Africa. The MS notes (.8946 *v.*) indicate that this plant grew at Gondar.

VERNACULAR NAME: Gumarro (MS). A similar name is recorded by Mooney (1963).
B 1977.14.8946 *r.* Outline drawing of flowering shoot with details of flowers, leaves, and dissected fruits.
Pencil; 302 × 199; inscr., top right, in pencil: Gumarro; and in pen and black ink, with letters against details; *v.* Extensive MS notes, inscr., top left, in pen and black ink: Gumarro... un albero chiamato Gumarrò: quest à ritrovato a Gonder circa la fine di Febraio... (A tree called Gumarrò which is found at Gonder about the end of February...). *Fig. 96.*

BOTANICAL REFERENCES: *R. albersii* Gilg in Engler, *Botanische Jahrbücher* (1903) 33, 208; Elfers et al. in *Flora Tropical East Africa*, Capparidaceae (1964) 21, fig. 5.
R. steudneri Gilg in Engler, *Botanische Jahrbücher* (1903) 33, 208; Cufodontis (1954) 125. The type specimen of this species also came from Gondar.

CELASTRACEAE Spindle-tree family

Catha edulis *Khath, Chaat*

A tree, or more usually seen as the cultivated "chaat" shrub, with leathery leaves and small, white flowers in axilliary inflorescences. It occurs throughout the mountains of eastern Africa in drier places. Its leaves are widely chewed and provide a mild narcotic effect. The plant was recorded at El Kaha on 10 Dec. (1770?).

VERNACULAR NAMES: Chaat, Ciath (MS). Similar Tigrinia and Amharic names are noted by Schweinfurth (1896) and Mooney (1963).

a. B 1977.14.8986 *r*. Rough sketch of a flowering stem, right; leafless stem, left; detail of leaf below. Pencil, indented for transfer; 222 × 210; inscr., top left: Dogmà; top right: Quest Albero/l'abbiamo pm̃a/Conosciuto sotto Nome/di Dogmà. Doppo/anno parlato p. Noi che era Ciath. (This tree we have first known under the name of Dogmà. Afterwards we found that it was spoken of as Ciath); *v*. MS notes only, inscr. in pen and black ink, with twenty-one lines of notes beginning: Dogmà. Nel rovesscio questo Foglio è rappresentato un ramo d'un Albero chiamato Dogmà. Quest è ritrovato al Kaha in Frutto circa alli 10 di Dicembre. (Dogmà. On the reverse of this sheet is represented a shoot of a tree called Dogma. This was found at El Kaha about the 10 December). *Fig. 97*.

b. B 1977.14.6467a. Finished drawing from a.
Watercolors and bodycolors over pencil outlines, within ruled pencil borders; 451 × 312; inscr. *r*., upper right, in pen and brown ink: Chaat; *v*., Chaat. *Fig. 98*.

BOTANICAL REFERENCES: *C. edulis* Forsskal, *Flora Aaegyptiaco-Arabica* (1775), CVII & 63; Cufodontis (1958) 481.

Maytenus arguta

A large shrub or small tree which occurs frequently in the undergrowth of montane forest and in secondary evergreen scrub of the Ethiopian highlands and in eastern Africa. It was recorded as "a large tree" at Geesh (*Travels* [1805] 7, 325; [1813] 7, 341).

VERNACULAR NAMES: Attat, Attath, Atatt (MS & *Travels*). Similar Tigrinia and Amharic names are recorded by Schweinfurth (1896) and Mooney (1963).

a. B 1977.14.8999 *r*. Outline drawing of fruiting stem with details of leaf and fruits.
Pencil, indented for transfer; 310 × 222; inscr., top left, in pen and black ink: Attath, with letters A to G against details; and with MS notes below beginning: Quest è un Picolo Albero ritrovato a Sakalla li 5 di Novembre e suo frutto non/ancora perfettamente matturo. (This is a small tree found at Sakalla on the 5th of November and its fruit not yet quite ripe). *Fig. 100*.

b. B 1977.14.6468i. Finished drawing from a.
Watercolors and bodycolors over pencil outlines, within ruled ink borders; 451 × 313; inscr. *r*., upper right, in pencil: 9; and *v*., in pen and brown ink: Attatt; in pencil: Barbary. *Fig. 101*.

BOTANICAL REFERENCES: *M. arguta* (Loesener) N. Robson in *Boletim Sociedade Broteriana*, Sér. 2, (1965) 39, 8.
M. ovatus (Wight & Arnott) Loesener var. *argutus* (Loesener) Blakelock-Cufodontis (1958) 479.

COMPOSITAE Daisy family

Carthamus tinctorius *Safflower*

A thistle-like annual herb with bright yellow flower heads. The safflower has been used since ancient Egyptian times as an orange-yellow dye for textiles, but it has also been known in Ethiopia to produce a good cooking oil. According to Bruce (*Travels* [1805] 7, 325; [1813] 7, 341) it was found on the way to Gondar, and it was used for dyeing cloth.

VERNACULAR NAMES: Samf, Somf, Suf, Sussa (MS). The same or similar name is recorded in Tigrinia and Amharic by Schweinfurth (1896) and Mooney (1963).

a. B 1977.14.9051 *r*. Slight sketch of habit, on left of juniper tree.
Pencil; 310 × 210; inscr.: Cardo Giallo/Chiamato Somf o suf (Yellow Thistle called Somf or Suf). *Fig. 326*.

b. B 1977.14.9100. Finished drawing of flowering plant.
Watercolors and bodycolors over pencil outlines, within ruled ink borders; 395 × 299; inscr. *v*., in pen and black in: Sussa Samf. *Fig. 7*.

BOTANICAL REFERENCES: *C. tinctorius* Linnaeus, *Species Plantarum* (1753) 830; Cufodontis (1967) 1178.

Echinops macrochaetus *Globe thistle*

A fiercely prickly globe thistle known only from the Ethiopian mountains. Flowers blue in spherical heads.

VERNACULAR NAMES: Coshillilla (MS and *Travels* [1805] 7, 325; [1813] 7, 341; Murray [1808] 450). Mooney (1963) records a similar Amharic name.

B 1977.14.9089i. Finished drawing of flowering plant.
Watercolors and bodycolors over pencil outlines; 533 × 373; inscr. *r*., upper right: 9. *Fig. 99*.

BOTANICAL REFERENCES: *E. macrochaetus* Fresenius in *Museum Senckenbergianum* (1840) 3, 69; Cufodontis (1967) 1171.

Guizotia abyssinica

An annual up to 3 ft. (1 m.) high with erect branches terminating in decorative yellow flower heads 3-5 cm across. Originating in Africa but cultivated elsewhere in warm countries for the sake of the oil from its seeds, which are nowadays often included in birdseed. In *Travels* (1805) 7, 325; (1813) 7, 341, it is recorded that this plant was drawn on the way to Gondar and that it was used for making oil.

VERNACULAR NAMES: Nuk, Nook (MS & *Travels*). Various similar names are noted by Schweinfurth (1896), Mooney (1963), and Cufodontis (1967).

a. B 1977.14.9000 *r*. Rough drawing of floral details. Pencil; 140 × 221; inscr. in black ink with letters against details and with two columns of MS notes below beginning: Le Foglie sono di un Verde Morto. (The leaves are a dull green). *Fig. 102*.

b. B 1977.14.9030 *r*. Very rough sketch of flowering shoot. *Fig. 104*.

Pencil; 222 × 172; inscr., top right, in pencil: Nuk.

c. B 1977.14.9030 *v*. Outline of leaf.

Pencil; inscr. in pen and black ink, full page of description beginning: Nuk. Quest è una pianta della quale i abbissini usano p. far oglio (Nuk. This plant the Abyssinians use to make oil.) *Fig. 105*.

d. B 1977.14.8891. Finished drawing by anonymous French artist of flowering plant and details of flowers. Watercolors and bodycolors over pencil outlines, within double ruled ink borders; 397 × 301; inscr. *r*., below borders, in pen and brown ink: Polymnia frondosa; *v*.: Polymnia Frondosa Nuk. *Fig. 103*.

BOTANICAL REFERENCES: *G. abyssinica* (Linnaeus fil.) Cassini in *Dictionnaire Sciences Naturelles* (1829) 59, 237 and 248; Cufodontis (1967) 1132; Bagø in *Botanisk Tidsskrift* (1974) 69, 20.

Polymnia abyssinica Linnaeus f., *Supplementum Plantarum* (1781) 383.

P. frondosa Bruce ex Murray in Bruce, *Travels* (1805 and 1813) 8, pl. 52; Murray (1808) pl. xv.

"Nuk" Bruce, *Travels* (1805) 7, 337–39 and 8, pl. 52; (1813) 7, 353 and 8, pl. 52; Murray (1808) 461–62.

Notes: Murray mentions that Bruce's notes on this plant were lost but this is incorrect (see .9030 *v*). He further suggested that the plate was done in London or Paris from a living plant grown from seed collected by Bruce—in fact, we know it was drawn in Paris. The specimen in Linnaeus's herbarium at Burlington House, London, may have been grown from Bruce's seeds.

Senecio gigas *A tree groundsel*

A small, soft-stemmed tree known only from the Ethiopian mountains with woolly leaves and inflorescences of yellow flower heads. It was found at Assua near the Blue Nile source.

VERNACULAR NAMES: Maraqua (MS).

a. B 1977.14.9031 *r*., right. Rough outline sketch of a leafy, budding shoot, with details of flowers.

Pencil; 202 × 257; inscr. right, in pen and black ink, with MS notes beginning: Questo rappresenta una Cima/di Albero Maraqua. (This represents a top shoot of the tree Maraqua); and: A, against flower detail. *Fig. 106*.

b. B 1977.14.9034 *r*. Outline sketch of a leafy stem, with detail of leaf.

Pencil, indented for transfer; 310 × 181; inscr., at the side and below, in pen and black ink, with MS notes beginning: Questo rappresenta un Ramo di un albero che si chiama Maraqua, quest è trovato a Sakala/a ¼ di miglio al SE delle Fontane de Nilo in Luogo detto Assua (This represents a shoot of a tree called Maraqua which was found at Sakalla, a quarter of a mile southeast of the source of the Nile in the place called Assua). *Fig. 107*.

c. B 1977.14.6467 d. Finished drawing of budding shoot from both a and b.

Watercolors and bodycolors over pencil outlines, within ruled ink borders; 451 × 310; inscr. *r*., upper right, in pen and black ink: 4; bottom: Maraqua; and *v*.: Maraqua; and, in pencil: Artemisia. *Fig. 8*.

BOTANICAL REFERENCES: *S. gigas* Vatke in *Linnaea* (1875) 39, 506; Cufodontis (1967) 1155.

Note: The specimen is depicted in young flower bud, but it is probably of this species.

Vernonia amygdalina *Bitter leaf*

A small tree or shrub forming thickets in upland in many parts of tropical Africa. The large heads of small, white flowers are sweetly scented in the evening. The tree is sometimes used as living compound stakes, and the wood is resistant to termites. Bruce (*Travels* [1805] 7, 325; [1813] 7, 341) recorded it at Gondar, and mentioned it as "a tree with the leaves of which they season beer."

VERNACULAR NAMES: Grawa (MS and *Travels*). The same Amharic name is noted by both Schweinfurth (1896) and Mooney (1963).

a. B 1977.14.9062 *r*. Outline sketch of flowering leafy shoot, with detail of leaf.

Pencil, indented for transfer; 222 × 199; inscr., top left, in pen and black ink, with MS notes beginning: Grawa Albero che/teniamo dentro la Casa vicino alla Cuc/ina (Grawa a tree which we found behind the house near the kitchen.) *Fig. 108*.

b. B 1977.14.6468n. Finished drawing from a.

Watercolors and bodycolors over pencil outlines, within ruled pencil borders; 451 × 313; inscr., upper right, in pencil: 14; *v*. in pen and brown ink: Grawa. *Fig. 111*.

BOTANICAL REFERENCES: *V. amygdalina* Delile,

Centuriae de plantes d'Afrique (1826) 41; Cufodontis (1966) 1066; Wilson and Mariam (1979) 33.

CUCURBITACEAE Marrow family

Coccinia abyssinica

A herbaceous climber from a stout, perennial root. The flowers are yellow and the fruits ovoid, green and orange, and longitudinally striped. The seeds are immersed in pink pulp. It was recorded (.9071*r*) to have mature fruits at Gondar in mid-October.

VERNACULAR NAMES: Uschis, Ueshish, Usheish, Usheest, Ushish (MS and *Travels* [1805] 7, 325; [1813] 7, 341). Similar names, which are also used for the pumpkin (*Cucurbita pepo*), are recorded by Schweinfurth (1896) and Mooney (1963).

a. B 1977.14.8963 *v*. Outline sketch of a plant with flowers and fruit, climbing a pole, with detail of tuber. Pencil, indented for transfer; 222 × 190. *Fig. 109.*

b. B 1977.14.9070 *r*. Outline sketch of a tuber.
Pencil; 172 × 110; inscr. above, in pen and black ink, with thirteen lines of MS notes beginning: Radice di Uschis... (Root of Uschis...). *Fig. 110.*

c. B 1977.14.9071 *r*. Outline sketch of details of fruit and leaves.
Pencil; 316 × 221; inscr., top, in pen and black ink: Ueshish; upper left: Useish; above with copious MS notes beginning: Questa Pianta è chiamata/in Amara [Amharic lettering] ed è ri/trovata a Gonder... (This plant was called in Amharic... and was found at Gondar...). *Fig. 112.*

d. B 1977.14.9071 *v*. Outlines of details of male flowers and tuber.
Pencil; inscr., top left, in pen and black ink: Fiori di Ueshish. Li fiori sono Biancastri nel calice, è Gialli nelle/Foglie striate di Verdastro (Flowers of Ueshish. The flowers are whitish in the calyx, and yellow in the leaves [petals] with greenish stripes); followed by six lines of MS notes, with separate notes against the tuber, left. The remaining drawing on the sheet is of a pelargonium. *Figs. 113, 131.*

e. B 1977.14.8883. Anonymous. Paris folder. Finished drawing of flowering shoot, with details below. Watercolors and bodycolors, within ruled ink borders; 406 × 308; inscr., *r*., in pen and brown ink: 1–9, against details; and *v*.: Bryonne d'abyssinie (Bryony of Ethiopia). *Fig. 9.*

f. B 1977.14.9101. Finished drawing from a.
Watercolors and bodycolors over pencil outlines, within ruled pencil borders; 406 × 306; inscr. *v*., in pen and brown ink: Usheest; and: No. 21 Ushesh. *Fig. 10.*

g. B 1977.14.9102. Finished drawing, the tuber from b, detail of fruit from c.
Watercolors and bodycolors over pencil outlines, within ruled pencil borders; 406 × 305; inscr. *v*., in pen and black ink: Usheest. *Fig. 116.*

h. an excellent dried specimen (the holotype) is present in Lamarck's herbarium in the Laboratoire de Phanérogamie, Paris (P–LA), grown from living material collected by Bruce.

BOTANICAL REFERENCES: *C. abyssinica* (Lamarck) Cogniaux in De Candolle, *Monographiae Phanergamarum* (1881) 3, 536; Cufodontis (1965) 1048. *Bryonia abyssinica* Lamarck, *Encyclopédie Méthodique Botanique* (1784) 1, 497.

Cucumis metuliferus *A wild cucumber*

A annual climbing or trailing herbaceous plant with fiercely prickly fruits that turn orange-red when mature. Occurring in wooded areas throughout much of tropical and southern Africa; it was found at the Tacazi River.

VERNACULAR NAMES: Cura-magiett, Kuaregh (MS). No similar names have been recorded by modern botanists.

a. B 1977.14.9054 *r*., left. Outline sketch of fruits and leaves.
Pencil; 302 × 397; inscr., along the side, above the drawing, in pen and brown ink; Balsamina; and: Cura-Màgiett; and along the right end, in Bruce's hand: Species of Balissan; *v*. inscr. below line in pen and brown ink, with copious notes beginning: Nel rovesscio di questo Foglio è disegnato una speccie di Balsamina che cresce nelle sponde del Tacazi. (On the back of this sheet is drawn a species of Balsam which grows at the source of the Tacazi). *Fig. 58.*

b. B 1977.14.9072 *r*. Sketch of fruiting plant climbing a pole.
Pencil, indented for transfer; 302 × 199. *Fig. 114.*

c. B 1977.14.9116. Fisnished drawing from b.
Watercolors and bodycolors over pencil outlines, within ruled pencil borders; 406 × 308; inscr. *v*., in pen and brown ink: Kuaregh. *Fig. 117.*

BOTANICAL REFERENCES: *C. metuliferus* Naudin in *Annales Sciences Naturelles*, sér. 4, (1859), 11, 10; Cufodontis (1965) 1043; Jeffrey in *Flora Tropical E. Africa*, Cucurbitaceae (1967) 98.

Lagenaria siceraria *Bottle gourd*

A densely hairy herbaceous climber of trailer producing bottle-shaped gourds that are often used as containers in Africa; it was found at Gondar.

VERNACULAR NAME: None recorded.

B 1977.14.8964 *v*., right. Outline sketch of a leafy fruiting stem.
Pencil; 409 × 308. *Fig. 79.*

BOTANICAL REFERENCES: *L. siceraria* (Molina) Standley in *Publications Field Museum Natural History Chicago*, Botany series (1930) 3, 435; Cufodontis (1965) 1046; Jeffrey in *Flora Tropical E. Africa*, Cucurbitaceae (1967) 51.

Momordica foetida

A perennial trailing or climbing herbaceous plant growing up to 14 ft. (4.5 m.) high, with an unpleasant smell when crushed. The egg-sized fruits are covered with soft spines, opening to expose scarlet interior and brown seeds. It is widespread in African thickets and was found at Gondar.

VERNACULAR NAME: Kuraregh (MS).

B 1977.14.9055 *r*. Sketches of details of leaves, flowers and fruits.

Pencil; 302 × 199; *v*. MS notes only.

Inscr., in pen and black ink, with extensive notes beginning: Nel rovesscio di questo foglio è rappresentato una Piant rampante ritrovata/a Gonder che in lingua Amara si chiama Kuraregh. (On the reverse of this sheet is represented a climbing plant found at Gondar which, in Amharic, is called Kuraregh). *Fig. 115.*

BOTANICAL REFERENCES: *M. foetida* Schumacher & Thonning, *Beskrivelse Guineiske Planter* (1827) 426; Cufodontis (1965) 1038; Jeffrey in *Flora Tropical E. Africa*, Cucurbitaceae (1962) 29.

Peponium vogelii

A robust climber or trailer with forked tendrils, pale yellow flowers and pendent fruits turning bright red. In upland und lowland forest, especially near water, throughout much of Africa; it was found at Maiagam (*Travels* [1805] 7, 325; [1813] 7, 341; Murray [1808] 450).

VERNACULAR NAMES: Ileef, Liff (MS), Leef (*Travels* [1805 & 1813]).

a. B 1977.14.8953 *v*. Outline sketch of fruiting plant climbing a pole.

Pencil, indented for transfer; 302 × 199; inscr. at the top: Liff. *Fig. 118.*

b. B 1977.14.6469l. Finished drawing from a.

Watercolors and bodycolors over pencil outlines, within ruled pencil borders; 451 × 310; inscr. *r*., upper right, in pen and brown ink: 12, and lower right: No. 1; *v*.: Ileef, *Fig. 119.*

BOTANICAL REFERENCES: *P. vogelii* (Hooker fil.) Engler, in Engler & Prantl, *Natürliche Pflanzenfamilien*, (1897), Nachtrage 1, 318; Cufodontis (1965) 1047; Jeffrey in *Flora Tropical E. Africa*, Cucurbitaceae (1967) 81, fig. 11 (1–7).

EBENACEAE Ebony family

Diospyros abyssinica *Ethiopian ebony*

A tall forest tree with a compact crown. The fragrant flowers are white and the spherical fruits turn black. It grows from the coastal forests of central Africa to the highlands of Ethiopia and its timber is useful for woodwork, such as loom shuttles for weaving sisal. It was found on the island of Mitraha in Lake Tana (*Travels* [1805] 7, 325; [1813] 7, 341).

VERNACULAR NAMES: Selcienn, Selchienn (MS & *Travels* [1805 & 1813]). Cufodontis (1963) also noted a similar name.

a. B 1977.14.9057 *v*. Outline sketch of short leafy shoot with fruit.

Pencil; 302 × 199; inscr., top right, in pencil: Selcienn. *Fig. 120.*

b. B 1977.14.9077 *v*. Outline sketch of fruiting shoots, similar to last, with details of fruit sections.

Pencil, indented for transfer; 302 × 199; inscr., top left: Selcienn followed by sixteen lines of MS notes beginning: Albero grande ritrovato Nell Isola di Mitrahà (Large tree found on the island of Mitrahà). *Fig. 121.*

BOTANICAL REFERENCES: *D. abyssinica* (Hiern) White in *Bulletin Jardin Botanique Bruxelles* (1956) 26, 241; Cufodontis (1960) 668.

Maba abyssinica Hiern in *Transactions Cambridge Philosophic Society* (1873) 12, 132.

Diospyros mespiliformis *African ebony*

A tall tree of forest and savanna. The white flowers are sweetly scented, and the round, yellow fruits have a four- or five-lobed calyx at their base. It is a very widely distributed lowland tropical African tree yielding useful timber.

VERNACULAR NAMES: Aijè, Ayè (MS). Schweinfurth (1896) and Mooney (1963) both record similar Tigrinia names.

a. B 1977.14.8957 *r*. Outline sketch of fruiting leafy shoot, with detail of leaf.

Pencil; 306 × 205; inscr., above left, in pencil: Aijé. *Fig. 122.*

b. B 1977.14.9078 *v*., left. Outline sketches of sections of fruit, within MS notes.

Pencil; 221 × 313; inscr., top left, in pen and black ink: Aijè; followed by seventeen lines of description beginning: Albero che porta frutti rottondi e chi si mangiano... (Tree which bears round fruits... which are eaten). *Fig. 124.*

c. B 1977.14.8983 *r*. Rough sketch of fruiting leafy stem.

Pencil, indented for transfer; 231 × 200. *Fig. 123.*

d. B 1977.14.6468h. Finished drawing from c.
Watercolors and bodycolors over pencil outlines, within ruled pencil borders; 451 × 313; inscr. *r.*, upper left, in pencil: 8; and *v.*, in pen and brown ink: Ayè. *Fig. 125.*

BOTANICAL REFERENCES: *D. mespiliformis* A. De Candolle, *Prodromus* (1844) 8, 672; Cufodontis (1960) 668.

EUPHORBIACEAE Spurge family

Croton macrostachyus

A medium-sized tree (in spite of being described as an "herb" by Bruce, because young plants have an herbaceous appearance). Its fruits are poisonous and used for a number of medicinal purposes in Ethiopia and, according to Bruce (*Travels* [1805] 7, 325), for "incantations" by the Falasha people.

VERNACULAR NAMES: Tambò, Mzenna (*Travels*). Similar Tigrinia and Amharic names are recorded by Schweinfurth (1896) and Mooney (1963).

B 1977.14.22702 *v.* Sketch of fruiting shoot.
Pencil; 204 × 173; inscr. in pen and black ink: Wanzie [deleted] Tambò Tigri/Bzena [deleted] Amara/ Mzenna in Amara ([called] Tambò in Tigrinia Mzenna in Amharic(?)). *Fig. 126.*

BOTANICAL REFERENCES: *C. macrostachyus* Hochstetter ex A. Richard, *Tentamen florae abyssinicae* (1851), 251; Cufodontis (1956) 419; Wilson and Mariam (1979) 30.

Euphorbia abyssinica *Ethiopian tree-spurge*

A cactuslike tree with numerous erect branches bearing sharp, recurved thorns. Flowers yellow, fruits red and three-lobed. The white latex of this and other species of *Euphorbia* causes wheals on the skin and severe eye irritation ending in blindness. Bruce (*Travels* [1790] 5, 41–44) reported that he found this tree covering Mount Taranta in West Samhar.

VERNACULAR NAMES: Kol-quall, Kol-qual (MS & *Travels*). Schweinfurth (1896) records similar Tigrinia names, while Mooney (1963) applies variants of the name to *Euphorbia candelabrum.*

a. B 1977.14.8949 *v.*, right. Outline sketches of details of flowers and fruits, one dissected, possibly of this species.
Pencil; 311 × 443. *Fig. 53.*

b. B 1977.14.22704 *r.* Outline of stems.
Pencil; 236 × 174; inscr. in pen and black ink: AM (?); *v.* list of plant names. *Fig. 127.*

c. B 1977.14.8910. Finished drawing of the tree's habit.
Watercolors and bodycolors over pencil outlines; 319 × 240; inscr. *v.*, in pen and brown ink: Kol-quall No. 14. *Fig. 11.*
Engr. in reverse by Heath, Bruce, *Travels* (1790) 5, pl. at p. 43; (1805) 8, pl. 11; (1813) 8, pl. 11.

d. B 1977.14.8911. Finished drawing of single trunk, with details of whole and dissected fruits.
Watercolors and bodycolors over pencil outlines; 318 × 247; inscr. *v.*, in pen and brown ink: Kol-quall No. 15. *Fig. 12.*
Engr. in reverse by Heath, Bruce, *Travels* (1790) 5, pl. at p. 42; (1805) 8, pl. 10; (1813) 8, pl. 10.

BOTANICAL REFERENCES: *E. abyssinica* J.F. Gmelin, *Systema Naturae* ed. 13 (1791) 2(1), 759; Cufodontis (1958) 441; J. A. Janse in *Succulenta* (1976) 55(11), 248–254, pls. at 252, 253; Bruce's plates are taken as the type of the species in lieu of an actual specimen.
'*Kol-qual*' Bruce, *Travels* (1790) 5, 41–44 and two pls.; IDC microfiche 5501–3: II.8, III.20; (1805) 7, 154–56 and 8 pls. 10 and 11; (1813) 7, 170–72 and 8, pls. 10 and 11.
E. officinarum Linnaeus var. *Kolquall* Willdenow, *Species Plantarum* (1799), 2, 884.
Note: Bruce (*Travels* 1790) described it as having "five thorns, four on the sides and one in the centre, scarce half-an-inch long, fragile, and of no resistance, but exceeding sharp and pointed", which does not accord with the *two* spines shown in the plate. N.E. Brown (in *Flora of Tropical Africa* (1912), 6(1), 588) therefore considered it an imperfectly known species, but Susan Carter of Kew informs us that there is no doubt about its identity since additional material with only two spines has been obtained from the same locality. The tree Bruce saw later at Geesh (*Travels*, loc. cit.) near the source of the Blue Nile by Lake Tana appears to have been the closely related species *E. candelabrum* Kotschy (non Welwitsch). Bruce's illustrations of "Kol-qual" show one plate of typical unbranched *E. abyssinica*, while the other one shows an older plant with the tree habit, but this should not be taken for *E. candelabrum.*

Euphorbia helioscopia *Sun spurge*

As this common European and Mediterranean weed has not been recorded from Ethiopia we presume that this watercolour was made of a plant found during the earlier part of Bruce's journey.

VERNACULAR NAME: None recorded.

B 1977.14.9125. Finished drawing of flowering plant with details of flowers and leaf and fruit.
Watercolors over pencil outlines; 375 × 540. *Fig. 130.*

BOTANICAL REFERENCES: *E. helioscopia* Linnaeus, *Species Plantarum* (1753) 459; Zohary, *Flora Palaestina* (1972) 2, 278; Tackholm, *Students Flora of Egypt* ed. 2, (1974) 330.

Euphorbia nubica *Nubian spurge*

A perennial herb or shrub with long, succulent stems branched irregularly, bearing yellowish flowers distally. It is widespread in eastern Africa.

VERNACULAR NAME: None recorded.

a. B 1977.14.8978 *r.*, right. Outline sketch of habit, with details of flowers.

Pencil; 295 × 180; *v.* inscr. only, detailed notes in pen and black ink. *Fig. 128.*

b. B 1977.14.9122. Finished drawing of flowering and fruiting shoot, with details of floral dissections. Watercolors and bodycolors over pencil outlines, within ruled pencil borders; 530 × 378. *Fig. 129.*

BOTANICAL REFERENCES: *E. nubica* N.E. Brown in *Flora Tropical Africa* (1911) 6(1), 554; Cufodontis (1958) 453.

Euphorbia ?tirucalli

A cactus-like tree often planted as a hedge, if this is the correct identification. *E. tirucalli* is actually an Indian species introduced into Africa. In *Travels* (1805) 7, 325; (1813) 7, 341, it is recorded as "a tree which produces a milky juice, said to be extremely hurtful to the eyes".

VERNACULAR NAMES: Cangieb, Canjeb (MS and *Travels* 1805 and 1813). Similar Amharic and Tigrinia names are recorded by Mooney (1963) and Cufodontis (1958); the latter stated (p. 458) that the names are used for *E. scoparia* N.E. Brown, a species he considered to be very near to *E. tirucalli*. The name is written on a drawing of *Juniperus procera* (q.v. p. 120).

BOTANICAL REFERENCES: *E. tirucalli* Linnaeus, *Species Plantarum* (1753) 452; Cufodontis (1958) 460.

GERANIACEAE Geranium family

Pelargonium alchemilloides subspecies **multibracteatum**

A soft-stemmed undershrub in the mountains of Ethiopia and East Africa, with white flowers.

VERNACULAR NAME: None recorded.

B 1977.14.9071 *v.* Faint sketch of leafy and flowering shoot, underlying drawing of *Coccinia abyssinica* (q.v.).

Pencil; 316 × 221. *Figs. 113, 131.*

BOTANICAL REFERENCES: *P. alchemilloides* (Linnaeus) Aiton subsp. *multibracteatum* (A. Richard) Kokwaro in *Kew Bulletin* (1969) 23, 530, and in *Flora Tropical E. Africa*, Geraniaceae (1971) 20.

P. multibracteatum A. Richard, *Tentamen florae abyssinicae* (1847) 1, 119; Cufodontis (1956) 351.

HYPERICACEAE St. John's-wort family

Hypericum quartinianum *A St. John's-wort*

A well-branched shrub with large, conspicuous yellow flowers. It occurs in upland evergreen bushland and at forest edges, in the mountains of eastern Africa as far south as Malawi. It was recorded at Adderghei near the River Lurni.

VERNACULAR NAMES: Feel-fetch, Fiel fetch (MS and *Travels* [1805] 7, 325; [1813] 7, 341).

a. B 1977.14.8993 *v.* Outline sketch of flowering shoot, with details of fruiting spray, leaf, and fruits. Pencil, indented for transfer; 302 × 199; inscr., top left, in pen and black ink: Fiel fetch; followed by copious MS notes beginning: Albero Picolo ritrovato a Adderghei in Vicinanza del rio Lurni... (Small tree found at Adderghei in the neighbourhood of the river Lurni...). *Fig. 132.*

b. B 1977.14.9103. Finished drawing from a, without additional details.

Watercolors and bodycolors over pencil outlines, within ruled pencil borders; 406 × 306; inscr. *v.*, in pen and brown ink: Feel Fetch. *Fig. 13.*

BOTANICAL REFERENCES: *H. quartinianum* A. Richard, *Tentamen florae abyssinicae* (1847) 1, 97; Cufodontis (1959) 589; Robson in *Flora Tropical E. Africa*, Guttiferae (1978) 29.

Note: Dr. N.K. Robson, the authority on this genus, points out that this drawing matches *H. quartinianum* rather than the similar *H. roeperianum* Schimper ex A. Richard, owing to the absence of obvious venation on the leaves, the acutely acuminated sepals, and the styles being considerably longer than the stamens.

LABIATAE Mint family

Coleus edulis *Ethiopian potato*

A soft-stemmed, fragrant-leaved herb with ascending branches and blue flowers. Widely grown in eastern tropical Africa for the tubers that are eaten like potatoes.

VERNACULAR NAMES: Denitch, Dinitch (MS). Similar Amharic names are noted by Cufodontis (1963).

a. B 1977.14.8939 *r.* Outline sketches of flowering stem, tubers, details of leaf and flowers.

Pencil; 302 × 185; inscr., upper left, in pencil: Dinitch; at the top: ogni rango di Fiori tiena 10 – o 12. (Every whore of flowers has 10 or 12); *v.* inscr. five

lines of MS notes: Nel rovescia di questo Foglio di desegneto la pianta di Denitch (on the back of this sheet is drawn the plant Denitch). *Fig. 133.*
b. B 1977.14.8892. Finished drawing by J.F. Miller, of leafy flowering shoot, with details of flowers.
Watercolors and bodycolors over pencil outlines, within two double ruled ink borders; 397 × 308; signed *r.*, lower left, between borders in pen and black ink: John Frederick Miller del 1775; and below, in pen and brown ink: Stachys resupinata, *v.*, in pen and brown ink: No. 2 Stachys Resupinata. *Fig. 14.*
One of the very rare signed watercolors by an artist outside the circle of Bruce and Balugani and having no connection with the preliminary drawing.

BOTANICAL REFERENCES: *C. edulis* Vatke in *Linnaea* (1871) 37, 319; Cufodontis (1963) 833.

Coleus species 1

VERNACULAR NAME: None recorded.
B 1977.14.8901. Anonymous. Paris folder. Finished drawing of habit of flowering plant.
Watercolors and bodycolors over pencil outlines, within two double ruled ink borders; 404 × 306. *Fig. 135.*
Note: The detail of this excellent watercolor is, however, insufficient for positive identification of the plant depicted. It could be either a *Coleus* or a *Plectranthus.*

Coleus species 2

VERNACULAR NAMES: Jajja diitch, Yeya deetch (MS), yeyadeetch (*Travels*, [1805] 7, 325; [1813] 7, 341; Murray [1808] 450).
B 1977.14.8991 *r.* Outline sketch of plant with variegated leaves, with details of dissected flowers and another, fainter sketch of a flowering plant.
Pencil; 334 × 229; inscr., top left, in pen and black ink: Jajja Diitch; top right: Yeya Deetch; and with copious MS notes beneath beginning: Quest è una pianta ritrovata a Coskam il pmo di Luglio (This is a plant found at Coskam on the first of July). *Fig. 134.*
Note: The identification of this plant has also presented a problem as there are numerous species in this genus.

Isodon ramosissimus

A tall, fragrant herb up to 13 ft. (4 m) high, straggling and much branched, with very glandular inflorescences and white and bluish flowers. It grows in streamside forest in upland through much of tropical Africa; the plant drawn was found at Sakalla near the source of the Blue Nile on 5 November 1770.

VERNACULAR NAME: None recorded.
B 1977.14.8945 *r.* Outline sketch of flowering stem, with details of flowers and leaves.
Pencil; 311 × 222; inscr., right and below, with extensive MS notes beginning: Quest è una pianta ritrovata a Sakalla li 5 Novembre 1770. Si chiama [space left] (This is a plant found at Sakalla on the 5th November 1770. It is called [no name given]); also with letters A to E against details. *Fig. 136.*

BOTANICAL REFERENCES: *I. ramosissimus* (Hooker fil.) Codd in *Taxon* (1968) 17, 239; Friis, Rasmussen & Vollesen in *Opera Botanica* (1982) no. 63, 48.
Plectranthus ramosissimus Hooker fil. in *Journal Linnean Society, Botany* (1862) 6, 17.
Homalocheilos ramosissimus (Hooker fil.) J.K. Morton in *Journal Linnean Society, Botany* (1962) 58, 268 t. 6, and in *Flora West Tropical Africa* ed. 2, (1963) 2,4 60.
Plectranthus schimperi Vatke in *Linnaea* (1871) 37, 317; Cufodontis (1963) 838.

Otostegia tomentosa subspecies **ambigens**

A branched herb with white flowers, growing on rocky hillsides in Ethiopia and found at La Malmon.

VERNACULAR NAMES: Tinguit (MS).
B 1977.14.8935 *r.* Outline sketch of flowering stem with details of flowers and leaf.
Pencil; 302 × 199; inscr. on both sides of central subject with extensive MS notes beginning: Queste è l'Idea di una Pianta ritrovata su la Malmon/si chiama Tinguit. (This is the sketch of a plant found at La Malmon called Tinguit). *Fig. 137.*

BOTANICAL REFERENCE: *O. tomentosa* A. Rich. subsp. *ambigens* (Chiov.) Sebald in *Stuttgarter Beiträger zur Naturkunde*, ser. A, (1973), no. 263, 42.

Phlomis herba-venti

A much-branched herb with pink mauve flowers, growing on rocky hillsides in Syria.

VERNACULAR NAME: (Tinguitt in pencil is misleading as that name applies to the Ethiopian *Otostegia*, see last entry).
a. B 1977.14.9108. Outline of habit of plant.
Pencil; 215 × 345. *Fig. 138.*
b. B 1977.14.9107. Finished drawing from a.
Watercolors and bodycolors over pencil outlines; 531 × 364; inscr. *v.* in pen and brown ink: Tinguitt. *Fig. 15.*

BOTANICAL REFERENCES: *P. herba-venti* Linnaeus, *Species Plantarum* (1753) 586; Dinsmore and Post, *Flora Syria, Palestine & Sinai* (1933) 2, 397.

Pycnostachys abyssinica

A soft-stemmed shrub 6–10 ft. (2–3 m) high in mountain forests of Ethiopia. The deep blue flowers are clustered in a dense, terminal inflorescence. It was found at Gondar.

VERNACULAR NAME: Kinchette (MS).

B 1977.14.9079 *r*. Outline sketch of part of a flowering plant with details of floral dissections, leaf, and flowering head.

Pencil; 316 × 221; inscr. with extensive MS notes to the right, in pen and black ink, with the name Kinchette (line 6) and beginning: Quest è un albero Aromatico ritro-/vato a Gonder li 24 Novbre (This is an aromatic tree found at Gondar on 24 November); and with letters against details; *v*. continued MS notes in pen and black ink. *Fig. 139*.

BOTANICAL REFERENCES: *P. abyssinica* Fresenius in *Flora* (1838) 2, 608; Cufodontis (1963) 830.

Salvia nilotica *Nile Sage*

A perennial herb with blue flowers growing in open places in the mountain forests of Ethiopia.

VERNACULAR NAME: None recorded by Bruce.

a. B 1977.14.8889. Anonymous. Paris folder. Finished drawing of a flowering shoot with details of floral dissections.

Watercolors and bodycolors over pencil outlines, within two double ruled ink borders; 406 × 308; inscr. *v*., in pen and brown ink: Sauge du Nil/No. 6. (Sage of the Nile No. 6). *Fig. 140*.

b. B 1977.14.8888. Anonymous. Paris folder. Finished drawing of a flowering shoot and root.

Watercolors and bodycolors over pencil outlines, within four ruled black and brown ink borders; 406 × 306; inscr. *v*., in pen and brown ink; Sauge du Nil/No. 5. *Fig. 16*.

BOTANICAL REFERENCES: *S. nilotica* Jussieu ex Jacquin, *Hortus Vindobensis* (1776) 3, 48 t. 92; Cufodontis (1962) 819. See note below.

S. abyssinica Linnaeus fil., *Supplementum* (1781) 88; Jacquin, *Icones Plantarum Rariorum* (1781) 2, t. 6 and *Collectanea* (1787) 1, 132.

Note: *Salvia nilotica*, though described by Jacquin, was named by Jussieu at Paris and grown in the royal gardens. It is reasonable to assume that it was grown from Ethiopian seeds brought back by James Bruce as the figures are in the Paris folder, and the French name "Sauge du Nil" fits in admirably with this assumption. There are dried specimens in the Natural History Museum, Vienna which are isotypes. Lamarck (*Encyc. Méth. Bot.* [1804] 6, 608) was evidently misled by the epithet into believing that it originated in Egypt. It is possible that Linnaeus filius described the same live plant under the name *Salvia abyssinica*. The above synonymy is that given by Cufodontis.

Labiatae indeterminate species

This plant is difficult to identify owing to the lack of flowers, but it could be *Salvia coccinea* (Etlinger, Salvia Diss [1777] 23) which is based on a plant grown from Bruce's seeds.

VERNACULAR NAME: Zatarindi (MS).

B 1977.14.9109. Finished drawing of a leafy shoot, without flowers.

Bodycolors over pencil outlines; 541 × 385; inscr. *r*., top right, in pen and brown ink, *Zatarindi*. *Fig. 144*.

LEGUMINOSAE Pea family

Acacia species *An acacia tree*

This tree is definitely not *Acacia seyal*, which is known by the same vernacular name, but the drawing is difficult to identify owing to lack of flowers and fruits; even the copious notes describe only the leaves and spines and provide no indication of habitat or locality, but according to information in *Travels* [1805] 7, 324; [1813] 7, 340, "the sunt and saiel acacias" were drawn in Egypt.

VERNACULAR NAME: Seyali Hindi (MS).

a. B 1977.14.8977 *r*. Outline sketch of a tree and, left, of bark.

Pencil; 296 × 181; inscr. above, right, in pencil: Accasia. *Fig. 142*.

b. B 1977.14.9035 *r*. Careful drawing of spiny shoots, with details of leaves.

Pencil; 450 × 342; inscr. above, in pen and brown ink: Seyali Hindi; below, twelve lines of notes beginning: Questo Alberetto di spini tiene al tiglia Color de Fuligine (This small, spiny tree has the branch the colour of soot). *Fig. 143*.

c. B 1977.14.9035 *v*. Pale unfinished outline of same shoot.

Pencil. *Fig. 144*.

Albizia gummifera *Gummy albizia tree*

A large forest tree with spreading crown; the leaves have many small leaflets and are deciduous. Flowers are numerous in heads, bearing white petals and long, white, staminal tube tipped with red; the thin pods are about 15 cm long and 2 cm wide. It occurs in the humid areas of both lowlands and uplands of eastern Africa and westwards to Nigeria. In Bruce MS (.8961 *v*.) it is stated that the trunk is used for building timber, and the properties of the soluble gum are described. The plate was drawn from a cultivated tree at

Enfras near the northeast bank of Lake Tana.

VERNACULAR NAME: Sassa (MS and *Travels*). Similar Amharic and Tigrinia names are noted by Mooney (1963).

a. B 1977.14.8940 *r*. Drawing of inflorescence and details of florets.

Pencil; 224 × 164. *Fig. 145*.

b. B 1977.14.8961 *r*. Outline drawing of a flowering leafy shoot, with detail of leaf.

Pencil, indented for transfer; 340 × 235; inscr. *v*. with extensive MS notes, in pen and black ink, beginning: Nel Retrodisegnato Foglio è Rappresentato un Ramo con fiori e foglie dell albero Sassa Ritrovato a Enfras (On the reverse is represented a shoot with flowers and leaves of the tree Sassa found at Enfras). *Fig. 146*.

c. B 1977.14.8906. Finished drawing from b with minor variations in flowering head.

Watercolors over pencil outlines; 317 × 246; inscr. *v*., in pen and brown ink: Sassa No. 18. *Fig. 152*.

Engr. in reverse, Bruce, *Travels* (1790) pl. opp. p. 29.

d. B 1977.14.8907. Finished drawing of flowering head of shoot, with details of inflorescence, florets, and leaf.

Watercolors and bodycolors over pencil outlines, within ruled pencil borders; 317 × 246; inscr. faintly, up the left and right sides, in pencil, in Bruce's hand: This [illegible] of leaves / and this parcel of flowers / are to be left out in the Engraving. *Fig. 148*.

Engr., without leaf and inflorescence, in Bruce, *Travels* (1790) 5, pl. opp. p. 28; (1805) 7, pls. 4, 5; (1813) 7, pl. 4, 5.

BOTANICAL REFERENCES: *A gummifera* (J.F. Gmelin) C.A. Smith in *Bulletin Miscellaneous Information Kew* (1930) 218; Cufodontis (1954) 184; Brenan in *Flora Tropical East Africa*, Leguminosae – Mimosoideae (1959) 157. Type: Bruce's two plates. "Sassa" Bruce, *Travels* (1790) 5, 27 & two pl. opp. p. 27; IDC microfiche 5501–2: III.8, IV.2; (1805) 7, 41–7, and 8, pls. 4 and 5; (1813) 7, 157–63, and 8, pls. 41 and 5.

Sassa gummifera J.F. Gmelin, *Systema Naturae*, (1791) 2(2), 1038. Type: as above.

Mimosa sassa Bruce, *Travels*, ed. Castera (1792) p. 6, pls. 4–5.

Albizia malacophylla *An albizia tree*

A medium-sized tree with white, fragrant flowers, but a shorter staminal tube than *A. gummifera*. It is distributed westward from Ethiopia to Guinée.

VERNACULAR NAME: None recorded.

a. B 1977.14.9089d. Finished drawing of a tree without flowers.

Watercolors and bodycolors over pencil outlines; 533 × 375; inscr. above right, in pencil: *4. Fig. 149*.

b. B 1977.14.9089e. Finished drawing of a flowering shoot.

Watercolors and bodycolors, over pencil outlines; 533 × 375; inscr. upper right, in pencil: *5. Fig. 150*.

BOTANICAL REFERENCES: *A. malacophylla* (A. Richard) Walpers, *Annales Botanices Systematicae* (1852) 2, 457; Brenan in *Flora Tropical East Africa*, Leguminosae – Mimosoideae (1959) 145.

Astragalus atropilosulus *A tragacanth vetch*

A perennial herb usually about 16 in. (40 cm) high with conspicuous stipules and purple flowers. It is a variable, widespread species of eastern Africa and the Yemen. It was found at Gondar.

VERNACULAR NAME: Allemitth (MS).

B 1977.14.9041 *r*., left. Sketch of infloresence and leaves, with details of flowers.

Pencil; 302 × 199; inscr. in pen and black ink: Allemitth; and below with copious MS notes beginning: Questa è una pianta ritrovata a Gonder. (This is a plant found at Gondar). *Fig. 147*.

BOTANICAL REFERENCES: *A. atropilosulus* (Hochstetter) Bunge, *Mémoires Académie Sciences St. Pétersbourg* (1868), 11(1), 6 and (1869), 15(2), 4; Gillett in *Flora Tropical East Africa*, Leguminosae – Papilionoideae (1971) 1054.

A. venosus sensu Cufodontis (1955), 289, non A. Richard.

Bauhinia farec

A shrub with white flowers known only from Bruce's description and illustrations (see notes below).

VERNACULAR NAME: Farek (*Travels*).

B 1977.14.8896. Anonymous. Paris folder. Finished drawing of a flowering shoot, with details of flowers below.

Watercolors and bodycolors over pencil outlines; within two double ruled ink borders; 317 × 247; inscr. above the top border, in Bruce's hand: No border but to be plain as the rest. *Fig. 151*.

Engr. Bruce, *Travels* (1790) 5, pl. 57; (1805) 8, pl. 18; (1813) 8, pl. 18.

BOTANICAL REFERENCES: *B. farec* Desv., *Journal de Botanique* (1814), 3, 74, as 'B. farek'; Cufodontis (1955) 212; Roti-Michelozzi in *Webbia* (1957) 13(1), 160. Type: Bruce's plate and description.

'*Farek*' or '*Bauhinia acuminata*' Bruce (non L.), *Travels* (1790) 57 and ('Farek' only) pl. IDC microfiche 5501–4: II, 7; (1805) 7, 167–174 and 8, pl. 18; (1813) 7, 183–90 and 8, pl. 18.

Note: Surprisingly, this plant has not been re-collected and it remains a mystery, although Bruce's plate is excellent and depicts the floral parts, if not the fruit. Bruce states in *Travels* (1790) that "this beautiful shrub was found on the banks of a brook, which, falling from the west side of the mountain of Geesh down the south face of the precipice where the village is situated, is the first water that runs southward into the lake Gooderoo, in the plain of Assoa." Such detailed directions should enable some botanist to find the spot, but Dr Mats Thulin, who visited the area in 1982, was unable to re-collect the plant and he fears extinction owing to destruction of the vegetation; on the other hand it might have been attributed to Bruce in error, and Bruce claimed it as his own.

One presumes that Bruce intended to provide it with the name *Bauhinia acuminata*, but this epithet was already used by Linnaeus for a different species and it was left to Desveaux to take up the vernacular name for the species.

Cadia purpurea

A shrub or small tree, with flexuous branches bearing pinnate leaves having numerous leaflets and decorative bell-shaped pink flowers.

BOTANICAL REFERENCES: *C. purpurea* (Picc.) Aiton, Hortus Kew. (1789) vol. 3, p. 492; Cufodontis (1955), p. 225.

Panciatica purpurea Piccivoli, Hortus Panciaticus (1783) p. 10.

Note: Although no drawing of this species occurs among Bruce's collection, there is a published engraving of the plant raised at the Panciatichi garden, Florence, from Bruce's seeds (see page 59 for discussion).

Cassia fistula *Purging cassia*

A medium-sized tree with very ornamental hanging inflorescences of yellow flowers. It is probably a native of tropical Africa but cultivated in warm countries throughout the world. It was found in a garden in Cairo.

VERNACULAR NAME: *Pudding pipe tree* (*Travels* 1805 and 1813).

a. B 1977.14.8978 *r.*, left. Outline sketch of open flowers and bud.

Pencil; 295 × 180; inscr. above, in pencil: Cassia. *Fig. 128.*

b. B 1977.14.8979 *v.*, right. Slight sketch of a shoot with flowers and fruit.

Pencil; 181 × 293; inscr. above, in pencil: Cassia. *Fig. 153.*

c. B 1977.14.9089a. Finished drawing of habit of flowering tree.

Watercolors and bodycolors over pencil outlines; 533 × 367; inscr. upper left, in pencil: Albero di Cassia (Cassia tree); upper right: 1. *Fig. 17.*

d. B 1977.14.9091. Finished drawing of leafy shoot with flowers.

Watercolors and bodycolors over pencil outlines; 486 × 345; inscr. lower left, in pencil, up the sheet: *Cassia fistula. Fig. 154.*

Engr. by Heath, in reverse, in Bruce, *Travels* (1805) 8, pl. 44; (1813) 8, pl. 44.

BOTANICAL REFERENCES: *C. fistula* L., *Species Plantarum* (1753) 337; Bruce, *Travels* (1805) 7, 327–330 and 8, pl. 44, as *Cassia;* (1813) 7, 343–346 and 8, pl. 44; Murray (1808) 452–454, pl. 7; Cufodontis (1955) 216.

Crotalaria pallida

A well-branched herb 3 ft. (1 m) or more high, with yellow flowers and inflated pods. Found on the riverside at Tobulachè.

VERNACULAR NAME: None recorded.

a. B 1977.14.8995 *r.*, right. Faint outline drawing of leafy shoot with flowers and seed pods.

Pencil; 302 × 397 (full sheet but lower right-hand half torn away with part of drawing); inscr. above, in pen and brown ink, MS notes headed: EE specie di biselli [ie. piselli], (EE species of ? pea—the lettering and detail drawing missing—and beginning: Pianta ritrovata a Canto un Rio a Tobulachè. (Plant found on the riverside at Tobulachè). *Fig. 74.*

b. B 1977.14.9039 *v.* Outline drawing of a leafy shoot with flowers and seed pods.

Pencil; 301 × 185; inscr. above, in pencil: Maraqua è il Nome di un Albero che tiene le foglie come/[?] Cardo ma senza spini ritrovato a Sachalla (Maraqua is the name of a tree which holds its leaves like a Thistle but without thorns. Found at Sachalla). Notes do not apply to this drawing. *Fig. 155.*

c. British Museum (Natural History) Type specimen prepared from a plant grown at Kew from seeds collected by Bruce.

BOTANICAL REFERENCES: *C. pallida* Aiton, *Hortus Kewensis* (1789) 3, 20; Cufodontis (1955) 235; Polhill in *Flora Tropical East Africa*, Leguminosae – Papilionoideae (1971) 905.

Dichrostachys cinerea *Marabou Thorn*

A spiny shrub or small tree sometimes forming thickets and colonizing farmland in dry areas. The hanging inflorescences are half purple and half yellow.

Widespread in tropical Africa and Asia. Bruce (*Travels* [1790]) records the locality as "the banks of the river Arno, between Emfras and the lake Tzana."

VERNACULAR NAMES: Goneck, Guoneck (MS), Ergett y dimmo, Ergett dimmo (MS and *Travels*). An Amharic name similar to the latter is recorded by Mooney (1963).

a. B 1977.14.9040 *v*. Uncompleted drawing of leafy spray.
Pencil; 336 × 153; inscr. above, in pen and black ink: Guoneck; and to the right: Goneck. *Fig. 156*.

b. Lord Elgin's collection, Broomhall. Finished drawing of flowering shoot, apparently from a.
Watercolors over pencil outlines; 317 × 247; inscr. *v*., in pen and brown ink: Ergett Dimmo No. 7. *Fig. 159*.
Engr., in reverse, Bruce, *Travels* (1790) 5, pl. opp. 34; (1805) 8, pl. 6; (1813) 8, pl. 6.

BOTANICAL REFERENCES: *D. cinerea* (L.) Wight & Arnott, *Prodromus florae peninsulae Indiae orientalis* (1834) 271; Brenan in *Flora Tropical East Africa*, Leguminosae – Mimosoideae (1959) 36.
"Ergett y dimmo" Bruce, *Travels* (1790) 5, 34 and pl. opp.; microfiche IDC 5501–3: I, 2, (1805) 7, 147–48 and 8, pl. 6; (1813) 7, 163–64 and 8, pl. 6.
Mimosa sanguinea J.F. Gmelin, *Systema naturae*, ed. 13 (1791) 2, 803. Type: Bruce's plate and description.
D. glomerata (Forsskal) Chiovenda – Cufodontis (1955) 208.

Erythrina abyssinica *Lucky bean*

A tree or even a shrub with a rounded, spreading crown and large three-foliolate leaves. The striking red flowers are pollinated by birds. The pods are constricted almost like a string of beads, containing poisonous seeds. It is widespread in eastern Africa and extends to Angola. The MS notes on the next species may also apply to this one.

VERNACULAR NAMES: Cuara (Amharic), Zuaua, (Tigrinia), Kuara (MS and *Travels*). Similar Tigrinia and Amharic names are noted by Mooney (1963).

a. B 1977.14.8998 *r*. Rough outline drawing of flowering and fruiting stem (the flowers are of *E. brucei*—see the following species).
Pencil indented for transfer; 254 × 199; inscr. top left, in pen and black ink: Zuaua Tigri (called Zuaua in Tigrinia); top center: Amhara Cuara (in Amharic Cuara); with MS notes, top right. *Fig. 157*.

b. Lord Elgin's collection, Broomhall. Finished drawing from a.
Watercolors and bodycolors over pencil outline; 320 × 249; inscr. *v*., in pen and brown ink: Kuara No. 4. *Fig. 160*.
Engr., in reverse, Bruce, *Travels* (1790) 5, pl. opp. 65; (1805) 8, pl. 19; (1813) 8, pl. 19.

c. B 1977.14.9065 *r*. Outline drawing of flowering shoot.
Pencil, indented for transfer; 308 × 204. *Fig. 158*.

d. B 1977.14.6467g. Finished drawing from c but with many minor details altered.
Watercolors and bodycolors over pencil outlines; 451 × 312; inscr., upper right, in pen and brown ink: 7; lower center: Kuara; and *v*.: Kuara. *Fig. 161*.

BOTANICAL REFERENCES: *E. abyssinica* De Candolle, *Prodromus* (1825) 2, 413; Cufodontis (1955): 316; Gillett in *Kew Bulletin* (1962) 15, 425–29, pl. 3; Verdcourt in *Flora Tropical East Africa*, Leguminosae – Papilionodeiae (1971) 555; Wilson and Mariam (1979) 31. Type: A plant that was cultivated at the Petit Trianon, Versailles, from seeds brought from Ethiopia by Bruce; a herbarium specimen prepared from this plant was sent by Thouin to De Candolle at Geneva, where it remains (as the holotype), hence no specimen is to be found at the Paris herbarium.
Note: Unfortunately, Balugani included the fruits of this species on the floral drawing of *Erythrina brucei* (q.v.), which long confused botanists until J. B. Gillett confirmed the identity of the specimen grown from Bruce's seeds. By studying the details of the sketch it is possible to explain the error: the superimposition of a fruiting shoot on one in flower were later assumed to be the same plant. It is likely that Balugani believed them to be the same species.

Erythrina brucei

This tree has a much smaller distribution than the last species, being restricted to Ethiopia. The local name *Kuara* is derived from the district west of Lake Tana where the tree was said to be very frequent and where it yielded an important timber. Its seeds (or those of *E. abyssinica*) were said (*Travels* [1790], vol. 5, pp. 65–66) to be used for weighing gold but Bruce evidently confused this with the true Carob (*Ceratonia siliqua*) of the Mediterranean area.

VERNACULAR NAMES: Cuara (Amharic), Zuara (Tigrinia), Kuara (MS and *Travels*).

a. B 1977.14.8998 *r*. Sketch of flowering shoot with fruits (which are of *E. abyssinica*, see previous species).
Pencil; 254 × 199; inscr. at top, in pen and black ink: Zuaua Tigri Amora Cuara (see previous entry). *Fig. 157*.

b. B 1977.14.8998 *v*. Details of dissected parts of a flower, some with numbers attached.
Pencil; inscr., in pen and black ink, with copious MS notes beginning: Qui a basso è rappresentato un Fiore

dell'Albero di Cuara, con tutte le sue parti in grandezza Naturale (Here below is represented a flower of the tree Cuara, with all its parts life size). See also above, *E. abyssinica*, a and b. *Fig. 162.*

BOTANICAL REFERENCES: *E. brucei* Schweinfurth emend Gillett in *Kew Bulletin* (1962) 15, 428, pl. 3 (excluding fruits) and fig. 1; Cufodontis (1955) 317. Type: Bruce's plate in *Travels*, excluding the fruits and seeds.
"Kuara" Bruce, *Travels* (1790) 5, 65–66 and pl.; IDC microfiche 5501–4, IV.1; (1805) 7, 174–175, and 8, pl. 19; (1813) 7, 190–191, and 8, pl. 19.

Desmodium species

The identification of this drawing is tentative. This is a leguminous herb with articulated pods, which places it in the tribe *Hedysareae*.

VERNACULAR NAME: Jadzeferi, Yad zeffere (MS).
a. B 1977.14.8952 *r.* Outline drawing of a leafy flowering spray.
Pencil; 302 × 200; inscr. above, in pen and black ink: M Albero Jadzeferi (M, tree called Jadzeferi); and: à Adderghei—Fa un fiore giallo chiaro piu/che paglia (at Adderghei—it bears a flower which is clear yellow rather than straw-coloured). *Fig. 163.*
b. B 1977.14.6467k. Finished drawing from a.
Watercolors and bodycolors over pencil outlines; 451 × 316; inscr. upper left, in pen and brown ink: II; and v.: Yad zeffere. *Fig. 166.*

Millettia ferruginea

A tree of mountain forests in Ethiopia, with beautiful purple flowers. The pods are used as fish poison.

VERNACULAR NAMES: Berberrà, Burberra Amkah Sassa (MS). Cufodontis (1955) notes the names Berbera, Bererebera and similar ones in Amharic and Tigrinia but the MS "Sassa" is misleading since this usually denotes Albizia.
a. B 1977.14.8947 *r.* Outline drawing of flowering shoot.
Pencil, indented for transfer; 302 × 198; inscr., above, in pen and black ink: Berberrà, Nella pronuncia di questo Nome si deve confondere il Suono / del B con il suono dell'U di modo che viene pronunciato in circa come il B greco (Berberrà. In the pronunciation of this name one should merge the sound of the B with the U sound in such a way that it comes to be pronounced something like a Greek B). *Fig. 164.*
b. B 1977.14.8947 *v.* Outline details of flowering stem, dissected flowers, and leaf.
Pencil, indented for transfer; inscr., top left, in pen and black ink: Fiore di Berberra (Flowers of Berberra); with, right, MS notes against the stamens. *Fig. 165.*
c. B 1977.14.9112. Finished drawing after a.
Watercolors and bodycolors over pencil outlines, within ruled ink borders; 406 × 305; inscr. *v.*, by Bruce in pen and brown ink: Sassa or Berberra No 1 the flower dissected its parts to be transferred from the / other leaf No 2 and distributed as gracefully as possible at the bottom of this... *Fig. 167.*
d. B 1977.14.9113. Finished drawing after b, a few details excluded.
Watercolors and bodycolors over pencil outlines, within ruled pencil borders; 406 × 308; inscr. *v.*, by Bruce, in pen and brown ink: Sassa or Burberra Amka / the parts of the flower as dissected here are to be transferred to the other drawing No / 1 of the larger drawing there to be omitted preserving only the parts that indicated its kind or class. *Fig. 18.*

BOTANICAL REFERENCES: *M. ferruginea* (Hochstetter) Baker in Oliver, *Flora Tropical Africa* (1871) 2, 130; Cufodontis (1955) 285; Wilson and Mariam (1979) 32.
Berrebera ferruginea Hochstetter in *Flora* (1846) 29, 597.

Mimosa pigra

A sharply armed shrub forming thickets along rivers in tropical Africa and America. The small, round flower heads contain numerous pink flowers. Bruce (*Travels* [1790]) records the locality as "the banks of the river Arno, between Emfras and lake Tzana."

VERNACULAR NAMES: Ergett el krone, Ergett el kroun (MS and *Travels*).
a. B 1977.14.9077 *r.* Outline drawing (unfinished) of stem with leaves, flower, and fruit.
Pencil, with touches of watercolors; 302 × 199. *Fig. 168.*
b. B 1977.14.8908. Finished drawing from a.
Watercolors and bodycolors over pencil outlines; 313 × 248; inscr. *v.*, in pen and brown ink: Ergett el kroun No 6. *Fig. 171.*
Engr. *Travels* (1790) 5, pl. opp. 35; (1805) 8, pl. 7; (1813) 8, pl. 7.

BOTANICAL REFERENCES: *M. pigra* Linnaeus, *Centuria I. plantarum* (1755) 1, 13; Cufodontis (1955) 207; Brenan in *Flora Tropical East Africa*, Leguminosae – Mimosoideae (1959) 43.
"Ergett el Krone" Bruce, *Travels* (1790) 5, 35–36 and pl. opp. p. 35; IDC microfiche 5501–3, I. 3; (1805) 7, 148–49, and 8, pl. 7; (1813) 7, 164–65, and 8, pl. 7.
Mimosa cornuta J.F. Gmelin, *Systema naturae*, ed. 13 (1791) 2, 804. Type: Bruce's plate and description.

Pterolobium stellatum

A straggly shrub with sharply reflexed prickles in dry woodland throughout eastern Africa and tropical Arabia. The sweetly scented white flowers occur on long inflorescences which become conspicuous as the scarlet pods develop. Bruce relates how this plant, with its sharp, hooked prickles, was a nuisance to Ethiopian horsemen because it caught and tore their clothes. The emperor always ordered it to be cleared from any path he was likely to travel, and Bruce once witnessed a local governor, a Shum, being instantly executed for failing to clear the path properly.

VERNACULAR NAMES: Cantaffa, Cantuffa, Kantaffa, Kantuffa (MS and *Travels*). Schweinfurth (1896) and Mooney (1963) both note several similar variants of these names in Amharic, Tigrinia, and Gallinia of Harar.

a. B 1977.14.9036 *r*. Faint outlines of flowering head and flowering stem, with details of flowers.
Pencil; 302 × 185; inscr., at the top, in pencil: 10 stamine con Testa gialla pistillo palloro... (10 stamens with yellow head, the pistil straw-colored...); beneath: Cantooffa (deleted); and Cantaffa; and beneath again: Cantuffa. *Fig. 169*.

b. B 1977.14.9068 *v*. Outline drawing of flowering, leafy shoot.
Pencil, indented for transfer; 221 × 203; inscr., above, in pen and black ink, with numbers and calculations. *Fig. 170*.

c. B 1977.14.8913. Finished drawing from b.
Watercolors over pencil outlines; 312 × 244; inscr. *v*., in pen and brown ink: Kantooffa No 8. *Fig. 19*.
Engr., *Travels* (1790) 5, pl. opp. 49; IDC microfiche 5501–3, III.7; (1805) 8, pl. 14; (1813) 8, pl. 13.

BOTANICAL REFERENCES: *P. stellatum* (Forsskål) Brenan in *Memoires New York Botanical Garden* (1954) 8, 425, and in *Flora Tropical East Africa*, Leguminosae – Caesalpinoidae (1967) 42; Cufodontis (1955) 221; Roti-Michelozzi in *Webbia* (1957) 13(1), 181.
"Kantuffa" Bruce, *Travels* (1790) 5, 49–51 and pl. opp. p. 49; IDC microfiche 5501–3: IV. 7; (1805) 7, 160–63, and 8, pl. 14; (1813) 7, 176–79, and 8, pl. 14.
Cantuffa exosa J.F. Gmelin, *Systema naturae*, ed. 13 (1791) 2(1), 677. Type: Bruce's plate.
Acacia kantuffa Poiret, *Encyclopédie, Supplementum* (1817) 5, 529. Type: Bruce's plate.

Trifolium polystachyum *A clover*

A perennial clover usually with rather long, creeping stems and oblong heads with many purplish flowers; found in damp places and margins of upland forest of northeast Africa. It occurred at Gondar.

VERNACULAR NAME: None recorded.

B 1977.14.8994 *r*., right. Slight outline of a leafy, flowering stem with details of leaves and flowers.
Pencil; 302 × 397; inscr., top left, in pen and black ink: III; with letters against details left and right, and with two columns of MS notes beginning: Quest è una specie di trefoglio / ritrovato a Gonder fiorita nel / principio di Ottobre (This is a species of clover found at Gondar flowering at the beginning of October). *Fig. 172*.

BOTANICAL REFERENCES: *T. polystachyum* Fresenius in *Flora* (1839) 22, 50; Cufodontis (1955) 252; Gillett in *Flora Tropical East Africa*, Leguminosae – Papilionoideae (1971) 1030.

Trifolium rueppellianum *A clover*

An annual clover of upland grassland in much of the northern portion of tropical Africa, with small, round heads of purple flowers, occurring at Gondar.

VERNACULAR NAMES: The MS name Gheccià on this sheet seems to relate to the space below, where Balugani evidently intended to draw another plant.

a. B 1977.14.8994 *r*., left. Careful outline drawing of flowering stem with leaves and details of leaves and flowers.
Pencil; 302 × 397; inscr., top, in pen and black ink: No. 38 Gheccià Giunco ritrovato a Gonder nel principio di Ottobre (No. 38 Gheccià a sedge which was found at Gondar at the beginning of Otober); a note with no relevance to the drawing below, the related lower note beginning: Il Trifoglio ritrovato a Gonder nel mese Settembre / Fiorito (Clover found in flower at Gonder in the month of September). *Fig. 172*.

b. B 1977.14.22701. (Drawing on same sheet as a bird) Outline of flowering shoot and leaves.
Pencil; 220 × 152. *Fig. 174*.

BOTANICAL REFERENCES: *T. rueppellianum* Fresenius in *Flora* (1839) 22, 51; Cufodontis (1959) 253; Gillett in *Flora Tropical East Africa*, Leguminosae – Papilionoideae (1971) 1030.

Trifolium schimperi *A clover*

A small clover occurring in the high country of Ethiopia.

VERNACULAR NAME: None recorded.

B 1977.14.8994 *v*., left. Careful outline drawing of plant with flowers and root with details of leaves and flowers.
Pencil; inscr., at the top, in pen and black ink: No. I; with a column of MS notes beginning: Quest e una Casta de Trefoglie assai particolare per la forma / delle Foglie e delli Frutti... ritrovato a Gonder... Semi

nel origio del mese di Ottobre. (This is a class of clover sufficiently distinguishable by the shape of the leaves and fruit... found at Gonder... In seed at the beginning of October). *Fig. 173.*

BOTANICAL REFERENCES: *T. schimperi* A. Richard, *Tentamen florae abyssinicae* (1847) 1, 173; Thulin in *Opera Botanica* (1983) 68, 200, fig. 47/3.

Trifolium usambarense *A clover*

An annual clover of wet upland throughout much of tropical Africa, with oblong heads of purple flowers. This identification is tentative owing to the absence of flowers from the drawing, especially as this species is at present not known from the Gondar area of Ethiopia but further south.

VERNACULAR NAME: None recorded.

B 1977.14.8994 *v.*, right. Outline sketch of a stem with leaves and details of leaves.
Pencil; inscr., at the top, in pen and black ink: IIII, and to the left of the central drawing, MS notes beginning: Quest' è un'altra specie di / Tre foglio ritrovato a Gonder / nel principio di Ottobre... (This is another species of clover found at Gondar at the beginning of October...). *Fig. 173.*

BOTANICAL REFERENCES: *T. usambarense* Taubert in Engler, *Pflanzenwelt Ost-Africa* (1895) C, 208; Cufodontis (1955) 255; Gillet in *Flora Tropical East Africa*, Leguminosae – Papilionoideae (1971) 1024.

LOGANIACEAE Strychnine family

Buddleia polystachya *An African buddleia*

A shrub or small tree occurring in the mountains of eastern Africa and Arabia; the leaves are conspicuously felted white on their undersurface; the inflorescences are long and narrow bearing orange flowers. In MS (.8938 *r.*) it is recorded from Addua, Gondar, and Saccala and it is mentioned that it flowers from October (at Gondar) to December (in Tigre).

VERNACULAR NAMES: Umfar, Amfar (MS and *Travels* [1805 and 1813]). Variations of the same Amharic name are recorded by Schweinfurth (1896), Cufodontis (1960) and Mooney (1963).

a. B 1977.14.8938 *r.* Outline drawing of flowering shoot, with detail of leaf.
Pencil, indented for transfer; inscr. around the drawing, with lengthy MS notes beginning at the top: Umfar Questo rappresenta L'estremita di un Ramo di un / albero chiamato Umfar. Questo si ritrova a Addua e a Gonder, e a Saccala... (This represents the end of a branch of a tree called Umfar. This is found at Addua, at Gonder and at Saccala...). *Fig. 177.*

b. B 1977.14.9105. Finished drawing from a.
Watercolors and bodycolors over pencil outlines, within ruled ink borders; 406 × 295; inscr., *v.*, in pencil: Umfar; and: Lentana [?]. *Fig. 20.*
Engr. in reverse, by Heath, Bruce, *Travels* (1805) 8, 53; (1813) 8, pl. 53; Murray (1808) pl. 16.

BOTANICAL REFERENCES: *B. polystachya* Fresenius in *Flora* (1838) 21, 605; Cufodontis (1960) 677; E. A. Bruce & Lewis in *Flora Tropical East Africa*, Loganiaceae (1960) 36.
"Umfar" or "Amfar" Bruce, *Travels*, (1805) 7, 339–40 and 8, pl. 53, as "Umfar" or "Lentana"; (1813) 7, 355–56, and 8, pl. 53; Murray (1808) 450, pl. XVI.
B. acuminata R. Brown in Salt, *Voyage to Abyssinia*, Appendix (1814) p. lxiii, non Poiret (1811). Type: Bruce's plate.

Nuxia oppositifolia

A tree of forest slopes in the mountains of tropical Africa. The fragrant white flowers occur in profusion in dense, spreading inflorescences at the ends of the shoots. It occurred by the river Lumi at Adderghei.

VERNACULAR NAMES: Atcarò, Hatcarù (MS).

a. B 1977.14.8937 *r.*, right. MS notes only, third item in column, inscr.: Hatcarù—Albero che tiene picoli semi come finochio (Hatcarù—Tree which bears small seeds like fennel).

b. B 1977.14.8993 *r.* Careful outline drawing of leafy, flowering shoot, with detail of leaf.
Pencil, indented for transfer; 302 × 199; inscr., above, with eleven lines of MS notes in pen and brown ink, beginning: Atcarò Albero che cresce a Canto il Rio Lumi a Adderghei. (Atcarò, a tree which grows by the side of the river Lumi at Adderghei). *Fig. 178.*

c. B 1977.14.64680. Finished drawing from b without detail of leaf.
Watercolors and bodycolors over pencil outlines, within ruled pencil borders; 451 × 313; inscr., upper right: 15; and *v.*: in pen and brown ink: Atcaro. *Fig. 175.*

BOTANICAL REFERENCES: *N. oppositifolia* (Hochstetter) Bentham in A. De Candolle, *Prodromus* (1846) 10, 455; Cufodontis (1960) 677; E. A. Bruce & Lewis in *Flora Tropical East Africa*, Loganiaceae (1960) 43; Leeuwenberg in *Mededeelingen Landbouwhoogeschool*, Wageningen (1975) 75–78, 52.

LORANTHACEAE Mistletoe family

Loranthus globiferus *An African mistletoe*

An African mistletoe with leathery leaves and shining red flowers, parasitic on the branches of trees and shrubs in tropical Africa. This is shown growing on *Gardenia ternifolia.*

VERNACULAR NAME: None recorded.
B 1977.14.8984 *r*. Rough sketch of flowering, leafy shoot.
Pencil; 231 × 200; inscr. top left, in pen and brown ink: Il fiore di questo aborto è / tutto rosso di Carmine... (The whole flower of this abortion is carmine red...). *Fig. 179.*

BOTANICAL REFERENCES: *L. globiferus* A. Richard, *Tentamen florae abyssinicae* (1847) 1, 341; Cufodontis (1953) 28.

Loranthus heteromorphus *An African mistletoe*

Another mistletoe that parasitizes acacia trees in eastern Africa.

VERNACULAR NAME: None recorded.
B1977.14.8996 *r*. Rough sketch of leaves and flowers, with details of the flower.
Pencil; 340 × 234. *Fig. 180.*

BOTANICAL REFERENCES: *L. heteromorpha* A. Richard, *Tentamen florae abyssinicae* (1847) 1, 340; Cufodontis (1953) 29.

LYTHRACEAE Purple loosestrife family

Lawsonia inermis *Henna*

A much-branched shrub of hot lowlands, with fragrant white flowers. This is the well-known henna that has long been grown in tropical and subtropical Africa for the sake of its fragrant flowers and the dye obtained from its leaves. The latter stains hair and nails an orange-brown color and it was used in the preparation of ancient Egyptians mummies. The Arabs also prepare a perfume from its flowers and it is surprising that Bruce and Balugani do not record a vernacular name, nor is there a field sketch or notes.

VERNACULAR NAME: None recorded.
B 1977.14.9124. Finished drawing of a flowering leafy shoot.
Watercolors over pencil outlines; 522 × 370. *Fig. 181.*

BOTANICAL REFERENCES: *L. inermis* Linnaeus, *Species Plantarum* (1753) 349; Keay in *Flora West Tropical Africa* ed. 2, (1954) 1, 163; Cufodontis (1959) 611.

Woodfordia uniflora

A much-branched shrub of seasonal stream beds in tropical Africa, with numerous narrow red flowers. Found at Mai Agam and Adderghei.

VERNACULAR NAMES: Aitan balalitti, Etam balalite, Etan balabelle, Etan balalli (MS). Cufodontis (1959) records variants of these names in Tigrinia and Amharic.

a. B 1977.14.8992 *r*. Outline sketch of a flowering plant, with details of flowers, fruits, and leaf.
Pencil, indented for transfer: 302 × 202; inscr. ,at the top, in pen and black ink: Aitan balalitti; and: Etan balaliti; with MS notes against details left, the lower: Quest albero è com mune a Mai Agam (this tree is common at Mai Agam); and with extensive notes below, beginning: Quest è un albero ritrov. a Adderghei; cresce in Mezzo un Rio / frà grosse Pietre. (This is a tree found at Adderghei. It grows in the middle of a river among large stones); *v*. inscr. in pen and ink: Assai commune a Mai Agam (Rather common at Mai Agam). *Fig. 182.*
b. B 1977.14.9111 *r*. Finished drawing of flowering shoot.
Watercolors and bodycolors over pencil outlines, within ruled pencil borders; 406 × 310; inscr. *v*., in pen and brown ink: Etan balabelle or Etan balalli. *Fig. 176.*

BOTANICAL REFERENCES: *W. uniflora* (A. Richard) Koehne in Engler, *Botanische Jahrbucher* (1881) 1, 334; Keay in *Flora West Tropical Africa* ed. 2, (1954) 1, 163; Cufodontis (1959) 610.

MALVACEAE Mallow family

Abutilon longicuspe *An abutilon*

A shrub up to 20 ft. (6 m) high with numerous branches and decorative, rather large mauve flowers. It occurs in the drier hills of N.E. Africa and Arabia. It was found at Gondar.

VERNACULAR NAMES: Neccilò, Nev, Netchilò (MS). (Mooney [1963] applies the latter name to members of other families).
a. B 1977.14.9037 *r*. Outline drawing of flowering shoot.
Pencil, indented for transfer; 222 × 195. *Fig. 183.*
b. B 1977.14.9038 *r*. Outline drawing of flowering shoot, with details of flowers and leaf.
Pencil; 316 × 228; inscr., on the right, in pen and black ink, with copious MS notes beginning, at the top, with the name in Amharic followed by: Netchilò Neccilò Quest è un / albero ritrovato a Gonder in fiore alli 21 di / Novembre. (This is a tree found at Gondar in flower on the 21st of November). *Fig. 184.*
c. B 1977.14.6467c. Finished drawing from a.
Watercolors and bodycolors over pencil outlines, within ruled pencil borders; 451 × 312; inscr. upper right, in pen and brown ink: 5; and *v*., in pen and brown ink over pencil: nev. *Fig. 186.*

BOTANICAL REFERENCES: *A. longicuspe* A. Richard, *Tentamen florae abyssinicae* (1847) 1, 69; Cufodontis (1959) 537.

Hibiscus cannabinus *Deccan hemp*

A tall, erect annual with a tough, prickly-hairy stem grown for fibre. The large yellow flowers have a crimson red centre.

VERNACULAR NAME: None recorded.

B 1977.14.9029 *v*. Outline drawing of flowering stem. Pencil; 302 × 200. *Fig. 185*.

BOTANICAL REFERENCES: *H. cannabinus* Linnaeus, *Systema naturae* ed. 10, (1759) 1149. Keay in *Flora West Tropical Africa* ed. 2 (1958) 2, 347; Cufodontis (1959) 550.

Hibiscus species

The drawings of two flowers lack an epicalyx and one flower has five petals while the other has six.

B 1977.14.9054 *v*., left. Outline drawing of two flowers.

Pencil; 302 × 397; inscr. below, in pen and brown ink, with seventeen lines of MS notes unconnected with the above, beginning: Nel Rovesscio di questo Foglio è disegnato una speccie di Balsamina... (On the reverse of this sheet is drawn a species of balsam... [see *Cucumis metuliferus* p. 85]). *Fig. 267*.

MELIANTHACEAE

Bersama abyssinica

A straggling tree with leaves poisonous to stock, occurring in the montane forests of tropical Africa. Inflorescences bearing numerous whitish flowers, developing into thick-walled capsules that ultimately split and expose the red arilled seeds. Found at Cioba.

VERNACULAR NAMES: Corzamà, Corzuma, Korsuma (MS). Schweinfurth (1896) records a similar Triginia name.

a. B 1977.14.8960 *r*. Outline drawing of a fruiting stem, with details of dissected fruits and leaf.

Pencil, indented for transfer; 302 × 200; inscr. at the top: Corzama Korsuma Ritrovato a Cioba nel piccolo Bosco / avanti il luogo dove abbiamo messo / la Tenda (Corzamà, Korsuma found at Cioba in a small wood in front of the place where we had pitched the tent). *Fig. 187*.

b. B 1977.14.6468g. Finished drawing, in part from a (the fruits).

Watercolors and bodycolors over pencil outlines, within ruled pencil borders; 451 × 316; inscr. upper right, in pencil: 7; and *v*., in pen and brown ink: Korsuma, and upper left, in pencil: Meba (inverted). *Fig. 21*.

BOTANICAL REFERENCES: *B. abyssinica* Fresenius in *Museum Senckenbergianum* (1837) 2, 281; Verdcourt in *Flora Tropical East Africa*, Melianthaceae (1958) 2; Cufodontis (1958) 495.

MORACEAE Mulberry family

Ficus vasta *A wild figtree*

A medium-sized figtree with widely spreading branches. It is native in dry parts of N.E. Africa and the Yemen where it is planted in public places for the sake of its shade. In *Travels* (1805) 7, 325; (1813) 7, 341, it is noted that it "stands in the public square at Gondar". The small fig fruits drawn presumably are this species since it is the only one mentioned.

VERNACULAR NAMES: Daroo (MS and *Travels* 1805 & 1813). Similar Tigrinia names are noted by Schweinfurth (1896) and Mooney (1963).

B 1977.14.8949 *v*., right. Outline drawing of figs, one in section, (and small flowers of *Euphorbia abyssinica* q.v.).

Pencil; 310 × 443. *Fig. 53*.

BOTANICAL REFERENCES: *F. vasta* Forsskål, Flora Aegyptiaco-Arabica (1775) 179; Cufodontis (1953) 16.

MYRSINACEAE Ardisia family

Myrsine africana *African myrsine*

An evergreen unisexual shrub with numerous small, leathery leaves, reddish flowers, and small, round, purple fruits. Common in the tropical African mountains extending eastwards to China and westwards to the Azores. In Bruce MS (.8956 *v*.) it is noted that the fruits are used as a purgative.

VERNACULAR NAMES: Cagiém, Cajem, Cajiem, Jujef (MS). Various spellings of the same Amharic names are given by Cufodontis (1960) and Mooney (1963).

a. B 1977.14.8956 *r*. Careful outline drawing of a leafy, fruiting stem, with detail of leaf.

Pencil, indented for transfer; 302 × 199; inscr., top left: Cagiém arbusto (Cagiém a shrub); *v*. inscr. with extensive notes in pen and black ink beginning: Cagiem Arbusto Nel rovescio del questo foglio (Cajiem a shrub on the back of this sheet). *Fig. 188*.

b. B 1977.14.6468e. Finished drawing from a.

Watercolors and bodycolors over pencil outlines, within ruled pencil borders; 451 × 313; inscr., upper right, in pencil: 5; and *v*, in pen and brown ink: Cajem; in pencil: Jujef. *Fig. 189*.

BOTANICAL REFERENCES: *M. africana* Linnaeus, *Species Plantarum* (1753) 196; Cufodontis (1960), 655; Halliday in *Flora Tropical East Africa*, Myrsinaceae (1984) 6, fig. 2.

OLEACEAE Olive family

Jasminum abyssinicum *Ethiopian Jasmine*

A climbing shrub of upland forest, with trifoliolate leaves and decorative, pink-tinged white flowers, occurring in the mountains of eastern Africa. Balugani found it at Dippabahar, but it has never been recorded again in Ethiopia.

VERNACULAR NAMES: Abbe zelim, Abbi zelim (MS), Abbeselim (*Travels*). Similar names in Amharic and Tigrinia are noted by Schweinfurth (1896) and Mooney (1963).

a. B 1977.14.8935 *v*. Quick outline sketch of flowering shoot, with details of leaves.

Pencil, indented for transfer; 302 × 199; inscr., above right, with lengthy MS notes beginning: Abbe Zelim Ritrovato a Dippabahar (Abbe Zelim found at Dippabahar). *Fig. 190*.

b. B 1977.14.6467h. Finished drawing from a.

Watercolors and bodycolors over pencil outlines, within ruled pencil borders; 451 × 313; inscr., upper right, in pen and brown ink: 8; and *v*., in pen and brown ink: Abbe Zelim. *Fig. 194*.

BOTANICAL REFERENCES: *J. abyssinicum* De Candolle, *Prodromus* (1844) 8, 311; Turrill in *Flora Tropical East Africa*, Oleaceae (1952) 18; Cufodontis (1960) 671.

Jasminum dichotomum *African jasmine*

A woody climber with simple leaves in whorls of three and fragrant white flowers with a red tube. It is widespread in tropical Africa, and Bruce recorded it at Addergai "on the way to Gondar."

VERNACULAR NAME: Terràh (MS and *Travels*, [1805] 7, 324; [1813] 7, 340).

a. B 1977.14.9056 *r*. Outline drawing of flowering shoot, with details of leaves and flowers.

Pencil, indented for transfer; 302 × 199; inscr., above and below, in pen and gray ink, with MS notes beginning at the top faintly: A Addergai (At Addergai); and: Terràh piccolo [deleted] albero che fa fiori Bianchi con 10 Foglie Ciàscuno... (Terràh small [deleted] tree which bears white flowers with ten petals each...). *Fig. 191*.

b. B 1977.14.9056 *v*. Small drawing of dissected fruits.

Pencil in second section; 302 × 199; inscr. with six lines of MS beginning: Il qui Sotto disegno rappresenta la Sezione del Frutto dell' Albero Terràh (The drawing here below represents the section of the fruit of the tree Terrah...).

c. B 1977.14.9104. Finished drawing from a.

Watercolors and bodycolors over pencil outlines, within ruled pencil borders; 406 × 306; inscr. *v*., in pen and brown ink: Terrah. *Fig. 22*.

BOTANICAL REFERENCES: *J. dichotomum* Vahl, *Enumeratio* (1804) 1, 26; Turrill in *Flora Tropical East Africa*, Oleaceae (1952) 23.

Olea europaea subspecies **africana** *African olive tree*

A medium-sized tree with a dense rounded crown of grey-green leaves similar to the cultivated olive. It occurs in the mountains of eastern Africa to south Africa, and the Middle East. In *Travels* (1805) 7, 325; (1813) 7, 341, it is noted that this tree "which seems to be the wild olive", was found "on Lamalmon", in Semien, at a place called Maccara.

VERNACULAR NAMES: Jauirà, Javeira (MS and *Travels* (1805 and 1813). Other variants of these names are recorded by Schweinfurth (1896) and Mooney (1963).

B 1977.14.9042 *r*. Outline of a flowering, leafy shoot, with floral details.

Pencil; 302 × 199; inscr. with extensive MS notes, in pen and black ink, beginning, top left: Jauirà Albero Ritrovato Alla Malmon dalla parte di Sud in luogo detto Maccara (Jauirà, a tree found in the south part of Lamalmon, in a place called Maccara). *Fig. 192*.

BOTANICAL REFERENCES: *O. europaea* L. subsp. *africana* (Miller) P. S. Green in *Kew Bulletin* (1979) 34, 69.

O. africana Miller, *Gardeners Dictionary* ed. 8, (1768) n. 4.

OXALIDACEAE Wood sorrel family

Oxalis corniculata *A wood sorrel*

A small, creeping herb common in warm countries, with yellow flowers and beaked pods, found at Sakalla.

VERNACULAR NAME: None recorded.

B 1977.14.8999 *v*., top. Outline drawing of shoot with leaves and buds.

Pencil; 310 × 222; inscr. at the top, in pen and black ink: Tré Foglio Ritrovato a Sakala si chiama (Trefoil found at Sakala called [name unstated]). *Fig. 193*.

BOTANICAL REFERENCE: *O. corniculata* L., *Species Plantarum* (1753) 435; Cufodontis (1956) 352; Kabuye in *Flora Tropical East Africa*, Oxalidaceae (1971) 3.

PHYTOLACCACEAE Pokeberry family

Phytolacca dodecandra *A pokeberry*

A straggly shrub or climber with long inflorescences of yellowish flowers. Widespread at forest edges in eastern Africa. Its poisonous properties are well known. In small quantities, however, the roots are used against malaria; in larger quantity as an abortifact, but the dosage has to be carefully regulated to avoid causing the death of the patient. Its fruits are used as a soap substitute and in MS (.8943 *v.*) it is noted that the Ethiopians used its seeds for washing their cotton material.

VERNACULAR NAMES: Endaud, Endood (MS), Endoud (*Travels* [1805] 7, 324; [1813] 7, 340); Murray (1808) 449.

a. B 1977.14.8943 *v.* Outline drawing of leafy, flowering shoot.

Pencil, indented for transfer; 302 × 199; inscr., at the top, in pen and brown ink: Endaud; and below, with thirteen lines of MS notes beginning: Questo Rapresenta un Ramo di un albero chiamato Endaud (This represents a shoot of a tree called Endaud). *Fig. 195.*

b B 1977 14.8979 *r.*, left. Outline drawing of the habit of a shrub.

Pencil; 181 × 294; inscr., above left, in pencil: Fitolacha. *Fig. 196.*

c. B 1977.14.6469m. Finished drawing from a.

Watercolors and bodycolors over pencil outlines, within ruled pencil borders; 451 × 318; inscr., upper right, in pen and brown ink: 13; and *v.*, in pen and brown ink: Endood. *Fig. 197.*

BOTANICAL REFERENCES: *P. dodecandra* L'Héritier, *Stirpes novae* (1791) t. 69; Cufodontis (1953) 82; Polhill in *Flora Tropical East Africa*, Phytolaccaceae (1971) 2, fig. 1; Wilson & Mariam (1979) 33. Type: Plant cultivated in Paris from seeds collected by Bruce "on the way to Gondar," but a specimen could not be found (in 1978) in either Lamarck's or Jussieu's herbaria in Paris.

PITTOSPORACEAE Parchment-bark family

Pittosporum viridiflorum *A pittosporum*

A medium-sized forest tree of tropical African mountains. The inflorescences bear creamy-white flowers and rounded fruits that split to expose the red seeds. It was found growing at Sakalla on 5 November 1770.

VERNACULAR NAMES: Lillahoo, Lillahù (MS). Similar names are noted by Cufodontis (1954) and Mooney (1963).

a. B 1977.14.8989 *r.* Outline drawing of leafy, fruiting shoot.

Pencil, indented for transfer; 222 × 172 (the sheet is unevenly mutilated). *Fig. 198.*

b. B 1977.14.8990 *r.* Rapid outline drawing of leafy, fruiting shoot with details of leaf and fruits.

Pencil, indented for transfer; 312 × 221; inscr., top left, in pen and black ink: Lillahù; and below, left, letters against details; and right, lengthy MS notes beginning: Lillahù quest è un Ramo di un albero di detto Nome Ritrovato a Sakalla li 5 Novembre... (Lillahù, this is a shoot of a tree with the said name found at Sakalla on 5 November...). *Fig. 199.*

c. B 1977.14.6468d. Finished drawing from a.

Watercolors and bodycolors over pencil outlines, within ruled pencil borders; 451 × 312; inscr., upper right, in pencil: 4; and *v.*, in pen and brown ink: Lillahoo. *Fig. 23.*

BOTANICAL REFERENCES: *P. viridiflorum* Sims in Curtis, *Botanical Magazine* (1814) 41, t. 1684; Cufodontis (1954) 174, and in *Flora Tropical East Africa*, Pittosporaceae (1966) 3.

PROTEACEAE Protea family

Protea gaguedi *A protea*

A shrub or small tree of rough mountainsides in eastern Africa. The large, rounded heads contain numerous individual white florets. According to MS notes, it was found at Dippabahar and Sagassa, while Bruce (*Travels* [1790]) notes it as occurring in Lamalmon.

VERNACULAR NAMES: Gaguedi (*Travels*), Gagudei (MS and *Travels* [1805] 7, 325; [1813] 7, 341; Murray (1808), p. 450) The same name in Amharic and Tigrinia is recorded by Schweinfurth (1896) and Mooney (1963).

a. B 1977.14.9058 *r.* Rapid sketch of flower-head from above, with details of floral parts.

Pencil, indented for transfer; 302 × 199; inscr., at the top, in pen and brown ink: A Dippabahar (At Dippabahar), and with MS notes against details. *Fig. 200.*

b. B 1977.14.9058 *v.* Rough outline drawing of flower-head and leafy stem, from the side.

Pencil. *Fig. 201.*

c. B 1977.14.9075 *r.* More careful drawing of a, without details.

Pencil, indented for transfer; 302 × 199. *Fig. 202.*

d. B 1977.14.9069 *r.* Simplified, more careful drawing of b.

Pen and brown ink over pencil, indented for transfer; 302 × 198; inscr., at the top, in pen and brown ink, Gagudei albero Ritrovato a / Dippabahar, si Ritrova ancora a Sagassa (Gagudei, tree found at Dippabahar, also at Sagassa) and with 14 lines of MS notes reading from bottom to top *Passato il Rio Gerama ed avanti di arrivare a Sembrasaghi / abbiamo ritrovato un grande cardo selvaggio...* (Moving from the River Gerama and before reaching Sembrasaghi we found a large wild thistle...). *Fig. 203.*

e. B 1977.14.9069 *v*. Outline of leaf, inflorescence from above, and stem from the side and in cross section to show spiral arrangement of leaf growth. Pencil and pen and black ink; inscr. above and to the right, with extensive notes beginning: Gagudei Albero ritrovato a due miglia di distanza di dippabahar... (Gagudei. Tree found two miles from Dippabahar...) and with letters and numbers against diagrams of stem. *Fig. 204*.

f. B 1977.14.8914. Finished drawing from b.

Watercolors and bodycolors over pencil outlines, within ruled pencil borders; 312 × 246; inscr. *v*., in pen and brown ink: Gaguedi No. 12.

Engr. in reverse, Bruce, *Travels* (1790) 5, pl. opp. p. 52; (1805) 8, pl. 15; (1813) pl. 15. *Fig. 24*.

g. B 1977.14.8915. Finished drawing from d, with floral details from a.

Watercolors and bodycolors, over pencil outlines; 321 × 249; inscr. *v*., in pen and brown ink: Gaguedi No. 13. *Fig. 206*.

Engr. Bruce, *Travels* (1790) 5, pl. opp. p. 53; (1805) 8, pl. 16; (1813) pl. 16.

BOTANICAL REFERENCES: *P. gaguedi* J.F. Gmelin, *Systema naturae*, ed. 13, (1791) 2(1), 225; Cufodontis (1953) 23. Type: Bruce's plates and description. "Gaguedi" Bruce, *Travels* (1790) 5, 52–53 and two pls.; IDC microfiche 5501–04; I.4, 6; (1805) 7, 163–64 and 8, pls. 15 & 16; (1813) 7, 179–80 and 8, pls. 15–16. *Protea abyssinica* Willdenow, *Species Plantarum*, ed. 4, (1798) 1, 522. Type as above.

Note: The vernacular name is consistently spelt *Gagudei* in the MS, but the form *Gaguedi* has been used for the plates in *Travels*, from where Gmelin adopted it as the epithet.

PUNICACEAE Pomegranate family

Punica granatum *Pomegranate*

The well-known cultivated pomegranate bush with scarlet flowers and rounded fruits full of seeds, each surrounded by juicy pulp. The fruits are eaten in Ethiopia, and the crushed leaves are used to expel tapeworm.

VERNACULAR NAME: None provided.

a. B 1977.14.9089 (Nos. 31 & 33). Subject incidental as bird perch.

Watercolors; 531 × 380.

Perhaps a copy by Bruce from Balugani; it certainly has less quality than the latter's work.

b. B 1977.14.22700. Drawing of above.

Pencil; 484 × 372. *Fig. 207*.

c. B 1977.14.22718. Drawing of fruit (with vine cluster).

Pencil; 237 × 197. *Fig. 208*.

BOTANICAL REFERENCES: *P. granatum* Linnaeus, *Species Platarum* (1753) 472; Wilson & Mariam (1979) 33.

RANUNCULACEAE Buttercup family

Clematis hirsuta *A clematis*

A climbing shrub covering trees and rocks, widespread in tropical Africa in savanna. It was found on the shore of Laka Tana near Meschelaxos. Flowers whitish and the small fruits are clustered and plumed. In *Travels* (1805) 7, 325; (1813) 7, 341; Murray (1808) 450, it is noted that with this plant "the women perfume their clothes".

VERNACULAR NAMES: Alzazo, Azazo, Azzo (MS and *Travels* 1805 and 1813). A similar Amharic name is noted by Mooney (1963).

a. B 1977.14.8982 *r*. Outline drawing of shoots, one with flowers without leaves, the other with terminal leaves only, with detail of stem nodes.

Pencil; 302 × 183; inscr., to the right, with lengthy MS notes beginning: Questa Pianta Rampante è Ritrovata Avanti di / Arrivare a Meschelaxos sù le sponde del Lago in luogo / pietroso si chiama Azzo. (This rampant plant was found before reaching Meschelaxos on the shores of the lake [Tana] in a stony place. It is called Azzo). *Fig. 205*.

b. B 1977.14.8937 *v*., right. MS note only, second item.

221 × 313; inscr. in pen and brown ink: Azazo—Grande Albero che fa un Seme ovale... (Azazo—Large tree which bears an oval seed).

BOTANICAL REFERENCES: *C. hirsuta* Guillemin & Perrottet, *Florae Senegambiae tentamen* (1830) 1, 1; Milne-Redhead and Turrill in *Flora Tropical East Africa*, Ranunculaceae (1952) 6; Cufodontis (1953) 107.

Delphinium wellbyi *Wild delphinium*

A tall, erect herb about 3 ft. (1 m) high with blue flowers, occurring in grassland restricted to the mountains of Ethiopia; found at Sakalla.

VERNACULAR NAME: Dorwan (MS and *Travels*, [1805] 7, 325; [1813] 7, 341; Murray [1808] 450).

a. B 1977.14.8966 *v*. Rough sketch of flowering plant, with floral details.

Pencil; 310 × 221; inscr., top left, in pen and black ink: Dorowan [deleted]; Dorwan; with fourteen lines of MS notes beginning: Quest è una pianta Ritrovata a Sakala. (This is a plant found at Sakalla); and with letters against details and scattered ink blots. *Fig. 209*.

b. B 1977.14.8955 *v*. Careful outline drawing of plant. Pencil, indented for transfer; 221 × 210; inscr., upper right, in pen and brown ink: Semezza. (This name refers to the drawing of *Ruttya speciosa* on the other side.) *Fig. 210.*

c. B 1977.14.9110. Finished drawing from b.
Watercolors and bodycolors over pencil outlines, within ruled ink borders; 406 × 306; inscr., lower left, in pen and brown ink: Dorwan; lower right: No 19 Larkspur Anemon. *Fig. 25.*

BOTANICAL REFERENCES: *D. wellbyi* Hemsley in *Kew Bulletin* (1907) 360; Milne-Redhead and Turrill in *Flora Tropical East Africa*, Ranunculaceae (1952) 20; Cufodontis (1952) 107.

RHAMNACEAE Buckthorn family

Rhamnus prinoides *African buckthorn*

An evergreen straggly shrub or small tree occurring in forests and thickets in upland areas of tropical Africa. The small, greenish flowers produce rounded red berries. The use of this plant for initiating the fermentation and the flavoring of mead (*tej*) and beer (*talla*) was well known to Bruce (MS .9061) and is still practiced in almost all parts of Ethiopia. It was recorded at Gondar on 28 June 1770.

VERNACULAR NAMES: Geshe, Ghesh (MS and *Travels*, [1805] 7, 325; [1813] 7, 341). Many Amharic and Tigrinia variants of this name are noted by Schweinfurth (1896), Cufodontis (1958), and Mooney (1963).

a. B 1977.14.9061 *r.* Outline drawing of flowering, leafy shoot, with details of leaf and flowers. Transection of flower marked *A*; diagrammatic sketch of inflorescence marked *B* (see comment on p. 67).
Pencil, indented for transfer; 302 × 199; inscr., with extensive MS notes surrounding drawing and beginning, at the top: Ghesh Albero con cui li Abbisini fanno uso per mettere nel vino di miele (Ghesh tree which the Abyssinians use for putting into honey wine). *Fig. 211.*

b. B 1977.14.9096. Finished drawing from a, without details and with berries added.
Watercolors and bodycolors over pencil outlines, within ruled pencil borders; 406 × 304; inscr. *v.*, in pencil: Geshe. *Fig. 212.*
Engr., in reverse, Bruce, *Travels* (1805) 8, pl. 50; (1813) 8, pl. 50; Murray (1808) pl. xiii.

BOTANICAL REFERENCES: *R. prinoides* L'Héritier, *Sertum anglicum* (1788) 6, pl. 9; Cufodontis (1958) 501; M.C. Johnston in *Flora Tropical East Africa*, Rhamnaceae (1972) 18.
"Gesh" Bruce, *Travels*,(1805) 7, 335 and 8, pl. 50, as "Gheshe"; (1813) 7, 351 and 8, pl. 50; Murray, (1808) 459, pl. xiii.

Ziziphus abyssinica *Ethiopian Christ-thorn*

A shrub or small tree with sharp axillary thorns of upland grassland throughout tropical Africa. The small flowers in leaf axils produce reddish marblelike fruits, which are sometimes eaten in Ethiopia.

VERNACULAR NAMES: Corzama, Habetteri, Korzama (MS). Similar names cited by Schweinfurth (1896) and Mooney (1963) refer to *Z. spina-christi.*

a. B 1977.14.8983 *v.* Careful outline drawing of shoot in fruit, with details of leaf.
Pencil, indented for transfer; 231 × 200; inscr. below, in pen and black ink, with eight lines of MS notes beginning: Questo Disegno Rappresenta Il Ramo di un albero che fa un frutto come una ceresa il quale si mangia: suo nome è Habetteri (This drawing represents a shoot of a tree which bears a fruit like a cherry which is edible; its name is Habetteri). *Fig. 212.*

b. B 1977.14.6468j. Finished drawing from a, without detail.
Watercolors and bodycolors over pencil outlines, within ruled pencil borders; 451 × 314; inscr., upper right, in pencil: 10; and *v.*, in pen and brown ink: Korzama. *Fig. 214.*

c. B 1977.14.9046 *v.* Thirty lines of MS only.
302 × 199; inscr. in pen and black ink beginning: Corzamà Albero grande che porta Semi racchiusi dentro un frutto / rottondo della Grandezza di un Cerasa (Carzamà a large tree which bears seeds enclosed inside a round fruit the size of a cherry).

BOTANICAL REFERENCES: *Z. abyssinica* A. Richard, *Tentamen florae abyssinicae* (1847) 1, 136; Cufodontis (1958) 497; M.C. Johnston in *Flora Tropical East Africa*, Rhamnaceae (1972) 27.

ROSACEAE Rose family

Hagenia abyssinica

A slender tree characteristic of eastern African forests and thickets above 7900 ft. (2400 m), with large drooping inflorescences of small, pink flowers. The widespread traditional use of this species as a worm medicine in Ethiopia is described by Bruce (*Travels* [1790] 5, 73–76).

VERNACULAR NAMES: Cusso, Cuzzo, Kusso (MS and *Travels*).

a. B 1977.14.9033 *r.* Fragment of a sketch of stem and leaf bases.
Pencil; 153 × 99; inscr., top left, in pen and black ink; Disposizione delle / Tiglie del Albero / Cuzzo. (Arrangement of the branches of the tree Cuzzo). *Fig. 94.*

b. B 1977.14.9032 *v.* Fragment with incomplete outlines of leaves.

Pencil; 153 × 100; probably originally part of a. *Fig. 215.*
c. B 1977.14.9067 *r.* Outline of sprig with leafy shoots, and flowering shoot, with some brushstrokes in watercolors.
Pencil and watercolors, stronger strokes over fainter outline, indented for transfer; 302 × 199; inscr. at the top in pen and brown ink: Idea di un Ramo di Cuzzo con le Tiglie, e Foglie, e Fiori. (Sketch of a sprig of Cuzzo with shoots, leaves and flowers). *Fig. 216.*
d. B 1977.14.9067 *v.* Left, careful outline drawing of tip of leafy shoot; right, sketch of flowering shoot with details of flowers and some brushstrokes in watercolors.
Pencil and watercolors, indented for transfer; inscr., top left, in pen and brown ink: Idea dell'estremità di una Tiglia vestita di Foglie di Cuzzo / in grandezza Naturale (Sketch of the end of a leafy shoot of Cuzzo, life size); top right: Idea di un Picolo Ramo di Fiori di Cuzzo (Sketch of a small flowering shoot of Cuzzo). *Fig. 217.*
e. B 1977.14.8920. Finished drawing from d (right).
Watercolors and bodycolors over pencil outlines, within ruled pencil borders; 386 × 293. *Fig. 26.*
Engr. Bruce, *Travels* (1790) 5, pl. before 75; (1805) 8, pl. 23; (1813) 8, pl. 23.
f. B 1977.14.8919. Finished drawing from c.
Watercolors and bodycolors over pencil outlines, within ruled pencil borders; 343 × 270; inscr., upper right, in pen and brown ink: Kusso UNS. *Fig. 27.*
Engr. Bruce, *Travels* (1790) 5, pl. opp. p. 74; (1805) 8, pl. 22; (1813) 8, pl. 22.

BOTANICAL REFERENCES: *H. abyssinica* (Bruce) J.F. Gmelin, *Systema naturae*, ed. 13, (1791) 2, 613; Cufodontis (1954) 181; R. A. Graham in *Flora Tropical East Africa*, Rosaceae (1960) 43, fig. 5; Wilson and Mariam (1979) 31; Jansen, Hepper and Friis in *Taxon* (1980) 29, 511.
Type: Bruce's plates and description.
"Cusso" or *Banksia abyssinica* Bruce, *Travels* (1790) 5, 73–75 and two pls. opp. p. 73, as *Bankesia;* IDC microfiche 5501–05: I. 8, II. 2; (1805) 7, 181–84 and 8, pls. 22 and 23; (1813) 7, 197–200 and 8, pls. 22 and 23; Murray (1808) 449.
Note: The genus *Hagenia* J.F. Gmelin (1791) was founded on Bruce's material.

RUBIACEAE Bedstraw family

Coffea arabica *Arabica coffee*

This well-known arabica coffee bush or small tree is now widely cultivated in many parts of the world, although its wild ancestors occur in the upland thickets of Northeast Africa.

VERNACULAR NAMES: Bun (MS). Similar Amharic names are recorded by Mooney (1963).
a. B 1977.14.9060 *v.* Outline drawing of stem with leaves and berries.
Pencil; 302 × 199; inscr., at the top and to the left, with MS notes, in pen and brown ink, beginning: Ramo di Caffè con frutto non ancora Maturo… In Abyssinia si chiama bun. (Stem of coffee with fruit not yet ripe. It is called in Abyssinia bun). *Fig. 218.*
b. B 1977.14.8518. Anonymous, Paris folder. Finished drawing of leafy shoot and berries.
Gray wash over pencil outlines, within double ruled pen and ink borders; 271 × 217; inscr., below bottom border, in pen and brown ink: Coffee No 1; and *v.*, in pencil: Coffee tree [?]. *Fig. 219.*
c. B 1977.14.8519. Anonymous, Paris folder. Finished drawing of leafy shoot, with details of flowers, leaf, and fruits.
Gray wash and pen and brown ink over pencil outlines, within ruled double pen and black ink borders; 277 × 217; inscr., below bottom border, in pen and brown ink: Coffee No 2; and with letters against details. *Fig. 220.*

BOTANICAL REFERENCES: *C. arabica* Linnaeus, *Species Plantarum* (1753) 172; Cufodontis (1965) 1012.

Gardenia ternifolia subspecies **jovis-tonantis**
Wild gardenia

A shrub with a rather gnarled appearance and often seen leafless in tropical African savanna, bearing large white flowers and woody fruits. In Bruce's *Travels* (1805) 7, 325; (1813) 7, 341; Murray (1808) 449, there are notes that the Ethiopians used this plant for dyeing their fingers and nails.

VERNACULAR NAMES: Aja aquarceti, Aquariti (MS and *Travels*).
a. B 1977.14.9043 *r.* Fragment with drawings of dissected fruits and seed arrangement.
Pencil; 199 × 99. *Fig. 221.*
b. B 1977.14.9043 *v.* Fragment of outline drawing of flowering shoot.
Pencil, indented for transfer. *Fig. 222.*
c. B 1977.14.9066 *r.* Outline drawing of leafy shoot with flowers and fruits.
Pencil; 308 × 203. *Fig. 224.*
d. B 1977.14.9066 *v.*, top. Outline drawing of details of leaf and flowers.
Pencil; inscr., against details: Periantion (i.e. calyx); and: B, Sezione del Fiore (Section of the flower); and: C, A. *Fig. 225.*
e. B 1977.14.9078 *v.*, right. MS notes only.

Inscr., at the top, in pen and black ink: Albero Aquariti. (Tree [called] Aquariti); and description beginning: Quest albero porta frutti come una noce... (This tree bears fruits like a nut...).
f. B 1977.14.9117. Finished drawing from b and/or c. Watercolors and bodycolors over pencil outlines, within ruled pencil borders; 406 × 310; inscr. *v.*, in pen and brown ink: Aja Aquareti; and: No 15 Aquareti. *Fig. 28.*
g. B 1977.14.9118. Finished drawing from d with additional detail of fruit.
Watercolors and bodycolors over pencil outlines, within ruled pencil borders; 406 × 310; inscr., *v.*, in pen and black ink: Aquariti; and: No 16 [inverted]. *Fig. 223.*

BOTANICAL REFERENCES: *G. ternifolia* Schum. & Thonn. subsp. *jovis-tonantis* (Welwitsch) Aubréville, *Flore Forestière Soudano-Guinéenne* (1950) 460; Verdcourt in *Kew Bulletin* (1979) 34, 354.

Mussaenda arcuata

A straggling shrub of upland streamside thickets in tropical Africa. The star center of the yellow flowers turns brown with age; the fruits eventually become orange-colored. It was found at Adderghei "on the way to Gondar" (*Travels*, [1805] 7, 325; [1813] 7, 341); Murray (1808) 449.

VERNACULAR NAME: Deh hack, Dehack (MS and *Travels*).

a. B 1977.14.9046 *r.* Careful outline drawing of flowering and fruiting shoot, with details of leaves and flowers.
Pencil, indented for transfer; 302 × 199; inscr., at the top: Albero Deh Hack a Adderghei (Tree [called] Deh Hack [found] at Adderghei). *Fig. 226.*
b. B 1977.14.9056 *v.* Seventeen lines of MS notes only, the drawing near the top unrelated to this plant. 302 × 199; inscr., below the first line of division, in pen and brown ink: A Adderghei (At Adderghei); and: Deh Hack Albero che fa fiori gialli... (Deh Hack [a] tree which bears yellow flowers...).
c. B 1977.14.9115. Finished drawing from a, without details.
Watercolors and bodycolors over pencil outlines, within ruled pencil borders; 406 × 306; inscr. *v.*, in pen and brown ink: Deh Hack. *Fig. 228.*

BOTANICAL REFERENCES: *M. arcuata* Lamarck ex Poiret, *Encyclopédie Méthodique, Botanique* (1797) 4, 392; Hepper in *Flora West Tropical Africa*, ed. 2, (1963) 2, 165; Cufodontis (1965) 999.

RUTACEAE Rue family

Clausena anisata *Mosquito plant*

A common small tree with odorous leaves, white flowers, and black fruits, widespread in tropical Africa and Asia; found at El Kaha.

VERNACULAR NAMES: Lambehk, Lembetch (MS). Similar Amharic names are recorded by Mooney (1963).

a. B 1977.14.8962 *r.* Careful outline drawings of flowering shoot and of a leaf, with details of leaflet and flowers.
Pencil; 222 × 210; inscr., at the top: Lembetch Albero Ritrovato al Kaha (Lembetch tree found at El Kaha); *v.* inscr. extensive notes, in pen and black ink, beginning: Lembetch Nel Rovescio di Questo Foglio (Lembetch. On the reverse of this sheet). *Fig. 227.*
b. B 1977.14.8965 *r.* Careful outline drawing of flowering shoot.
Pencil, indented for transfer; 222 × 190. *Fig. 229.*
c. B 1977.14.6468k. Finished drawing from b.
Watercolors and bodycolors over pencil outlines, within ruled pencil borders; 451 × 313; inscr., upper right: 11; and *v.*, in pen and brown ink: Lambehk. *Fig. 29.*

BOTANICAL REFERENCES: *C. anisata* Hooker fil. ex Bentham in Hooker, *Flora Nigritana* (1849) 256; Cufodontis (1956) 371.

SALICACEAE Willow family

Salix subserrata *Willow tree*

A small, much-branched willow tree of riversides in tropical Africa and the Middle East. The greenish catkins produce an abundance of small, downy seeds. In Ethiopia, it is an extremely common species along rivers and smaller streams, often forming dense thickets; it was found at Gutta on the banks of the Blue Nile (Abbai).

VERNACULAR NAMES: Hà (MS and *Travels* [1805] 7, 325; [1813] 7, 341; Murray [1808] 450). Similar Amharic names are noted by Mooney (1963).

a. B 1977.14.9039 *r.* Outline drawing of leafy shoot with inflorescences and details of leaves and flowers.
Pencil, indented for transfer; 310 × 185; inscr., at the top, in pen and black ink: Hà Quest è una speccie di Salice Ritrovata su li Bordi dell' / Abbai nel Luogho dove l'abbiamo passato a Guttà. (Hà. This is a species of willow found on the banks of the Abbai [Blue Nile] at the place where we crossed to Guttà [Gouto?]). *Fig. 230.*
b. B 1977.14.6468 m. Finished drawing from a, without details.
Watercolors and bodycolors over pencil outlines,

within ruled pencil borders; 451 × 314; inscr., upper right, in pencil: 13; and *v.*, in pen and brown ink: Ha; and in pencil: Salix [inverted]. *Fig. 233.*

BOTANICAL REFERENCES: *S. subserrata* Willdenow, *Species Plantarum*, ed. 4, (1806) 4, 671; Cufodontis (1953) 4.

SANTALACEAE Sandalwood family

Osyris lanceolata *African sandalwood*

A shrub with waxy green leaves and short inflorescences of yellow flowers; fruits bright red. Common at the edges of upland forest in eastern Africa. It was seen at Sagassa.

VERNACULAR NAMES: Carratt, Carrat (MS). Similar names in Tigrinia and Amharic are recorded by Schweinfurth (1896) and Mooney (1963).

a. B 1977.14.8959 *r.* Careful outline drawing of fruiting shoot, with details of leaves.
Pencil, indented for transfer; 302 × 199; inscr., top left: Carratt. Arbusto (Carratt, shrub); *Fig. 232*; *v.* inscr. with detailed notes beginning: Carratt arbusto ritrovato a Sagassa (Carratt, shrub found at Sagassa).

b. B 1977.14.6468c. Finished drawing from a, without details.
Watercolors and bodycolors over pencil outlines, within ruled pencil borders; 451 × 312; inscr., upper right, in pencil: 3; and *v.*, in pen and brown ink: Carratt. *Fig. 231.*

BOTANICAL REFERENCES: *O. lanceolata* A. De Candolle, *Prodromus* (1857) 14, 633. *O. abyssinica* Hochstetter ex A. Richard (1851); Cufodontis (1953) 23.

SAPINDACEAE Litchi family

Cardiospermum halicacabum *Balloon vine*

A slender climber with tendrils on the inflorescences, occurring widely in tropical African thickets. The fruits are inflated bladders. Found at Emfras.

VERNACULAR NAMES: Semec, Semeck (MS and *Travels* [1805] 7, 325; [1813] 7, 341; Murray [1808] 450).

a. B 1977.14.8944 *r.* Lower part of drawing of a plant with fruits, growing up a stick.
Pencil; 151 × 198. *Fig. 234.*

b. B 1977.14.8944 *v.* Outline drawing of part of stem of plant with leafy and fruiting shoots, more visible in fainter outlines.
Pencil. *Fig. 235.*

c. B 1977.14.9001 *r.* Careful outline drawings of leaf and details of fruits with dissections.
Pencil; 221 × 190; inscr. *r.*, with six lines of MS notes in pen and black ink, beginning: Pianta Rampante chiamata Semeck che porta un frutto come / bulle unite insieme... (Climbing plant called Semeck which bears a fruit like round beads pressed together). *Fig. 236.*

d. B 1977.14.9001 *v.* Two unconnected outline drawings, one showing fruits, the other a leafy spray.
Pencil. *Fig. 237.*

BOTANICAL REFERENCES: *C. halicacabum* Linnaeus, *Species Plantarum* (1753) 366; Cufodontis (1958) 490; Keay in *Flora West Tropical Africa*, ed. 2, (1958) 1, 711.

SAPOTACEAE Chicle family

Mimusops kummel

A large tree of eastern African forests. The fragrant white flowers occur in the axils of upper leaves, and the pointed fruits are yellow. The leaves and fruits of this tree have been recorded from the tombs of ancient Egypt.

VERNACULAR NAMES: Cumel, Kumel, Kummel, Kummell—Tigre (MS and *Travels* [1805] 7, 325; [1803] 7, 341; Murray [1808] 450). Scie—Amhara (MS). The same Tigrinia and Amhara names are given by Cufodontis (1960).

a. B 1977.14.9028 *r.* Sketchy outline of shoot with leaves and fruits.
Pencil, rubbed; 233 × 171; inscr., top left, in pen and brown ink: Tigre Amara (i.e., Tigre and Amhara languages); in pencil: Cumèl Tigre / Cumèl, Tigrè (i.e., called Cumèl in Tigre); and right: Sciè Amara / Sciè, Amharic (i.e., scie in Amhara); beneath, in pen and brown ink: Alla Fine di decembre, ma soupratutto in Gennaio / Li frutti di questo albero Cumel sono Maturi (At the end of December, but above all in January, the fruits of this tree Kummel are ripe). *Fig. 238.*

b. B 1977.14.9076 *r.* Shoot with leaves and fruits, closely similar design to a.
Pen and black ink over pencil outlines, indented for transfer; 217 × 310; inscr. below, in pen and black ink, with MS notes beginning: Quest è l'Idea di una Branca di cumel con suoi frutti (This is a sketch of a shoot of Kummel with its fruits). *Fig. 241.*

c. B 1977.14.9026 *r.* Careful drawing of details of buds, flowers, and leaf.
Pencil; 302 × 199; inscr., below the subjects at the top with extensive notes in pen and brown ink beginning: Questo Rappresenta il Fiore di Cumel (This represents the flower of Kummel); *Fig. 239*; *v.* continuation of 7 lines of MS notes.

d. B 1977.14.9106. Finished drawing from b with some of the flowers from c.

Watercolors and bodycolors over pencil outlines, within ruled ink borders; 406 × 305; inscr. *v.*, in pencil: Kummell. *Fig. 240.*

Engr. in reverse, by Heath, in Bruce, *Travels* (1805) 8, pl. 54; (1813) 8, pl. 54; Murray (1808) pl. XVII.

e. B 1977.14.22702 *r.* Drawing of whole tree.

Pen and black ink; 237 × 173; inscr. in pen and black ink: Camel/scie in... (Kummel, scie in...). *Fig. 242.*

BOTANICAL REFERENCES: *M. Kummel* A. De Candolle, *Prodromus* (1844) 8, 203; Cufodontis (1960) 665; Hemsley in *Flora Tropical East Africa*, Sapotaceae (1968) 54.

"*Kummel*" Bruce, *Travels*, (1805) 7, 341–342 and 8, pl. 54; (1813) 7, 357–358 and 8, pl. 54; Murray (1808) 450, 464–465, pl. XVII.

SCROPHULARIACEAE Figwort family

Hebenstreitia dentata

A perennial herb of grassy hillsides in tropical Africa, with white flowers conspicuously marked orange in the throat. Found at Sakalla on 5 November 1770.

VERNACULAR NAMES: None recorded.

B 1977.14.9053 *v.* Rough sketch of shoot and flowering head and root, with details of inflorescence and flowers.

Pencil; 310 × 221; inscr., left of central drawing, in pen and black ink: Quest e una Pianta Ritrovata a Sakalla / li 5 Novbre 1770. Questa Cresce all'altezza / di 1 Piede e ½, o 2 Piedi in Circa (This is a plant found at Sakalla on November 1770. This grows to a height of about 1½ or 2 feet). *Fig. 243.*

BOTANICAL REFERENCES: *H. dentata* Linnaeus, *Species Plantarum* (1753) 629; Cufodontis (1963) 912.

Verbascum sinaiticum *Mullein*

A stout herb, one of the mulleins, with yellow flowers and woolly leaves, occurring in dry upland in northeast Africa and Arabia. The crushed leaves are a commonly used remedy against excessive bleeding. Found at Lamalmon in Semien, among scattered rocks not far from a stream.

VERNACULAR NAME: Cottina (MS). Cufodontis (1963) notes a similar name for a related species.

a. B 1977.14.9049 *v.* Outline drawing of flowering shoot, detail of leaf and floral parts.

Pencil; 302 × 199; inscr., right, with extensive notes in pen and black ink, beginning, top: Quest è una Pianta Ritrovata su la Malmon. Si chiama Cottinà (This is a plant found at Lamalmon. It is called Cottinà). *Fig. 244.*

b. B 1977.14.9049 *r.* Careful outline drawing of habit of plant in flower.

Pencil; inscr., below, with seven lines of MS notes in pen and black ink, continuing from other side of sheet: Una mezza linea di d^{e}. Stamine a basso è liscea (Half a line [about 1 mm] of the said stamens at the base is smooth). *Fig. 246.*

c. B 1977.14.8884. Anonymous, Paris folder. Finished drawing of flowering shoot and root.

Watercolors and bodycolors over pencil outlines, within two double ruled ink borders; 406 × 305; inscr. *v.*, in pen and brown ink: Bouillon Blanc d'abissinie. (Mullein of Abyssinia). *Fig. 245.*

d. B 1977.14.8885. Anonymous (same hand as previous drawing), Paris folder. Finished drawing of section of stem and leaves, with floral details.

Watercolors and bodycolors over pencil outlines, within two double ruled ink borders; 406 × 305; inscr., below details at foot, numbers 1–12, and *v.*, in pen and brown ink: Bouillon blanc d'abissinie (Mullein of Abyssinia). *Fig. 30.*

BOTANICAL REFERENCES: *V. sinaiticum* Bentham in De Candolle, *Prodromus* (1846) 10, 236; Cufodontis (1963) 884; Wilson and Mariam (1979) 33.

Note: Bruce (*Travels* [1790] 5, 60) refers to "Verbascum abyssinicum" described by de Jussieu from a plant grown from Bruce's Ethiopian seeds. This name was not published and the plant itself has not been traced in Paris, but it was painted and the painting sent to Bruce in the "Paris folder," in which it is still to be found in the Yale collection.

SIMAROUBACEAE Tree of heaven family

Brucea antidysenterica *James Bruce's tree*

A small tree with pinnate leaves and narrow inflorescences of yellow flowers; the small, rounded fruits are bright red. It occurs on tropical African mountains, where it was found at Sagassa and elsewhere; its leaves and bark are used medicinally by local people. (See remarks below.)

VERNACULAR NAMES: Avallò, Woginoussi, Woginus, Wooginoos, Wryginous (MS and *Travels* [1805] 7, 325; [1813] 7, 341). Similar Amharic names are recorded by Schweinfurth (1896) and Mooney (1963).

a. B 1977.14.9031 *v.* Outline drawing of top part of stem of shrub, with young leaves and inflorescences in fruit, with detail of dissected fruit.

Pencil; 257 × 203; inscr., top, in pen and black ink: Woginus A Saccalla si chiama Avallò (Woginus, at Saccalla called Avallò); and with note against detail below, with eight lines of MS notes beginning: In

questo Stato l'abbiamo Ritrovato a Saccalla alli 5 di Novembre (We have found it in this condition at Saccalla on the 5th of November). *Fig. 247.*
b. B 1977.14.9047 *r.* Rapid outline sketch of habit of shrub bearing fruits.
Pencil; 302 × 199; inscr. above, in pencil: Woginus arbusto Velenoso (Woginus, poisonous shrub); and below this, in Bruce's hand: Roots dried and reduced to powder good against the bloody flux. *Fig. 248.*
c. B 1977.14.9047 *v.* Rapid outline sketch of leaf supporting fruiting inflorescence, with detail of leaf tip.
Pencil; inscr., above and below, with extensive MS notes in pen and brown ink, beginning: Ritrovato a Sagassa Waigin, Wyginous [both last names deleted] Wryginous in Amara / Woginus Arbusto che gl'Abbissini Reputano per Veleno (Found at Sagassa, Wryginous in Amhara, Woginus, shrub which the Abyssinians hold to be poisonous). *Fig. 249.*
d. B 1977.14.9048 *r.* Similar bold outline of part of stem with flowering shoots and leaves.
Pen and black ink over weak pencil outlines, indented for transfer; 302 × 199. *Fig. 251.*
e. B 1977.14.8918. Finished drawing from d, with stem of fruits added.
Watercolors and bodycolors over pencil outlines, within ruled pencil borders; 364 × 271; inscr. *v.*, in pen and black ink: Woginoussi 1 / or Brucea the flowers to be added from the folio Drawing No. 2. *Fig. 250.*
Engr. Bruce, *Travels* (1790) pl. opp. p. 69, with floral details from f following; (1805) pl. 21; (1813) pl. 21.
f. B 1977.14.8882. By F. P. Nodder. Finished drawing of part of stem with flowering shoots and leafy shoots.
Watercolors and bodycolors over pencil outlines, on vellum; 478 × 344; inscr., at the bottom, in pen and black ink: Brucea antidysenterica; and in lower left-hand corner: Fred^{k}. Polydore Nodder Pinx^{t}. 1777; *v.*, in pen and brown ink: Wooginous or Brucea No. 2, the flower here to be copied into the / smaller drawing No. 1. *Fig. 31.*
g. British Museum (Nat. Hist.) Dried specimen (holotype) grown at Kew in 1775 from seed brought back from Ethiopia by Bruce.
Note: A comparison of Nodder's and Balugani's drawings of the same plant, though not the same specimen, the former grown from seed in England, the latter found in Ethiopia at Sakalla, reveals that although Nodder's is the more elegant drawing (the parchment encourages this quality), Balugani's more weatherworn portrait is at least equally convincing.

BOTANICAL REFERENCES: *B. antidysenterica* J. Miller, *Icones* (1779) 25; Bruce, *Travels* (1790) 5, 69–73 and pl.; IDC microfiche 5501–05: I. 1; (1805) 7, 178–81 and 8, pl. 21; (1813) 7, 194–97, 341 and 8, pl. 21; Britten in *Journal of Botany* (1913) 51, 255–57 and (1914) 57, 353; Cufodontis (1956) 374.
Note: Bruce (*Travels* [1790] 5, 72) claimed that J. Miller drew the shrub on the order of Sir Joseph Banks, who had grown it at Kew and provided it with the name in his honor. However, the credit for the name must actually go to John Miller (the father of J. F. Miller, to whom it is usually attributed), as he first published it. Bruce would have been gratified to know that the generic name *Brucea* has been officially conserved. The type specimen is now housed in the General Herbarium of the British Museum (Natural History). Bruce's attempt to reciprocate with the genus *Banksia* come to nought since Linnaeus fil. had already published the name for a proteaceous plant. The drawing published in the *Travels*, though said to have been made at Hor Cacamoot, in the province of Ras el Feel in western Ethiopia, was actually drawn from sketches by Balugani of trees found at Sakalla.

SOLANACEAE Potato family
Discopodium penninervium
A soft-stemmed shrub occurring in shady upland forests in tropical Africa. The clustered greenish flowers produce orange-yellow berries. It was found at Lamalmon.

VERNACULAR NAMES: Almit, Almol (MS). Mooney (1963) notes similar in Amharic.
a. B 1977.14.8987 *r.* Outline drawing of leafy shoot bearing flowers and fruits, with details of leaves and flowers.
Pencil, indented for transfer; 302 × 199; inscr., top, in pen and black ink: Almit Albero Ritrovato su la Malmon (Almit, tree found at Lamalmon); and with words and letters against details; *v.* extensive MS notes, inscr. in pen and black ink beginning: Nel rovescio di questo Foglio è rappresentato un Ramo di un Albero / retrovato su La Malmon, che si chiama Almit. (On the back of this sheet is shown a shoot of a tree found at La Malmon called Almit). *Fig. 252.*
b. B 1977.14.9114. Finished drawing from a.
Bodycolors and watercolors over pencil outlines, within ruled pencil borders; 406 × 308; inscr., *v.*, in pen and brown ink: Almol. *Fig. 32.*

BOTANICAL REFERENCES: *D. penninervium* Hochstetter in *Flora* (1844) 22; Heine in *Flora West*

Tropical Africa, ed. 2 (1963) 2, 328; Cufodontis (1963) 856.

Solanum adoense *A nightshade*

A prickly undershrub occurring only in northeast Africa. The blue flowers produce numerous orange-yellow berries used for tanning leather at Sakalla.

VERNACULAR NAMES: Mergiombe, Mergombey, Merjombey, Ombuai, Ombuay (MS and *Travels*). The Amharic name Embwai with several variant spellings, used for various spiny species of *Solanum*, is noted by Mooney (1963).

a. B 1977.14.8945 *v.*, top. Outline drawing of shoot bearing small fruits, with details of leaf and flower. Pencil; 311 × 222; inscr., left, with MS notes beginning: Quest e una Foglia ed un / Ramo di Frutti di Ombuai Ri-/ trovato a Sakalla (This is a leaf and fruiting shoot of Ombuai found at Sakalla). *Fig. 253.*

b. B 1977.14.9068 *r.* Outline drawing of shoot with leaves, flowers, and fruits.
Pencil: indented for transfer; 221 × 203; inscr., top right, by Bruce in pencil: 1 Tree, Lamaln [for Lamalmon?] Merjombey. *Fig. 254.*

c. B 1977.14.9097. Finished drawing from b.
Watercolors and bodycolors over pencil outlines, within ruled pencil borders; 406 × 310; inscr., *v.*, in pencil: Merjombey. *Fig. 33.*
Engr. by Heath, Bruce, *Travels* (1805), pl. 51; (1813), pl. 51.

BOTANICAL REFERENCES: *S. adoense* Hochstetter ex A. Richard, *Tentamen florae abyssinicae* (1851) 2, 105; Wright in *Flora Tropical Africa* (1906) 4, 233. "Merjombey" Bruce, *Travels*, (1805) 7, 336–37 and 8, pl. 51; (1813) 7, 352–53 and 8, pl. 51; Murray (1808) 460–61, pl. XIV.

Solanum capsicoides *A nightshade*

A very prickly undershrub with white flowers and tomato-like fruits.

VERNACULAR NAME: None recorded.

B 1977.14.8900. Anonymous, Paris folder. Finished drawing of flowering and fruiting plant, with details of fruit and flowers.
Watercolors and bodycolors over pencil outlines, within two double ruled ink borders; 404 × 308. *Fig. 255.*

BOTANICAL REFERENCES: *S. capsicoides* Allioni, *Auctuarium synopsim methodicam stirpium horti regii taurinensis* (1773) 64.
S. ciliatum Lamarck, *Tableau Encyclopédie Méthodique, Botanique* (1797) 2, 21.
Note: Although this watercolor is among Bruce's material it is not certain that the plant was grown from his seed brought back from Ethiopia. The species originates from tropical America, and it is now widely naturalized in tropical Africa and Asia, but it would be surprising if it was in Ethiopia by the 1770s. Even Cufodontis (1963) does not list it, and there is a possibility that by some confusion the painting was attributed to Bruce's collection and included in the Paris folder (see note on *Bauhinia acuminata*, p. 86).

Solanum incanum *Wild egg plant*

A rather prickly shrub of open pastures in the drier parts of tropical Africa and the Middle East, with blue flowers and yellow fruits.

VERNACULAR NAMES: Umboy—Amhara (MS), Angoule, Angule (MS and *Travels* [1805] 7, 325; [1813] 7, 341; Murray [1808] 449); Schweinfurth (1896) notes similar Tigrinia names.

a. B 1977.14.8936 *v.*, right. Rough sketch of leafy, flowering shoot with two fruits.
Pencil; 222 × 316; inscr., top left, in pencil: Angule. *Fig. 310.*

b. B 1977.14.8937 *v.*, left. Drawings of dissected fruits and flowers.
Pencil, indented for transfer; 221 × 313; inscr., top, in pencil: Sezione transversale (transverse section); above the central drawings: Sezione per lungo (longitudinal section). *Fig. 257.*

c. B 1977.14.6469n. Finished drawing from b.
Watercolors and bodycolors over pencil outlines, within ruled pencil borders; 451 × 319; inscr., upper right, in pen and brown ink: 14, *v.*: Angoule In Amhara Umboy. *Fig. 34.*

BOTANICAL REFERENCES: *S. incanum* Linnaeus, *Species Plantarum* (1753) 188; Cufodontis (1963) 868; Wilson and Mariam (1979) 33.
Note: The large, round fruits shown in the drawing are typical of the species, but the conspicuous spines on the surface of the leaves are odd and the leaves are not shown as pubescent.

Solanum marginatum *A nightshade*

A prickly shrub covered with white, woolly hairs, occurring in waste places in Ethiopia. Pendent white flowers produce tomatolike fruits.

VERNACULAR NAME: None recorded.

a. B 1977.14.8886. Anonymous, Paris folder. Finished drawing of whole plant with root.
Watercolors and bodycolors over pencil outlines, within two double ruled ink borders; 407 × 308; inscr. *v.*, in pen and brown ink: Solanum d'Abissinie. *Fig. 258.*

b. B 1977.14.8887. Anonymous (same hand as previous drawing?), Paris folder. Finished drawing of plant in flower with floral details.
Watercolors and bodycolors over pencil outlines, within two double ruled ink borders; 406 × 305; inscr. below details, in pen and brown ink, with numbers 1–6; and *v.*: Solanum d'Abissinie. *Fig. 35.*

BOTANICAL REFERENCES: *S. marginatum* Linnaeus fil., *Supplementum* (1781) 147; Lamarck, *Tableau Encyclopédie Méthodique, Botanique* (1797) 2, 22; Cufodontis (1963) 871.
Note: Linnaeus the younger must have used the specimen in his herbarium, now at the Linnean Society, but he gave no indication of the source except that it came from "Abyssinia." Since the above watercolors are included in the Paris folder it appears that the plants were grown in Paris, as Lamarck refers to seeds presented by Bruce. We also note that *Solanum abyssinicum* Jacquin ex Vitman (*Summa Plantarum* [1789] 1, 492) is a synonym of *S. marginatum*, but we have been unable to trace type material, although it is presumably based on Bruce's plants (see p. 62).

Solanum piperiferum *A nightshade*

A large, prickly shrub or small tree with small, variable leaves and small orange fruits containing seeds like those of black pepper. Found at Maccara, Lamalmon. In MS (.9060 *r.*) it is described as an ingredient in purgatives.

VERNACULAR NAMES: Mergiombei, Mergombey (MS). See also *S. adoense* above.
a. B 1977.14.9060 *r.* Outline drawing of leafy shoot in fruit.
Pencil, indented for transfer; 302 × 199; inscr. with extensive MS notes surrounding the drawing, in pen and brown ink, beginning: Mergiombei Albero Piccolo Ritrovato su la Malmon in luogo dello Maccara (Mergombey, small tree found at Lamalmon in a place called Maccara). *Fig. 256.*
b. B 1977.14.64690. Finished drawing from a.
Watercolors and bodycolors over pencil outlines, within ruled pencil borders; 451 × 318; inscr., upper right, in pen and brown ink: 15; and *v.*: Mergombey. *Fig. 259.*

BOTANICAL REFERENCES: *S. piperiferum* A. Richard, *Tentamen florae Abyssinicae* (1851) 2, 106; Cufodontis (1963) 875.
Note: The small leaves and few-flowered inflorescences as well as the straight spines on the stem serve to distinguish this from *S. adoense*. The small, entire leaves also distinguish it from *S. coagulans* Forsskal (Syn. *S. dubium* Fresenius).

STERCULIACEAE Cocoa family

Dombeya torrida

A small tree with conspicuous clusters of white flowers. It is common in the montane forests of eastern Africa; found at Sakalla.

VERNACULAR NAMES: Valkoffa, Walkaffa, Walkoffa, Wala kuffa, Walkuffa (MS and *Travels*). Mooney (1963) cites two similar Amharic names.
a. B 1977.14.8938 *v.* Sketch of flowering shoot.
Pen and black ink over pencil outlines, indented for transfer; 302 × 199; inscr. above in pen and black: Kantuffa [deleted] disegnato in gran/dezza Naturale (Kantuffa drawn natural size); and below in Bruce's hand, in pen and brown ink: Walkaffa. The extensive MS notes written from the bottom upwards do not relate to this drawing and are continued from the *recto*. *Fig. 260.*
b. B 1977.14.8966 *r.* Rough outline drawing of budding shoot with details of bud clusters, leaf, and enlarged stellate hairs.
Pencil; 310 × 221: inscr., right and below, with MS notes in pen and black ink beginning: Questo Rappresenta un Ramo di un Albero chiamato / Valkoffa è Ritrovato a Sakalla. (This represents a shoot of a tree called Valkoffa found at Sakalla); and with letters against details. *Fig. 261.*
c. B 1977.14.8990 *v.* Outline drawing of flowering head with details of flowers.
Pencil; 312 × 221; inscr., below with seventeen lines of MS notes in pen and black ink beginning: Questo Rappresenta un Ramo di Fiori de Walkoffa disegnato in grandezza Naturale... (This represents a shoot of flowers of Walkoffa drawn life-size...); and with letters against details. *Fig. 265.*
d. B 1977.14.8917. Finished drawing of leafy flowering shoot, from a. with floral details from b.
Watercolors and bodycolors over pencil outlines; 312 × 246; inscr. *v.*, in pen and brown ink: Walkuffa No. 5. *Fig. 36.*
Engr. Bruce, *Travels* (1790), pl. opp. p. 67; (1805), pl. 20; (1813), pl. 20.

BOTANICAL REFERENCES: *D. torrida* (J.F. Gmelin) Bamps in *Bulletin Jardin Botanique Bruxelles* (1962) 32, 170; Hepper & Friis in *Botaniska Notiser* (1979) no. 132, 397–98.
Walcuffa torrida J.F Gmelin, *Systema naturae* ed. 13, (1791) 2, 1029. Type: Bruce's plate and description.
Dombeya bruceana A. Richard, *Tentamen florae abyssinicae* (1847) 1, 77; Cufodontis (1959) 580.
'*Walkuffa*' Bruce, *Travels* (1790) 5, 67–68 and pl. opp. p. 67; (1805) 7, 176–77, and 8, pl. 20; (1813) 7, 192–93 and 8, pl. 20.

Sterculia africana

A small, spreading tree with pale, smooth bark inhabiting rocky places and streamsides in drier parts of tropical Africa. The flowers are not very conspicuous, but the fruits are distinctive with three to five spreading follicles that split open to reveal seeds with red arils.

VERNACULAR NAME: Anguah No. 2 (MS and *Travels* but probably applies to *Boswellia papyrifera* and not to the fruits illustrated).

B 1977.14.9095. Finished drawing of stem with seed pods, some open showing seeds.

Watercolors and bodycolors over pencil outlines, within ruled pencil borders; 406 × 305, inscr. *v.*, in pen and brown ink: No. 14; and, in pencil: Anguah. *Fig. 263.*

Engr. by Heath, in reverse, Bruce, *Travels* (1805) 8, pl. 49; (1813) 8, pl. 49; Murray (1808), pl. XII.

BOTANICAL REFERENCES: *S. africana* (Loureiro) Fiori in *Agricoltura Coloniales Italia* (1912) 5 suppl., 37; Cufodontis (1959) 584; H. Wild in *Flora Zambesiaca* (1961) 1 (2), 553.

"Anguah No. 2" Bruce, *Travels* (1805) 7, 334 and 8, pl. 49; (1813) 7, 350 and 8, pl. 49 (1813); Murray (1808) 458, pl. XII.

TILIACEAE Linden tree family

Grewia ferruginea

A large, straggly shrub with white flowers, inhabiting thickets and streamside forests in the mountains of northeast Africa. In MS (.9078 *v.*) it is stated that the plant was found on rocky ground at a stream near Meisbenni.

VERNACULAR NAMES: Logheba, Logheta (MS & *Travels* [1805] 7, 325; [1813] 7, 341). Mooney (1963) records a similar Amharic name.

a. B 1977.14.9078 *r.*, left. Careful outline drawing of leafy sprays beginning to flower, with detail of leaf.

Pencil, indented for transfer; 221 × 313; inscr., top left: Logheta; and right, with MS notes beginning: Il dissegno qui a Canto Represcenta un Ramo di un Arbusto / chiamata Logheta e ritrovato a Meisbenni. (The drawing alongside represents a sprig of a shrub called Logheta and found at Meisbenni). *Fig. 262.*

b. B 1977.14.6468l. Finished drawing from a, including detail of leaf.

Watercolors and bodycolors over pencil outlines, within ruled pencil borders; 451 × 314; inscr., upper right: 12; and *v.*, in pen and brown ink: Logheba, in pencil: Ipusa (inverted). *Fig. 264.*

BOTANICAL REFERENCES: *G. ferruginea* A. Richard, *Tentamen florae abyssinicae* (1847) 1, 87; Cufodontis (1958) 520.

UMBELLIFERAE Parsley family

Alepidea peduncularis

An erect herb with conspicuous white bracts around the inflorescence. It is found in grasslands on the mountains of eastern Africa.

VERNACULAR NAMES: None recorded.

B 1977.14.9005 *r.* Light outline of habit of plant, with details of leaves and inflorescence.

Pencil; 340 × 236; inscr., top right, with six lines of MS notes in pen and black ink, beginning: Il dritto del Fiore e bianco con una linea / quasi negra in Ciascuna Foglia... (The front side of the flower is white with an almost black line on each petal [i.e., bract]). *Fig. 266.*

BOTANICAL REFERENCES: *A. peduncularis* Steudel ex A. Richard, *Tentamen florae abyssinicae* (1847) 1, 320; Cufodontis (1959) 639.

Coriandrum sativum *Coriander*

An annual herb with umbels of white flowers and small, round fruits. This is the commonly cultivated coriander, which is used in the characteristically spicy Ethiopian and Arabian dishes. Bruce records this from Gondar.

VERNACULAR NAME: Dembelal (MS and *Travels* [1805] 7, 325; [1813] 7, 341). A similar Amharic name is noted by Mooney (1963).

B 1977.14.8949 *r.* Careful outline drawings of habit of plant, with root, and details of fruits and foliage.

Pencil; 311 × 443; inscr., at the top, with eleven lines of MS notes in pen and black ink, beginning: Dembelal Quest è una Pianta alquanto arromatica Ritrovata a Gonder alla metta di Novembre in Frutto... (Dembelal. This is a somewhat aromatic plant found at Gondar in the middle of November in fruit...). *Fig. 49.*

BOTANICAL REFERENCE: *C. sativum* Linnaeus, *Species Plantarum* (1753) 256; Cufodontis (1959) 640.

Diplolophium africanum

An erect perennial herb with finely cut leaves and umbels of yellowish flowers. Common in wooded grassland of eastern African mountains; found between Meisbenni and Dagashea.

VERNACULAR NAMES: None recorded.

B 1977.14.9054 *v.*, right. Careful drawing of flowering heads with floral detail.

Pencil; 302 × 397; inscr., top, with ten lines of MS notes in pen and brown ink, beginning: Disegno di una Pianta di Finochio Ritrovata fra Meisbenni, e Dagasheà (Drawing of a fennel plant found between Meisbenni and Dagasheà). *Fig. 267.*

BOTANICAL REFERENCES: *D. africanum* Turczaninow in *Bulletin Societé Imperiale Naturalistes Moscow* (1847) 20(1), 173.

Steganotaenia araliacea *The carrot tree*

A small tree, which is an unusual habit for the parsley and carrot family. Widespread in tropical African savanna; it was found at Zinghetch.

VERNACULAR NAMES: None recorded.

B 1977.14.8971 *r.* Outline of stem without leaves showing inflorescence and some fruits, with detail of flowering shoot.

Pencil; 301 × 199; *v.* extensive MS notes only; inscr. in pen and black ink, beginning: Nel Rovesscio di questo Foglio e rappresentato il Ramo di un Albero ritrovato a Zinghetch... (On the reverse of this sheet is illustrated the shoot of a tree found at Zinghetch...). *Fig. 268.*

BOTANICAL REFERENCES: *S. araliacea* Hochst. in *Flora* (1844) 27 Bes. Beil., p. 4; Cufodontis (1959) 650.

VERBENACEAE Vervain family

Clerodendrum myricoides *A clerodendron*

A laxly branched shrub of montane forest in eastern tropical Africa, with blue flowers on slender stalks, found at Meisbenni.

VERNACULAR NAMES: None recorded.

B 1977.14.8984 *v.* Outline of flowering shoot.

Pencil; 231 × 200; inscr. above and to the left, with MS notes in pen and black ink, beginning: Disegno di una Pianta ritrovata a Meisbenni cresce in luogo umido / fra le pietre (Drawing of a plant found at Meisbenni. It grows in a humid situation among stones). *Fig. 269.*

BOTANICAL REFERENCES: *C. myricoides* (Hochstetter) R. Brown ex Gürke in Engler & Prantl, *Natürliche Pflanzenfamilien* IV. (1895) 3a, 176; Wilson & Mariam (1979), 30.

VITIDACEAE Vine family

Vitis vinifera *Grapevine*

This familiar vine yields grapes for eating fresh and dried as raisins, and for fermented wine.

VERNACULAR NAME: None recorded.

B 1977.14.22718 *r.* Rough sketch of cluster of grapes.

Pencil; 237 × 197. *Fig. 272.*

BOTANICAL REFERENCES: *V. vinifera* Linnaeus, *Species Plantarum* (1753) 202; Cufodontis (1958) 503.

MONOCOTYLEDONS

AGAVACEAE Sisal family

Dracaena steudneri *A dragon tree*

A slender tree with a conical base to the trunk and a few erect branches bearing tufts of stiff leaves. The fragrant, creamy flowers are borne on a large inflorescence. It occurs in the wetter parts of secondary forests in the uplands of eastern tropical Africa. (The drawings show the leaves to be unusually narrow for this species.)

VERNACULAR NAMES: Gibara, Jebaara (MS).

a. B 1977.14.8941 *r.* Outline of habit of tree.

Pencil; 243 × 199; inscr., top right, in pen and black ink: Gibara; bottom right, with numbers in a calculation. *Fig. 271.*

b. B 1977.14.8941 *v.* Outline detail of inflorescence among leaves.

Pencil, indented for transfer; inscr., top left, in pencil: Gibara: [deleted]; and on all sides, in pen and black ink, with numbered calculations. *Fig. 270.*

c. B 1977.14.8942 *r.* Outline detail of inflorescence.

Pencil, indented for transfer; 241 × 192; inscr., top right, in pencil: Gibara; and lower left: Sassa [deleted]. *Fig. 273.*

d. B 1977.14.6467c. Finished drawing from c, with floral details below.

Watercolors and bodycolors over pencil outlines, within ruled pencil borders; 451 × 310; inscr., upper right, in pen and brown ink: 3; bottom center: Jibara; and *v.*, in pen and brown ink: Jebaara parts; and Dracina. *Fig. 37.*

e. B 1977.14.6467b. Finished drawing from b.

Watercolors and bodycolors over pencil outlines, within ruled pencil borders; 451 × 312; inscr., upper right, in pen and brown ink: 2; lower center: Jebaara; and *v.*, in pen and brown ink: Jebaara. *Fig. 275.*

BOTANICAL REFERENCE: *D. steudneri* Engler, *Pflanzenwelt Ost-Afrika* (1895) c, 43; Cufodontis (1971) 1569.

AMARYLLIDACEAE Daffodil family

Crinum schimperi *Crinum lily*

A bulbous plant with large, white- or pink-striped flowers topping a two-foot-high (sixty-centimeter) stalk.

VERNACULAR NAMES: None recorded.

B 1977.14.9064 *r.* Outline of inflorescence.

Pencil; 302 × 199. *Fig. 274.*

BOTANICAL REFERENCES: *C. schimperi* Vatke ex K. Schumann in Gartenflora (1889) t. 1309, 561; Cufodontis (1971) 1576.

Crinum zeylanicum *Crinum lily*

A bulbous plant with a large inflorescence of decorative pink-striped, white flowers. It grows in wet places in both tropical Africa and Asia; found on the island of Mitradh in Lake Tana.

VERNACULAR NAME: Scioneurth (MS).

a. B 1977.14.8967 *r*. Outline of inflorescence, with detailed dissections of flowers.

Pencil on vellum; 283 × 171; inscr., top left, in pen and black ink: Ritrovato a Mitradh (Found at Mitradh); and with words against details; *v*. MS notes only, inscr. in pen and black ink: Nel Rovesscio di questo Foglio è Rappresentato una Specie di Lillea / che in abbissinia chiamano Scioneurth... (On the reverse of this sheet is represented a species of lily which in Abyssinia they call Scioneurth...). *Fig. 276.*

BOTANICAL REFERENCES: *C. zeylanicum* (Linnaeus) Linnaeus, *Systema Naturae*, ed. 13, (1770) 2, 236; Nordal in *Norwegian Journal of Botany* (1977) 24, 188.

C. ornatum (Linnaeus fil. ex Aiton) Bury, *Selection of Hexandrian plants* (1831–34) t. 18; Cufodontis (1971) 1575.

Scadoxus puniceus *Blood lily*

A bulbous plant with a ball of scarlet flowers. It occurs in woods throughout much of the eastern upland of tropical Africa. In Ethiopia it is presumably quite rare and was only rediscovered at the end of the 1930s.

VERNACULAR NAME: Scionkust (MS).

a. B 1977.14.8943 *r*. Drawings of inflorescence, with details of parts of florets, and diagrams of the inflorescence.

Pencil, some parts indented for transfer; 199 × 302; inscr. with extensive MS notes in pen and black ink, against details, that to the right of the plan of the inflorescence beginning: Pianta della parte inferiora del Fiore... (Plan of the lower part of the flower...). *Fig. 278.*

b. B 1977.14.9072 *v*. Outline of base of stem and root, with cross-section of lower stem.

Pencil; 302 × 199; inscr. extensively, mainly to the left, in pen and black ink, beginning: Questo Rappresenta la Radice ed / una porzione della Tiglia di una Specie di Cipolla che si chia -/ma Scionkust questa / cresca nella Montagna in luogo asciutto. (This represents the root and a portion of the stem of a species of onion which is called Scionkust and grows on the mountain in dry places). *Fig. 277.*

c. B 1977.14.9057 *r*. Careful drawing of whole plant. Pencil, indented for transfer; 302 × 199. *Fig. 279.*

d. B 1977.14.6469h. Finished drawing from c, with some details from a.

Watercolors and bodycolors over pencil outlines, within ruled pencil borders; 451 × 316. *Fig. 38.*

BOTANICAL REFERENCES: *S. puniceus* (Linneaus) Friis & Nordal in *Norwegian Journal of Botany* (1976) 26, 64.

Haemanthus puniceus Linnaeus, *Species Plantarum* (1753) 325; Bjørnstad & Friis in *Norwegian Journal of Botany* (1974) 21, 244.

H. multiflorus sensu Cufodontis (1971) 1574.

CANNACEAE Canna lily family

Canna bidentata *Canna lily; Indian shot plant*

A perennial herb with an inflorescence of bright red flowers. It is of American origin but is now widely cultivated as an ornamental plant in Egypt and Ethiopia.

VERNACULAR NAME: In MS notes it is called Guinco or Giunco but this is an Italian word for jonquil or reed.

a. B 1977.14.8951 *r*. Outline of habit of whole plant, without flowers.

Pencil, bolder outlines over lighter strokes, indented for transfer; 302 × 200. *Fig. 280.*

b. B 1977.14.8951 *v*. Outline of habit closely similar and largely from a (or vice-versa) but with flowering head.

Pencil, bolder outlines over lighter strokes, indented for transfer; inscr., top left: Giunco di/Rossetto (Rush or Jonquil of Rossetto [Egypt?]). *Fig. 281.*

c. B 1977.14.9089f. Finished drawing largely from a (?).

Watercolors and bodycolors over pencil outlines; 533 × 367; inscr., upper right, in pencil: 6; and upper left, very lightly: Giunco (Jonquil or reed). *Fig. 283.*

d. B 1977.14.9089g. Finished drawing of end of stem and flowering head from b.

Watercolors and bodycolors over pencil outlines; 533 × 375; inscr., upper right, in pencil: 7, and, upper left, very lightly: Parte del Giunco (Part of the jonquil or reed). *Fig. 284.*

e. B 1977.14.6469g. Finished drawing from b.

Watercolors and bodycolors over pencil outlines, within ruled pencil borders; 451 × 313; inscr., upper right, in pen and brown ink: 7. *Fig. 39.*

BOTANICAL REFERENCE: *C. bidentata* Bertoloni in *Memorie Accademia Scienze Bologna* (1859) 10, 33; Cufodontis (1972) 1596.

COMMELINACEAE Spiderwort family

Commelina africana

A common tropical African perennial herb with semiprostrate branches and bright blue flowers. It was found at Gondar.

VERNACULAR NAME: Lanbutt (MS).

B 1977.14.9041 *v.*, right. Careful outline of flowering plant, with roots and with details of flowers.

Pencil; 302 × 199; inscr., top left, in pen and black ink, the name in Amharic script followed by: Lanbutt; and right and left, extensive notes beginning: Quest è una / Pianta Ritrovata a Gonder e sue vi / cinanze. Nel principio / di Ottobre... non ancor maturo (This is a plant found at Gonder and in its environs. At the beginning of October not yet ripe). *Fig. 147.*

BOTANICAL REFERENCE: *C. africana* Linnaeus, *Species Plantarum* (1753) 45; Cufodontis (1970) 1424; Brenan in *Flora West Tropical Africa* ed. 2 (1972) 3, 45.

CYPERACEAE Sedge family

Cyperus esculentus *Tiger nut*

An aggressive, weedy sedge of wet places in the tropics, such as rice fields, with nutlike tubers which are edible as "tiger nuts."

VERNACULAR NAMES: Albasis, Abdelasic, Abdelasis (MS).

a. B 1977.14.6469f. Finished drawing of whole plant with roots.

Watercolors over pencil outlines, within ruled pencil borders, and with scale of Italian measurement below; 451 × 313; inscr., upper left, in pencil: 38; upper right: 6.

b. B 1977.14.8980 *r.*, left. Very rough outline of habit of flowering plant.

Pen and brown ink; 293 × 181; inscr., top left in pen and brown ink: Abasis ò Abdelasis (Abasis or Abdelasis); and to the left and right, 24 lines of MS notes beginning: Il fruto che si trova nella / radice tiene 3 linea e mezzo di diametro (The fruit of which is found in the root is 3½ lines [about 7 mm] in diameter). *Fig. 282.*

c. B 1977.14.9089h. Finished drawing of whole plant, with detail of inflorescence.

Watercolors and bodycolors over pencil outlines; 533 × 371; inscr., upper left, in pencil: Albasis opure Abdeladic [?] (Albasis or Abdeladic), upper right: 8. *Fig. 285.*

BOTANICAL REFERENCES: *C. esculentus* Linnaeus, *Species Plantarum* (1753) 45; Cufodontis (1970) 1424; Hooper in *Flora West Tropical Africa*, ed. 2, (1972) 3, 286.

Note: While .9089h shows the root tubers characteristic of *C. esculentus*, .6469f does not show them and it is therefore tentatively assigned to this species.

Cyperus papyrus *Papyrus sedge*

The famous papyrus sedge forming massive beds in lakes and at river margins with moplike inflorescences up to 13 ft. (4 m) high. It is native in tropical Africa and Huleh Lake in the upper Jordan valley. Bruce gives a long description of his attempts to make papyrus writing material.

VERNACULAR NAMES: Papyrus (MS and *Travels*), Babeer – Syrian Arabic (*Travels*), *el Berdi* – Egyptian Arabic (*Travels* [1790] 5, 15).

a. B 1977.14.8903. Finished drawing of stem and flowering head, with details of inflorescence left and right.

Watercolors and bodycolors over pencil outlines; 523 × 347; inscr., *v.*, in pen and brown ink: --- [illegible] Papyrus; and with Bruce's instructions: This drawing as it cannot be reduced without making its head too small is therefore to be divided in the stalk & the sections placed side by side as gracefully as possible. *Fig. 40.*

Engr. in reverse, in Bruce, *Travels* (1790), pl. opp. p. 1; IDC microfiche 5501–1: III.7; (1805), pl. 1; (1813), pl. 1.

b. B 1977.14.9119. Finished drawing of inflorescence. Bodycolors and watercolors over pencil outlines; 523 × 357. *Fig. 41.*

BOTANICAL REFERENCES: *C. papyrus* Linnaeus, *Species Plantarum* (1753) 47; Cufodontis (1970) 1433. Note: The second illustration is said (*Travels* [1805] 7, 324; [1813] 7, 340) to have been painted at Sidon on 29 July 1767, but on that day Bruce was at Paneas, Upper Jordan (Murray [1808] 52) where the papyrus grew nearby.

Cyperus species 1

It is not certain that the drawings represent a *Cyperus* species as it is sterile, but someone has written on its reverse "Caryx," i.e., *Carex*. The MS (.9041 *V.*) indicates that it came from Gondar and it was also found in the lowlands of Dembia; it has clusters of perennial roots which are yellow inside. The whole plant is said to be fragrant and dries out completely after the rains.

VERNACULAR NAMES: Tzadge saar, Tsadg-saar (MS), Tsadjessar, Tsadjesar (*Travels* [1805] 7, 325; [1813] 7, 341), see note below.

a. B 1977.14.9041 *v.* MS notes only.

302 × 199; inscr. extensively, in pen and black ink,

beginning at the top with name in Amharic and: Tsadg = ssar, Quest è una Casta di Erba odorosa ritrovata / a Gonder e nelle Vicinanze sue (Tsadg – ssar. This is a type of scented grass found at Gondar and in its environs).
b. B 1977.14.6469b. Finished drawing of sedge-like plant with roots.
Watercolors over pencil outlines; 451 × 312; inscr., upper right, in pen and brown ink; 2; and *v.*, in pen and brown ink: Tzadge Saar (?); and Caryx. *Fig. 286.*
Note: Mooney (1963) quotes the Amharic name *Tej-sar* for the citron grass *Cymbopogon citratus*, so there remains some doubt as to its identity.

Cyperus species 2
VERNACULAR NAME: None recorded.
B 1977.14.6469d. Finished drawing, of flowering plant and root.
Watercolors over pencil outlines; 451 × 312; inscr., upper right, in pen and brown ink: 4; and beneath: cinque piedi Inglesi (Five English feet); *v.*, in pen and brown ink: *Cyperus. Fig. 287.*

Mariscus cyperoides *A sedge*
An erect perennial sedge 1–2 ft. (30–60 cm) high. A common plant of African woodland savanna. In *Travels* (1805) 7, 325 and (1813) 7, 341, it is recorded as "a rush found at Adua".
VERNACULAR NAMES: Dangheeh or Dangheck (MS), Dangheeli (*Travels* 1805 and 1813).
B 1977.14.6469e. Finished drawing of inflorescence.
Watercolors and bodycolors over pencil outlines; 451 × 313; inscr., upper right, in pen and brown ink: 5; *v.*, in pencil: Dangheeh or Dangheck. *Fig. 288.*
BOTANICAL REFERENCES: *Mariscus cyperoides* (Linnaeus) Urban, *Symbolae Antillanae* (1900) 2(1), 164.
M. sieberianus Nees ex Steudel (1854); Cufodontis (1970) 1456.
Note: It is difficult to be sure of the identity of this drawing, but it seems to match this species.

GRAMINEAE Grass family
Arundinaria alpina *African bamboo*
A bamboo forming dense forests in the tropical Africain mountains at high altitude. In Ethiopia it is not common in the central inhabited highland, but it is often grown to provide building materials, particularly for supporting the roof, as mentioned by Bruce, who recorded it in great abundance at the precipice at Sakalla. It was worshipped by the Agow people of that area.

VERNACULAR NAMES: Krhaha, Krehaha, Krihàhà (MS and *Travels* [1805] 7, 325; [1813] 7, 341). Similar Tigrinia & Amharic names are recorded by Schweinfurth (1896) and Mooney (1963).
a. B 1977.14.9025 *v.* Drawing of leafless habit with, at right, detail of stem.
Pencil; 302 × 199; inscr., with extensive notes in pen and black ink, beginning, top left: Quest è un Casta di Canna Ritrovata a Sackalla che si chiama Krihàhà (This is a type of bamboo found at Sackalla called Krihàhà). *Fig. 289.*
b. B 1977.14.9025 *r.* Drawing of base of stem and roots, with detail of leaves on shoot.
Pencil; inscr., left, in pen and black ink, with eighteen lines of MS notes beginning with two asterisks referring overleaf and: Per quanto / io posso pensare mi / sembra che le sude / grandi Foglie... (So far as I can judge it seems that the above-mentioned large leaves...). *Fig. 290.*
c. B 1977.14.9073 *r.* Drawing of base of stem and roots, closely similar to those shown in b.
Pencil, indented for transfer; 298 × 199; the sheet mutilated along the edges and creased. *Fig. 291.*
d. B 1977.14.9073 *v.* Details of largely leafless shoot, stem, and shoot with leaves, closely similar to that shown in b.
Pencil, indented for transfer. *Fig. 292.*
e. B 1977.14.9094. Finished drawing of sections of stem from d and c, and of shoot with leaves from d.
Watercolors over pencil outlines; 361 × 262; inscr. *v.*, in pen and brown ink: Krihaha No. 2. *Fig. 42.*
Engr. by Heath, in reverse, Bruce, *Travels* (1805), pl. 47; Murray (1808), pl. X.
f. B 1977.14.9093r. Finished drawing of habit with roots.
Watercolors over pencil outlines, within ruled pencil borders; 362 × 263; inscr., lower right, in pen and brown ink: ID; lower center, in pencil: No. 1; *v.*, inscr. in pen and brown ink: Krihaha No. 1. *Fig. 293.*
Engr. by Heath, in reverse, in Bruce, *Travels* (1805), pl. 46; Murray (1808), pl. IX.
BOTANICAL REFERENCES: *A. alpina* K. Schumann in Engler, *Pflanzenwelt Ost-Afrika* (1895) C, 116; Hubbard in *Flora Tropical East Africa*, Gramineae (1970) 9; Cufodontis (1970) 1413.
"Krihaha" Bruce, *Travels*, (1805) 7, 332–34, and 8, pls. 46, 47; (1813) 7, 348–50 and 8, pl. 46.

?Chloris species *A Rhodes grass*
The drawing represents an erect grass, but in this species, if the identity is correct, the stems should be

horizontal and mat-forming. It was seen at Massawa on the coast (*Travels* 1805 and 1813).

VERNACULAR NAMES: Muem il msalib (MS), Meim mesalib, or grass of the cross (*Travels* [1805] 7, 324; Murray [1813] 449.

B 1977.14.6469c. Finished drawing of flowering grass.

Watercolors over pencil outlines, within ruled pencil borders; 451 × 312; inscr., at the foot, in pencil, against a scale of measurement: sei piedi Inglesi (six English feet); *Fig. 294; v.*, in pen and brown ink: Muem il Msalib (?).

Eragrostis tef *Teff*

An annual cereal with very small grains cultivated in Ethiopia and to a lesser extent in southwest Arabia. It is used for baking the characteristic Ethiopian pancakelike bread *Enjera*, which is much larger and more pliable than the well-known *chapati* from India or *pitta* of the Middle East. Teff is also the most important ingredient in making Ethiopian beer, *talla*. It is grown in a number of varieties: dark teff, white teff, etc.

VERNACULAR NAME: Teff (*Travels*). The same well-known Amharic name is cited by Schweinfurth (1896) and Mooney (1963).

a. B 1977.14.8893. By A. Fyfe, Paris folder. Finished drawing of flowering grass.

Watercolors ,within double ruled ink borders; 406 × 300; inscr., bottom left, within the border, by the artist, in pen and brown ink: A Fyfe Edinb[h]. Botanick Garden 1775. *Fig. 295.*

One of only two signed drawings in the botanical collection preserved by Bruce. Alexander Fyfe (1754–1824) was an anatomist at the University of Edinburgh and for about forty years superintended dissections, gave dry lectures, and made good anatomical drawings. As a young man in 1775, the very year in which the above drawing was made, he was awarded the annual prize medal by the commissioners for improvements in Scotland for the best drawing in the academy.

b. 1977.14.22699. Finished drawing of flowering grass.

Watercolors, within single ink borders; 541 × 378 (lower right corner bevelled). This plant was probably grown at Paris or Versailles. *Fig. 296.*

c. Vienna – Watercolor drawing presumably of the cultivated *Poa abyssinica* (see below). *Fig. 298.*

BOTANICAL REFERENCES: *E. tef* (Zuccagni) Trotter in *Società Botanica Italiana* (1918) 62, in observation; Cufodontis (1968) 1258. Type: plants grown in the Botanical Garden of Florence from seeds collected by Bruce in Ethiopia (see note below). "Teff" Bruce, *Travels* (1790) 5, 76–80, excl. pl. opp. p. 76; (1805) 7, 184–87, excl. 8, pl. 24; (1813) 7, 200–03, excl. 8, pl. 24.

Poa tef Zuccagni, Dissertazione... Tef: 21 (1775). Type as above.

P. abyssinica Jacquin, Miscellanea (1781) 2, 364. Type: specimen at Vienna Natural History Museum herbarium indicated by B. de Winter as the probable holotype.

Note: Unfortunately, the published plate supposed to illustrate this grass represents *Panicum deustum* Thunb., or at least that is the impression, for it is difficult to be certain owing to inaccurate delineation of the glumes. Even Desvaux fell into this trap in supposedly correcting an error of those who had placed teff in *Poa* by calling it *Panicum teff* ("Erreur des agriculteurs et des botaniste sur le Teff des Abyssins," *Opuscules* [1831] 3, 43–46 and 97). In Bank's library at the British Museum and in the Botanical Library of the University of Copenhagen there are copies of A. Zuccagni's "Dissertazione concernente l'istoria di una pianta panizzabile dell'Abisinia consciuta da quei popoli sotto il nome di Tef." 45 pp., t. 1 (Firenze 1775), with a plate depicting the true teff. Evidently the wrong illustration was selected, but no herbarium specimens have been traced in the British Museum (Natural History), or elsewhere.

Ischaemum afrum

A perennial grass with slender inflorescences. It prefers black clay soils and is widespread throughout Africa.

VERNACULAR NAMES: Gir gir, Geshe el aube (*Travels*).

a. B 1977.14.8894. Anonymous, Paris folder. Finished drawing of whole plant, with inflorescence.

Watercolors and bodycolors over pencil outlines, within two double ruled ink borders; 322 × 251; inscr. by Bruce above the top borders, in pen and brown ink: No border but to be plain as the rest. *Fig. 297.*

b. B 1977.14.8895. Anonymous, Paris folder. Finished drawing of inflorescence and leaves, with fifteen separate details below.

Watercolors and bodycolors, within two double ruled ink borders; 317 × 247; inscr. by Bruce above the top borders, in pen and brown ink: No border but to be plain like the rest; and numbers: 1–15 against details; *v.*, in pen and brown ink: Geshe el Aube No. 16. *Fig. 43.*

Engr. Bruce, *Travels* (1790), pl. opp. p. 47; (1805), pl. 13; (1813), pl. 13.

BOTANICAL REFERENCES: *I. afrum* (J.F. Gmelin) Dandy in F.W. Andrews, *Flowering Plants Sudan* (1956) 3, 476; Cufodontis (1976) 1364.
Andropogon afer J.F. Gmelin, *Systema naturae*, ed. 13, (1791) 2, 166.
"Gir gir" or "Geshe elaube" Bruce, *Travels* (1790) 5, 47–48, pl. opp. p. 47; IDC microfiche 5501–3: IV. 3; (1805) 7, 159–60 and 8, pl. 13; (1813) 7, 175–76 and 8, pl. 13.

Panicum deustum *A wild millet*

A tufted grass with a loose inflorescence. It is widespread in Africa from Ethiopia to South Africa. As the drawing is not quite accurate the identification is tentative.

VERNACULAR NAME: Teff (MS and *Travels*, in error).

a. B 1977.14.8897. Anonymous, Paris folder. Finished drawing of inflorescence. Watercolors over pencil outlines; within two double ruled ink borders; 343 × 279; inscr. by Bruce below lower border, in pencil: Teff; and left and right, above top border: No border. *Fig. 291.*
Engr. in reverse, Bruce, *Travels* (1790), pl. opp. p. 76; (1805), pl. 24; (1813), pl. 24.
b. B 1977.14.8902. Anonymous, Paris folder. Finished drawing of whole plant in flower.
Watercolors and bodycolors over pencil outlines, within two double ruled ink borders; 404 × 305; inscr. *v.*, in pen and brown ink: No. 1. *Fig. 300.*

BOTANICAL REFERENCES: *P. deustum* Thunberg, *Prodromus* (1794) 1, 19; Stapf in *Flora Tropical Africa* (1920) 9, 651; Cufodontis (1969) 1306.
"Teff" Bruce, *Travels*, (1790) 5, pl. opp. p. 76; (1805) 7, pl. 24; (1813) 7, pl. 24. See notes above under *Eragrostis tef.*

HYDROCHARITACEAE Waterweed family

Ottelia ulvifolia

A water plant with submerged leaves and white or yellow flowers held above the water level. It is widespread in tropical Africa; found at Tobulachè.

VERNACULAR NAMES: None recorded.

a. B 1977.14.8958 *r.* Details of flowers and leaves.
Pencil; 302 × 197. *Fig. 301.*
b. B 1977.14.8958 *v.* Detail of leaf.
Pencil; inscr., at the top, in pen and black ink, with five lines of MS notes beginning: La Rettro disegnata Pianta Rappresenta una Ninfa Aquatica / Ritrovata a Tobulachè (The plant drawn overleaf represents a water-lily found at Tobulachè). *Fig. 302.*

BOTANICAL REFERENCES: *O. ulvifolia* (Planchon) Walpers, *Annales* (1852) 3, 510; Cufodontis (1968) 1206.

IRIDACEAE Iris family

Oenostachys abyssinica *Ethiopian gladiolus*

A perennial herb endemic to Ethiopia with a graceful inflorescence of red flowers. It occurs in pastures at higher altitudes; found at Coskam.

VERNACULAR NAMES: Azerazaei, Enzerazai or Enzerazei (MS).

a. B 1977.14.9031 *r.*, left. Outlines of base of stem and corms.
Pencil; 203 × 257; inscr. extensively with MS notes in pen and black ink, beginning, top right: Questo Rappresenta / la Radice della pianta / Enzerazai [or Enzerazei]... (This represents the roots of the plant Enzerazai [or Enzerazei]). *Fig. 305.*
b. B 1977.14.9045 *r.* Drawing of habit and outline of flowering stem, with details of flowers.
Pencil, indented for transfer; 310 × 221; inscr., top right, in pencil, Enzerazei; below this, the name in Amharic script, in pen and black ink; and, left and center, with extensive MS notes beginning: Azerazaei Pianta Ritrovata in Fiore, e frutto nel principio di ottobre / Sopra la Montagna di Coskam (Azerazaei, plant found in flower and fruit at the beginning of October on the mountain of Coskam); also with letters against details. *Fig. 303.*
c. B 1977.14.9127. Finished drawing of plant in flower.
Watercolors over pencil outlines, within ruled pencil borders; 406 × 310. *Fig. 44.*
d. Lord Elgin's collection, Broomhall. Finished drawing of flowering stem only, from a.
Watercolors over pencil outlines; 406 × 315; inscr. *v.*, in pen and brown ink: Ath Antheoliza. *Fig. 305.*

BOTANICAL REFERENCES: *O. abyssinica* (A. Richard) N.E. Brown in *Transactions Royal Society South Africa* (1932) 20, 280; Cufodontis (1972) 1591.

LILIACEAE Lily family

Albuca abyssinica *Ethiopian squill*

A bulbous plant having a rather wide distribution in hilly savanna across Africa to Arabia. The inflorescence is tall, with large, pendulous, yellow and green flowers.

VERNACULAR NAMES: None recorded.

B 1977.14.8890. Anonymous, Paris folder. Finished drawing of flowering plant, with floral details.

Watercolors and bodycolors over pencil outlines, within two double ruled ink borders; 404 × 306; inscr. *v.*, in pen and brown ink: Ornithogalum des sources du Nil (Ornithogalum from the springs of the Nile). *Fig. 306.*

BOTANICAL REFERENCES: *A. abyssinica* Murray, *Systema Vegetabile* (May–June 1784) 326; Dryander in *Kongl. Vetenskap Academiens Handlingar, Stockholm* (Oct.–Dec. 1784) 298; Jacquin, *Collectanea* (1787) 1, 55. Note: Since this watercolor is to be found in the Paris folder the plant depicted was probably raised at Paris from seed or a bulb brought back by James Bruce. Unfortunately, none of the authors gives actual details, and Dryander even doubted the provenance of "Abyssinia." The usual authority given for this species is Dryander, but since he refers to Murray's earlier publication there is now no doubt about Murray having priority.

Allium species *Onion*

VERNACULAR NAMES: Shongourt (*Travels*).

Note: "In the island of Mitraha, in the lake of Dembea (i.e., Lake Tana), several kinds of shongourt (allium)" (*Travels*, [1805] 7, 325; (1813] 7, 341; Murray [1808] 450). We have seen no drawings, and the identity remains in doubt, unless it is the ordinary culinary onion, *Allium cepa*, but the vernacular name is often used for bulbs in general.

Aloe macrocarpa *Aloe*

A perennial succulent plant with long, sharply toothed mottled leaves and a tall inflorescence of red flowers. It inhabits rocky places in dry country from Eritrea to West Africa.

VERNACULAR NAMES: Errett, Erret (MS and *Travels*, [1813] 7, 341). Similar names are noted by Schweinfurth (1896) and Mooney (1963).

a. B 1977.14.8936 *r.*, left, outline of habit, right, rough outline of inflorescence, with floral details.

Pencil, indented for transfer; 222 × 316; inscr., top and left with extensive notes, in pen and brown ink, beginning: Ques [sic] e una speccie di Aloè che chiamano Erret (This is a species of aloe which they call Erret). *Fig. 304.*

b. B 1977.14.6469j. Finished drawing from a (right) without details.

Watercolors and bodycolors over pencil outlines, within ruled ink borders; 451 × 310; inscr., upper right in pen and brown ink: 10; *v.*: Errett?. *Fig. 307.*

BOTANICAL REFERENCE: *A. macrocarpa* Todaro, *Hortus Botanicus Panormitanus* (1875–76) 1, 36, t. 9; Cufodontis (1971) 1546.

Aloe species 1 *An aloe*

VERNACULAR NAME: Erret (MS).

B 1977.14.8936 *v.*, left. Outline drawings of inflorescence with flowers open and with closed buds on the stem, together with details of flowers and leaves, the latter not of aloe.

Pencil; inscr., top, with five lines of MS notes in pen and brown ink, headed: Aloè – Pianta (Aloe plant); and beginning: Questo è un altro Aloe che parimente chiamano Erret... (This is another aloe which they likewise call Errett). *Fig. 310.*

Aloe species 2 *An aloe*

B 1977.14.8937 *r.* Rough outline of seed head, with cross-sections of fruits.

Pencil; 221 × 313; inscr., top, in pen and brown ink: Aloè Pianta 2ª (Second aloe plant); and with extensive MS notes, right, beginning: Quando l'Aloè à fatto, il fiore, è che / è Maturo, Sua tiglia si va Indurendo / e prende Consistenza per erigere in alto l'ovario... (When the aloe has flowered and when it is ripe its stem becomes hardened and acquires strength to hold up the ovary...). *Fig. 309.*

Note: This sketch is inadequate for certain identification to species rank.

Asparagus species *Wild asparagus*

A wiry-stemmed, rambling herb or small shrub with reflexed prickles. The roots of asparagus are eaten in Ethiopia either grilled or fried.

VERNACULAR NAMES: None recorded.

B 1977.14.9075 *v.* Drawing of the plant's habit, largely without detail, with, above, a smaller shoot with leaflike stem clusters.

Pencil; 302 × 199; inscr., at the top, in pen and brown ink: Pianta Ritrovata a Cioba (Plant found at Cioba); and with a column of MS notes to the right, beginning: Quest è una Pianta Lignosa che cresce / in luogho di Montagna fra le Pietre... (This is a woody pant which in mountainous places grows among the stones...). *Fig. 308.*

Kniphofia foliosa *Redhot poker*

A robust pernnial forming a clump with masses of long leaves. Inflorescences 2–3 ft. (60–100 cm) high. The yellow- or orange-colured flowers group in a dense cylinder in the upper part. It is known only from Ethiopia at altitudes between 6,700 ft. and 13,100 ft. (2,050 m and 4,000 m) and was found on the mountains at Geesh.

VERNACULAR NAME: Effarazengh—Agau (MS and *Travels* [1805] 7, 325; [1813] 7, 341).

B 1977.14.9053 *r*. Careful outlines of whole plant and single flowering stem, with details of inflorescence, florets, cross-sections of root, leaf, etc.

Pencil; 310 × 221; inscr., center, in pen and black ink: Effarazengh / nome suo in Agau (Effarazengh, its name in Agau); and with extensive MS notes beginning top center: Questo Rappresenta un Ramo de una speccie / di Aloè Ritrovato supra la Montagna Gheesh a Sakalla / li fiori erano in suo perfetto Stato era li 5 Novembre 1770 (This represents a stem of a species of aloe found on Mount Geesh at Sakalla. It was the 5th of November 1770 when the flowers were in perfect condition). *Fig. 311*.

BOTANICAL REFERENCES: *K. foliosa* Hochstetter in *Flora* (1844) 31; Cufodontis (1971) 1538.

Kniphofia pumila *Redhot poker*

A perennial with a small cormlike rootstock and narrow leaves up to 2½ ft. (80 cm) long. The inflorescence is usually 2–4½ ft. (60–150 cm) high with a very dense cylindrical mass of yellow flowers toward its top. The species is most frequent in Ethiopia, with outlying stations in Uganda, Sudan, and Zaire.

VERNACULAR NAMES: None recorded.

a. B 1977.14.8898. Anonymous, Paris folder. Finished drawing of flowering plant.

Watercolors and bodycolors over pencil outlines; within two double ruled ink borders; 407 × 310. *Fig. 314*.

b. B 1977.14.8899. In the same hand as previous drawing, Paris folder. Finished drawing of leaves and inflorescence, with floral details.

Watercolors and bodycolors over pencil outlines; within two double ruled ink borders; 404 × 305. *Fig. 313*.

BOTANICAL REFERENCES: *K. pumila* (Aiton) Kunth, *Enumeratio plantarum* (1843) 552; Cufodontis (1971) 1541.

Kniphofia probably **pumila × schimperi**

A perennial herb with small flowers that are probably red or yellow. It grows commonly in the uplands of East Africa.

VERNACULAR NAME: Ansceck-due (MS). No modern botanist has recorded a similar name.

B 1977.14.9045 *v*. Outline of habit, left, and right, details of inflorescence and fruits.

Pencil; 310 × 221; inscr., top left, in pencil: Ansceck = due; with below, letters in Amharic script showing through from reverse. *Fig. 312*.

Note: Mr W. Marais, who is a specialist on the genus *Kniphofia*, suggests that the plant illustrated belongs to *K. pumila*, but the much elongated inflorescence could indicate that it was a hybrid between *K. pumila* and *K. schimperi*, which is common in Ethiopia.

Sansevieria abyssinica *Ethiopian bowstring hemp*

A stout perennial with fleshy leaves, inhabiting dry shady places in Northeast Africa. The strongly scented white flowers open during the night, the fruits are round, orange-colored berries. In MS (8960 *v*) it is recorded that the leaves are left to rot in water forming long tough fibres which can be used to make ropes. It was found at Adderghei.

VERNACULAR NAME: Hecà (MS).

B 1977.14.8960 *v*. Part of inflorescence, left, and across the bottom, outline of habit.

Pencil; 302 × 200; inscr., top left, in pen and black ink: Hecà / Aloè Ritrovato à / Adderghei (Hecà, aloe found at Adderghei); with extensive MS notes beginning: Questo Rappresenta una Speccie di Aloè che gl' / Abbissini chiamano Hecà... (This represents a species of aloe which the Abyssinians call Hecà...). *Fig. 315*.

BOTANICAL REFERENCES: *S. abyssinica* N.E. Brown in *Kew Bulletin* (1913) 306; Cufodontis (1971) 1569.

MUSACEAE Banana family

Ensete ventricosum *African wild banana*

This wild banana inhabits moist forests in eastern Africa. Its fruits are small and full of large, black seeds and are inedible, but the young leaves are used for meal like cereals.

Bruce notes (*Travels* [1790], vol. 5, p. 36–41) that "it grows and comes to great perfection at Gondar, but it most abounds in that part of Maitsha and Goutto west of the Nile, where there are large plantations of it, and is there, almost exclusive of anything else, the food of the Galla inhabiting that province."

VERNACULAR NAMES: Ensete, Ensett, Ensette (*Travels*), Enset (*Travels* [1805] 7, 325; [1813] 7, 341). The last spelling was that used by Mooney (1963).

a. B 1977.14.9063 *r*. Drawing of habit with inflorescence showing folding of leaves, with details of leaves in cross-sections, flowers, stem in cross-section, roots, etc.

Pencil; 302 × 199; inscr., top left, in pen and black ink: Æ; with the name in Amharic script beneath, followed by: Ensett Albero Ritrovato a Gonder (Ensett, tree found at Gondar); and with notes on dimensions and letters against details. *Fig. 317*.

b. B 1977.14.9063 *v*. Rough outline of stem with

root and inflorescence, the latter, in part, diagrammatic.
Pencil; inscr., top, in pen and gray ink: Figura J. [or I. (?)]; with notes in pencil against details and numbers giving dimensions. *Fig. 316.*
c. B 1977.14.9059 *r.* Habit with inflorescence, closely similar to a, but without details.
Pen and black ink over pencil outlines; 217 × 188. *Fig. 321.*
d. Lord Elgin's collection, Broomhall. Finished drawing from c.
Watercolors; 313 × 245; inscr. *v.* in pen and brown ink: Ensete No. 9. *Fig. 318.*
Engr. by Heath, Bruce, *Travels* (1790) first pl. opp. p. 36; (1805), pl. 8; (1813), pl. 8.
e. B 1977.14.8985 *r.* Outline drawing of inflorescence, with details of floral dissections.
Pencil with touches of watercolor; 222 × 317; inscr., prolifically, with MS notes in pen and black ink, beginning, top left: Questi sono li fiori Rappresentati per quella parte / che riguarda il centro, e sono del rango interiora (These are the flowers represented by that part which faces the centre and are from the inner row). *Fig. 319.*
f. B 1977.14.8985 *v.* Two outlines of inflorescence. Pencil; inscr., in pen and black ink, above the left-hand figure: Parte interiora / delle Foglie con li fiori (The interior part of the 'bud' with the flowers). *Fig. 320.*
g. B 1977.14.8948 *r.* Details of transverse sections of stem and longitudinal section of base.
Pencil; 212 × 317; inscr. with MS notes down the left, in pen and black ink, beginning: Qui a Basso e Rappresentato un Tronco d'Ensette con sua Radice... (Here at the bottom is represented a trunk of Ensette with its root...); and with letters and numbers, in ink and pencil, against details. *Fig. 323.*
h. B 1977.14.8948 *v.* Extensive MS notes, the right-hand section written over an outline detail of the inflorescence.
Pencil (for the drawing); inscr., at the top, in pen and black ink: Ritrovato a Gonder li 15 ottobre 1770 (Found at Gondar on the 15th of October, 1770); and continuing: L'Albero Ensett è Rappresentato nel Foglio marcato / Æ ma il disegno è Rilevato da una pianta Imperfatta cioè / che non teneva ne Radice grande ne fiore ne frutto, ma unica / mente le foglie in perfezione... (The Ensett tree is represented on the sheet marked Æ [see a] but the drawing is taken from an imperfect plant, that is to say, it did not have a large root or flower or fruit but only leaves in perfection...). *Fig. 324.*
i. B 1977.14.9074 *r.* Careful drawing of inflorescence as in d, with details of fruit and floral parts and, at the foot, scales of measurement.
Pencil, indented for transfer; 316 × 220; inscr., above the scales, in pencil: Polici (thumb width [i.e., contemporary Italian units of about one inch]; and: Piedi (feet). *Fig. 322.*
j. B 1977.14.8909. Finished drawing from i.
Watercolors over pencil outlines: 310 × 245; inscr., *v.*, in pen and brown ink: Ensete No. 10. *Fig. 45.*
Engr. by Heath, in reverse, Bruce, *Travels* (1790), pl. before p. 37; (1805), pl. 9; (1813), pl. 9.

BOTANICAL REFERENCES: *E. ventricosum*(Welw.) Chees. in Kew Bull. 2: 101 (1948). *M. ensete* J.F. Gmelin, *Systema Naturae*, ed. 13 (1791) 2(1), 567. Type: Bruce's plate and description of "Ensete." "Ensete" Bruce, *Travels* (1790) 5, 36 and pl. opp. pp. 36 and 37; IDC microfiche 3: I. 6, 8; (1805) 7, 149–53 and 8, pls. 8 and 9; (1813) 7, 165–66 and 8, pls. 8 and 9; Murray (1808) pls. 8 and 9.

Musa × sapientum *Banana*

This is the well-known cultivated sweet banana of Asiatic origin. It was drawn at Sidon on 31 July 1767 (*Travels* [1805] 7, 324; Murray, [1808] 449).

VERNACULAR NAMES: None recorded by Bruce.

a. B 1977.14.9121. Preliminary outline drawing of habit of tree with fruit.
Pencil; 529 × 359; inscr., lower left, in pencil against the base of the trunk: Circonferenza d34; and: P9:8 and P2, vertically against sections of the main trunk indicating measurements in feet. *Fig. 325.*
b. 1977.14.9120. Finished drawing from a, with details of inflorescence. Watercolors and bodycolors over pencil outlines; 521 × 365. *Fig. 46.*

BOTANICAL REFERENCES: *M. × sapientum* Linnaeus, *Systema* ed. 10 (1763) 1303; Tackholm & Drar, *Flora of Egypt* (1954) 3, 553.
M. × paradisiaca subsp. *sapientum* K. Schumann—Cufodontis (1972) 1593.

PALMAE

Hyphaene thebaica *Dom palm*

A palm tree with characteristically forked branches and fan leaves. It is widely distributed across the southern Sahara and into Arabia. It was recorded (*Travels* [1790] 5, 45; [1805] 7, 324; [1813] 7, 340; Murray [1808] 449) that this was found at Sibt in Egypt, and the implication is that a drawing was made, but we have not seen it.

VERNACULAR NAMES: Doom, palma cuciofera (*Travels*).

BOTANICAL REFERENCES: *H. thebaica* (Linnaeus) Martius, *Historia Naturalis Palmarum* (1839) 3, 226, tt. 131–33; Tackholm & Drar, *Flora of Egypt* (1950) 2, 273.

PANDANACEAE

Pandanus probably **odoratissimus** *Screwpine*

A small tree with very sharply toothed leaves that arise from the trunk in twisted ranks, hence the common name of screwpine. The green fruits are reminiscent of large pinecones. It has not been recorded from Ethiopian but this Asian tree is cultivated in valleys in the Yemen for its fragrant male inflorescences.

VERNACULAR NAMES: None recorded.

B 1977.14.9126. Finished drawing for male inflorescence.

Watercolors; 176 × 134. *Fig. 47.*

BOTANICAL REFERENCE: *P. odoratissimus* Linnaeus f., *Supplementum* (1781) 424.

GYMNOSPERMS

CUPRESSACEAE Cypress tree family

Juniperus procera *African pencil cedar*

A medium-sized to large coniferous tree that forms forests on many African mountains between 5000 and 10,000 ft. (1500–3000 m). It is characteristic of the drier African mountain forest. In his account of his travels Bruce often referred to this tree as the 'cedar', but of course it is more correct to say 'juniper', although the name 'cedar' is still current for it in East Africa.

VERNACULAR NAME: Cedar, Arze (*Travels*).

a. B 1977.14.9051 *r.* Rough sketch of tree's habit, with slight sketch of thistle, left (see *Carthamus tinctorius* p. 77).

Pen and brown ink over pencil outlines, indented for transfer (the thistle in pencil); 310 × 210; inscr. above, in pen and brown ink: Congieb. *Fig. 326.*

b. B 1977.14.9128. Finished drawing of tree's habit. Watercolors and bodycolors over pencil outlines, within ruled pencil borders; 419 × 329. *Fig. 48.*

BOTANICAL REFERENCES: *J. procera* Endlicher, *Synopsis coniferarum* (1847) 26; Melville in *Flora Tropical East Africa*, Gymnospermae (1958) 16, fig. 5; Cufodontis (1953) 1.

PODOCARPACEAE Podo tree family

Podocarpus gracilior *Podo tree*

A coniferous tree of upland forest in eastern Africa.

VERNACULAR NAME: Zegvà (MS). Cufodontis (1953) and Mooney (1963) note similar names.

B 1977.14.8997 *r.* Outline of leafy shoot, with details of leaves.

Pencil: 222 × 210; inscr., top left: Zegvà; *v.* MS notes only; inscr., in pen and black ink: Nel Roverscio di questo Foglio è Rappresentato un picolo Ramo di un Albero / Ritrovato in Decembre al Kaha senza Fiore, e senza Frutto... (On the reverse of this sheet is represented a small shoot of a tree found in December at El Kaha (?) without flower or fruit...). *Fig. 327.*

BOTANICAL REFERENCES: *P. gracilior* Pilger in Engler, *Pflanzenreich* (1903) IV.5, 71; Cufodontis (1953) 1; Melville in *Flora Tropical East Africa*, Gymnospermae (1958) 21.

ALGAE

Alga indeterminate species *Seaweed*

It has not been possible to identify this drawing of a slender seaweed.

VERNACULAR NAME: Yesser (MS).

B 1977.14.6469a. Finished drawing of habit of weed. Watercolors over pencil outlines, within ruled ink borders; 312 × 451; inscr., upper right, in pen and brown ink: 1; lower right, vertically: Yesser and *v.*: Yesser; also: Gorgonia. *Fig. 330.*

Unidentified species of flowering plant

VERNACULAR NAME: Yesser (MS).

B 1977.14.9050 *r.* Outline sketch of habit.

Pencil; 213 × 231; inscr. above in pen and ink: Idea della pianta Cadi La piu giusta che si è descritta / che sarà a disegnose alorche non / sia stato possibile di avere una mi- / glior descrizione oppore che non / la potiamo Vedere noi stessi (Sketch of a plant Cadi just as it was described, to be drawn when it has not been possible to get a better description or rather when we have not been able to see it ourselves). *Fig. 329.*

Unidentified species of flowering plant

VERNACULAR NAME: Sciaura (MS).

B 1977.14.8981 *v.* Outline sketches of seed.

Pencil; 293 × 181; inscr. above in pen and ink: Coccia esteriore e Mandola (outer shell and almond); and below with 11 lines of MS notes beginning: Idea del Frutto che fo la Pianta Sciaura / Il frutto esteriormente è della forma di un Mandola (Sketch of the fruit of the plant Sciaura. The fruit has on the outside the shape of an almond...) *Fig. 328.*

· Appendix I ·

List of Accession Numbers on Drawings, with Identifications Yale Center for British Art

11 Watercolors hardbound in dark green boards
B 1977.14.
6467 a *Catha edulis* Forssk.
b *Dracaena steudneri* Engl.
c *Dracaena steudneri* Engl.
d *Senecio gigas* Vatke
e *Abutilon longicuspe* A. Rich.
f *Capparis tomentosa* Lam.
g *Erythrina abyssinica* DC.
h *Jasminum abyssinicum* DC.
i *Carissa edulis* (Forssk.) Vahl
j *Acokanthera schimperi* (DC) Schweinfurth
k *Desmodium* species

15 Watercolors hardbound in dark green boards
B 1977.14.
6468 a *Rhus ? retinorrhoea* Oliver
b *Ehretia cymosa* Thonning
c *Osyris lanceolata* A.DC.
d *Pittosporum viridiflorum* Sims
e *Myrsine africana* L.
f *Rhus glutinosa* A. Rich.
g *Bersama abyssinica* Fresen.
h *Diospyros mespiliformis* A. DC.
i *Maytenus arguta* (Loes.) N. Robson
j *Ziziphus abyssinica* A. Rich.
k *Clausena anisata* (Willd.) Benth.
l *Grewia ferruginea* A. Rich.
m *Salix subserrata* Willd.
n *Vernonia amygdalina* Del.
o *Nuxia oppositifolia* (Hochst.) Benth.

15 Watercolors hardbound in dark green boards
B 1977.14.
6469 a *Alga* indeterminate
b *Cyperus* species 1
c *? Chloris* species
d *Cyperus* species 2
e *Mariscus cyperoides* (L.) O. Kuntze
f *Cyperus esculentus* L.
g *Canna bidentata* Bertol.
h *Scadoxus puniceus* (L.) Friis & Nordal
i *Impatiens rothii* Hook. f.
j *Aloe macrocarpa* Todaro
k *Dregea abyssinica* (Hochst.) K. Schum.
l *Peponium vogelii* Engl.
m *Phytolacca dodecandra* L'Hérit.
n *Solanum incanum* L.
o *Solanum piperiferum* A. Rich.

Loose watercolors and drawings
B 1977.14.
8518 *Coffea arabica* L.
8519 *Coffea arabica* L.
8693 *Commiphora gileadensis* (L.) C. Chr.
8882 *Brucea antidysenterica* J. Miller
8883 *Coccinia abyssinica* (Lam.) Cogn.
8884 *Verbascum sinaiticum* Benth.
8885 *Verbascum sinaiticum* Benth.
8886 *Solanum marginatum* Linn.f.
8887 *Solanum marginatum* Linn. f.
8888 *Salvia nilotica* Juss. ex Jacq.
8889 *Salvia nilotica* Juss. ex Jacq.
8890 *Albuca abyssinica* Murray
8891 *Guizotia abyssinica* (Linn. f.) Cass.
8892 *Coleus edulis* Vatke
8893 *Eragrostis tef* (Zuccagni) Trotter
8894 *Ischaemum afrum* (J.F. Gmel.) Dandy
8895 *Ischaemum afrum* (J.F. Gmel.) Dandy
8896 *Bauhinia farec* Desv.
8897 *Panicum deustum* Thunb.
8898 *Kniphofia pumila* (Ait.) Kunth
8899 *Kniphofia pumila* (Ait.) Kunth
8900 *Solanum capsicoides* Allioni
8901 *Coleus* species 1
8902 *Panicum deustum* Thunb.
8903 *Cyperus papyrus* L.
8904 *Commiphora gileadensis* (L.) C. Chr.
8905 *Commiphora gileadensis* (L.) C. Chr.
8906 *Albizia gummifera* (J.F. Gmel.) C.A. Sm.
8907 *Albizia gummifera* (J.F. Gmel.) C.A. Sm.
8908 *Mimosa pigra* L.

8909 *Ensete ventricosum* (Welw.) Cheesman
8910 *Euphorbia abyssinica* J. F. Gmel.
8911 *Euphorbia abyssinica* J. F. Gmel.
8912 *Avicennia marina* (Forssk.) Vierh.
8913 *Pterolobium stellatum* (Forssk.) Brenan
8914 *Protea gaguedi* J. F. Gmel.
8915 *Protea gaguedi* J. F. Gmel.
8916 *Cordia africana* Lam.
8917 *Dombeya torrida* (J. F. Gmel.) Bamps
8918 *Brucea antidysenterica* J. Miller
8919 *Hagenia abyssinica* (Bruce) J. F. Gmel.
8920 *Hagenia abyssinica* (Bruce) J. F. Gmel.
(8921–8934 nonbotanical)
8935 *v.* *Jasminum abyssinicum* DC.
r. *Otostegia tomentosa* A. Rich. subsp. *ambigens* (Chiov.) Sebald
8936 *r.* *Aloe macrocarpa* Todaro, *Aloe* species 1 & *Solanum incanum* L.
8937 *r.* *Aloe* species 2
r. *Nuxia oppositifolia* (Hochst.) Benth.
r. *Clematis hirsuta* Guill. & Perr.
v. *Rhus glutinosa* A. Rich.
v. *Solanum incanum* L.
v. *Rhus? retinorrhoea* Oliv.
8938 *v.* *Dombeya torrida* (J. F. Gmel.) Bamps
r. *Buddleia polystachya* Fresen.
8939 *r.* *Coleus edulis* Vatke
8940 *r.* *Albizia gummifera* (J. F. Gmel.) C. A. Sm.
8941 *r.*, *v.* *Dracaena steudneri* Engl.
8942 *r.* *Dracaena steudneri* Engl.
8943 *v.* *Phytolacca dodecandra* L'Hérit.
r. *Scadoxus puniceus* (L.) Friis & Nordal
8944 *r.* *Cardiospermum halicacabum* L.
8945 *v.* *Solanum adoense* Hochst. ex Rich.
r. *Isodon ramosissimus* (Hook. f.) Codd.
8946 *r.* *Ritchiea albersii* Gilg
8947 *r.* *Millettia ferruginea* (Hochst.) Baker
8948 *r.*, *v.* *Ensete ventricosum* (Welw.) Cheesman
8949 *r.* *Coriandrum sativum* L.
r. *Acanthus sennii* Chiov.
v. *Hygrophila auriculata* (Schumach.) Heine
v. *Euphorbia abyssinica* J. F. Gmel.
v. *Ficus vasta* Forssk.
8950 *r.*, *v.* *Acanthus sennii* Chiov.
8951 *r.*, *v.* *Canna bidentata* Bertol.
8952 *v.* *Dregea abyssinica* (Hochst.) K. Schum.
r. *Desmodium* species
8953 *v.* *Peponium vogelii* Engl.
r. *Dregea abyssinica* (Hochst.) K. Schum.
8954 *r.* *Rhus ? retinorrhoea* Oliver
8955 *v.* *Delphinium wellbyi* Hemsley
r. *Ruttya speciosa* (Hochst.) Engl.
8956 *r.* *Myrsine africana* L.
8957 *r.* *Diospyros mespiliformis* A. DC.
8958 *r.* *Ottelia ulvifolia* (Planch.) Walp.
8959 *r.*, *v.* *Osyris lanceolata* A. DC.
8960 *r.* *Bersama abyssinica* Fresen.
v. *Sansevieria* prob. *abyssinica* N. E. Br.
8961 *r.* *Albizia gummifera* (J. F. Gmel.) C. A. Sm.
8962 *r.*, *v.* *Clausena anisata* (Willd.) Benth.
8963 *r.* *Impatiens rothii* Hook. f.
v. *Coccinia abyssinica* (Lam.) Cogn.
8964 *r.* *Cynanchum altiscandens* K. Schum.
r. *Tacazzea venosa* Hochst. ex Decne.
v. *Lagenaria siceraria* (Molina) Standl.
v. *Tacazzea galactagoga* Bullock
8965 *r.* *Clausena anisata* (Willd.)Benth.
v. ? indeterminate
8966 *v.* *Delphinium wellbyi* Hemsley
r. *Dombeya torrida* (J. F. Gmel.) Bamps
8967 *r.*, *v.* *Crinum zeylanicum* (L.) L.
8968 *r.* *Annona squamosa* L.
8969–8970 nonbotanical
8971 *r.*, *v.* *Steganotaenia araliacea* Hochst.
8972–8976 nonbotanical
8977 *v.* *Annona squamosa* L.
r. *Acacia* species
8978 *r.* *Cassia fistula* L.
r., *v.* *Euphorbia nubica* N. E. Br.
8979 *r.* *Phytolacca dodecandra* L.Hérit.
v. *Cassia fistula* L.
8980 *r.* *Cyperus esculentus* L.
8981 *v.* Unidentified species 2
8982 *r.* *Clematis hirsuta* Guill. & Perr.
8983 *r.* *Diospyros mespiliformis* A. DC.
v. *Ziziphus abyssinica* A. Rich.
8984 *v.* *Clerodendrum myricoides* (Hochst.) R. Br. ex Vatke
r. *Loranthus globiferus* A. Rich.
8985 *Ensete ventricosum* (Welw.) Cheesman
8986 *r.*, *v.* *Catha edulis* Forssk.
8987 *r.* *Discopodium penninervium* Hochst.
8988 *r.* *Rhus glutinosa* A. Rich.
8989 *r.* *Pittosporum viridiflorum* Sims
8990 *r.* *Pittosporum viridiflorum* Sims
v. *Dombeya torrida* (J. F. Gmel.) Bamps
8991 *r.* *Coleus* species 2
8992 *r.* *Woodfordia uniflora* (A. Rich) Koehne
8993 *r.* *Nuxia oppositifolia* (Hochst.) Benth.
v. *Hypericum quartinianum* A. Rich.

8994 r. *Trifolium polystachyum* Fresen.
r. *Trifolium rueppellianum* Fresen.
v. *Trifolium usumbarense* Taub.
v. *Trifolium schimperi* A. Rich.
8995 v. *Capparis tomentosa* Lam.
r. *Kanahia laniflora* (Forssk.) R. Br.
r. *Crotalaria pallida* Ait.
8996 r. *Loranthus heteromorphus* A. Rich.
8997 r., v. *Podocarpus gracilior* Pilger
8998 r. *Erythrina abyssinica* DC.
v. *Erythrina brucei* Schweinf. emend Gillett
8999 r. *Maytenus arguta* (Loes.) N. Robson
v. *Oxalis corniculata* L.
9000 r. *Guizotia abyssinica* (Linn. f.) Cass.
9001 r. *Cardiospermum halicacabum* L.
9002–9004 nonbotanical
9005 r. *Alepidea peduncularis* Steud. ex A. Rich.
9006–9024 nonbotanical
9025 r. *Arundinaria alpina* K. Schum.
9026 r., v. *Mimsuops kummel* A. DC.
9027 r., v. *Boswellia papyrifera* (Del.) Hochst.
9028 r. *Mimusops kummel* A. DC.
v. *Carissa edulis* (Forssk.) Vahl
9029 r. *Boswellia papyrifera* (Del.) Hochst.
v. *Hibiscus cannabinus* L.
9030 r., v. *Guizotia abyssinica* (Linn. f.) Cass.
9031 v. *Brucea antidysenterica* J. Miller
r. *Oenostachys abyssinica* (A. Rich.) N.E. Br.
r. *Senecio gigas* Vatke
9032 r. *Michauxia campanuloides* L'Hérit. ex Ait. and *Hagenia abyssinica* (Bruce) J. F. Gmel.
9033 r. *Hagenia abyssinica* (Bruce) J. F. Gmel.
9034 r. *Senecio gigas* Vatke
v. *Impatiens rothii* Hook.f.
9035 r., v. *Acacia* species
9036 r. *Pterolobium stellatum* (Forssk.) Brenan
9037 r. *Abutilon longicuspe* A. Rich.
9038 r., v. *Abutilon longicuspe* A. Rich.
9039 r. *Salix subserrata* Willd.
v. *Crotalaria pallida* Ait.
9040 r. *Capparis tomentosa* Lam.
v. *Dicrostachys cinerea* (L.) Wight & Arn.
9041 v. *Cyperus* species 1
r. *Astragalus atropilosulus* (Hochst.) Bunge
v. *Commelina africana* L.
r. *Gomphocarpus fruticosus* (L.) Ait. f.
9042 r. *Olea africana* P. Mill.
v. *Tacazzea galactagoga* Bullock
9043 r., v. *Gardenia ternifolia* Schum. & Thonn. subsp. *jovis-tonantis* (Welw.) Verdc.
9044 r. *Ruttya speciosa* (Hochst.) Engl.
9045 r. *Oenostachys abyssinica* (A. Rich.) N.E. Br.
v. *Kniphofia* prob. *pumila* × *schimperi*
9046 r. *Mussaenda arcuata* Lam. ex Poir.
v. *Ziziphus abyssinica* A. Rich.
9047 r., v. *Brucea antidysenterica* J. Miller
9048 r. *Brucea antidysenterica* J. Miller
9049 v. *Verbascum sinaiticum* Benth.
9050 r. Unidentified species 1
9051 r. *Juniperus procera* Endl.
9052 r. *Cordia abyssinica* R. Br.
9053 r. *Kniphofia foliosa* Hochst.
v. *Hebenstreitia dentata* L.
9054 r. *Rhus? retinorrhoea* Oliver
r. *Cucumis metuliferus* Naud.
v. *Hibiscus* species
v. *Diplolophium africanum* Turcz.
9055 r. *Cucumis metuliferus* Naud.
9056 r. *Jasminum dichotomum* Vahl
v. *Dregea abyssinica* (Hochst.) K. Schum.
v. *Mussaenda arcuata* Lam. ex Poir.
v. *Cardiospermum halicacabum* L.
9057 r. *Scadoxus puniceus* (L.) Friis & Nordal
r. *Diospyros abyssinica* (Hiern) White
9058 v. *Protea gaguedi* J. F. Gmel.
9059 r., v. *Ensete ventricosum* (Welw.) Cheesman
9060 r. *Solanum piperiferum* A. Rich.
v. *Coffea arabica* L.
9061 r. *Rhamnus prinoides* L'Hérit.
9062 r. *Vernonia amygdalina* Del.
9063 r., v. *Ensete ventricosum* (Welw.) Cheesman
9064 r. *Crinum schimperi* Vatke ex K. Schum.
v. *Capparis tomentosa* Lam.
9065 r. *Erythrina abyssinica* DC.
9066 r., v. *Gardenia ternifolia* Schum. & Thonn. subsp. *jovis-tonantis* (Welw.) Verdc.
9067 r., v. *Hagenia abyssinica* (Bruce) J. F. Gmel.
9068 r. *Solanum adoense* Hochst. ex Rich.
v. *Pterolobium stellatum* (Forssk.) Brenan
9069 r. *Protea gaguedi* J. F. Gmel.
9070 r. *Coccinea abyssinica* (Lam.) Cogn.
9071 r., v. *Coccinea abyssinica* (Lam.) Cogn.
v. *Pelargonium alchemilloides* (L.) Ait. subsp. *multibracteatum* (A. Rich.) Kokwaro
9072 r. *Cucumis metuliferus* Naud.
r. *Scadoxus puniceus* (L.) Friis & Nordal
9073 r., v. *Arundinaria alpina* K. Schum.
9074 r. *Ensete ventricosum* (Welw.) Cheesman

9075 *r.* *Protea gaguedi* J. F. Gmel.
v. *Asparagus* species
9076 *r.* *Mimusops kummel* A. DC.
9077 *r.* *Mimosa pigra* L.
v. *Diospyros abyssinica* (Hiern) White
9078 *r.* *Grewia ferruginea* A. Rich.
v. *Diospyros mespiliformis* A. DC.
v. *Gardenia ternifolia* Schum. & Thonn. subsp. *jovis – tonantis* (Welw.) Verdc.
9079 *r.* *Pycnostachys abyssinica* Fresen.
(9080–9088 nonbotanical)
9089 *Punica granatum* L. (Nos. 31 & 33)
9089 a *Cassia fistula* L.
b *Annona squamosa* L.
c *Annona squamosa* L.
d *Albizia malacophylla* (A. Rich.) Walp.
e *Albizia malacophylla* (A. Rich.) Walp.
f *Canna bidentata* Bertol.
g *Canna bidentata* Bertol.
h *Cyperus esculentus* L.
i *Echinops macrochaetus* Fresen.
9090 nonbotanical
9091 *Cassia fistula* L.
9092 *Saba comorensis* (Bojer) Pichon
9093 *Arundinaria alpina* K. Schum.
9094 *Arundinaria alpina* K. Schum.
9095 *Sterculia africana* (Lour.) Fiori
9096 *Rhamnus prinoides* L'Hérit.
9097 *Solanum adoense* Hochst. ex A. Rich.
9098 *Ruttya speciosa* (Hochst.) Engl.
9099 *Acanthus sennii* Chiov.
9100 *Carthamus tinctorius* L.
9101 *Coccinia abyssinica* (Lam.) Cogn.
9102 *Coccinia abyssinica* (Lam.) Cogn.
9103 *Hypericum quartinianum* A. Rich.
9104 *Jasminum dichotomum* Vahl
9105 *Buddleia polystachya* Fresen.
9106 *Mimusops kummel* A. DC.
9107 *Phlomis herba-venti* L.
9108 *Phlomis herba-venti* L.
9109 *Labiatae* indeterminate species
9110 *Delphinium wellbyi* Hemsley
9111 *Woodfordia uniflora* (A. Rich.) Koehne
9112 *Millettia ferruginea* (Hochst.) Baker
9113 *Millettia ferruginea* (Hochst.) Baker
9114 *Discopodium penninervium* Hochst.
9115 *Mussaenda arcuata* Lam. ex Poir.
9116 *Cucumis metuliferus* Naud.
9117 *Gardenia ternifolia* Schum. & Thonn. subsp. *jovis–tonantis* (Welw.) Verdc.
9118 *Gardenia ternifolia* Schum. & Thonn. subsp. *jovis–tonantis* (Welw.) Verdc.
9119 *Cyperus papyrus* L.
9120 *Musa sapientum* L.
9121 *Musa sapientum* L.
9122 *Euphorbia nubica* N.E. Br.
9123 *Calotropis procera* (Ait.) Ait. f.
9124 *Lawsonia inermis* L.
9125 *Euphorbia helioscopia* L.
9126 *Pandanus* prob. *odoratissimus* L.
9127 *Oenostachys abyssinica* (A. Rich.) N.E. Br.
9128 *Juniperus procera* Endl.
22699 *Eragrostis tef* (Zucc.) Trotter
22700 *Phlomis herba–venti* L.
22701 *Trifolium rueppellianum* Fresen.
22702 *r.* *Mimusops kummel* A. DC.
v. *Croton macrostachyus* Hochst. ex A. Rich.
22703 *Saba comorensis* (Bojer) Pichon
22704 *Euphorbia abyssinica* J. F. Gmel.
22718 *r.* *Vitis vinifera* L.
v. *Punica granatum* L.

· Appendix II ·

Summary of James Bruce's Itinerary

June	1762	Left Britain for France and Italy
20 March	1763	Arrived in Algiers
Late March	1765	Luigi Balugani joins Bruce in Algiers
25 August	1765	Bruce and Balugani leave for Tunis
Sept. 1765–Feb.	1766	Excursions in Tunisia and Algeria to Tripoli
25 Oct.	1766	Left Tripoli, to Benghazi, shipwrecked later
29 Dec.	1766	From Benghazi to Crete and Syria
29 July	1767	Near Jordan R. source
Sept.–Oct.	1767	Journeys to Baalbek and Palmyra
15 June	1768	Sailed from Sidon to Alexandria, thence to Cairo
12 Dec.	1768	Sailed south from Cairo to Aswan
16 Feb.	1769	Left Qena for Quseir across desert
5 April	1769	Sailed from Quseir for Jidda via Tor and Rabigh
8 July	1769	Left Jidda for Bab el Mandeb
6 August	1769	Left Al Luhayya
15 Sept.	1769	Dahalac Ils.
19 Sept.	1769	Arrived at Massawa
15 Nov.	1769	Left Arkeeko, near Massawa
23 Nov.	1769	At Dixan
25 Nov.	1769	Left Dixan
26 Nov.	1769	Left Hadadid
1 Dec.	1769	Arrived at Kellah, near Debra Damo
6 Dec.	1769	Arrived at Adowa, stayed until 17 Jan. 1770
18 Jan.	1770	Reached Axum
22 Jan.	1770	In the province of Shire (Sire)
26 Jan.	1770	Crossed the river Tacazze
31 Jan.	1770	Adderghei, near the river Mai-Lumi
8 Feb.	1770	At Lamalmon
11 Feb.	1770	At Mackara, at Lamalmon, near the pass in the Semien mountains
15 Feb.	1770	Camped at Angareb River just N of Gondar
3 March	1770	At Gondar
4 Apr.	1770	Left Gondar
5 Apr.	1770	Arrived at Emfras where Bruce stayed, with occasional visits to Gondar, until 17 May
13 May	1770	The Emperor's army left Gondar, marching southwards
18 May	1770	Bruce followed the army out of Emfras
May	1770	Visited the Tissisat Falls, and crossed the Blue Nile where it flows out of Lake Tana. Continued to Dingleber in the direction of Gondar
3 June	1770	Arrived at Gondar
5 June	1770	Fled from a rebellion to Koscam. Gondar captured by rebels. Received Geesh and the source of the Nile as endowment.
28 Oct.	1770	Left Gondar on a journey to the source of the (Blue) Nile
30 Oct.	1770	Arrived at Fasil's Court at Bamba. Bruce is granted an escort and guide by Fasil who is governor of Gojam and Agow-Midre.
4 Nov.	1770	Arrived at Geesh, Sacala and Blue Nile source
10 Nov.	1770	Left Geesh to return to Gondar
19 Nov.	1770	Reached Koscam The Emperor on his way back from Tigré. Bruce set out to meet the Emperor
24 Dec.	1770	The Emperor marched into Gondar. Revenge executions started

After 14 Feb.	1771	Luigi Balugani died at Gondar
19 May	1771	The rebels' army met the Emperor's in the battles of Serbraxos, just outside Gondar. Bruce took part in the battles. The victory went to the rebels, and the Emperor reigned on as puppet. Bruce returned to Koscam
Aug.-Oct.	1771	In ill health at Koscam
26 Dec.	1771	Left Koscam to return home via Sennar
2 Jan.	1772	Arrived at Cherkin
26 Jan.	1772	Arrived at Hor Cacamoot
18 March	1772	Left Hor Cacamoot
25 Mar.	1772	Arrived at Teawa
17 Apr.	1772	Left Teawa
29 Apr.–5 Sept.	1772	At Sennar
6 Sept.	1772	Left Sennar
23 Sept.	1772	Reached the meeting of the Blue Nile and the White Nile
29 Nov.	1772	Reached Aswan after perilous journey through the Nubian desert
11 Dec.	1772	Left Aswan for Cairo
25 Mar.	1773	Reached Marseilles from Alexandria, to Paris
end July	1773	Left France for Italy
Aug.	1773	Arrived in Bologna, stayed at Poretta, on to Rome etc.; back to France
21 June	1774	Finally reached London

· Appendix III ·

The Early Illustrated Editions of Bruce's "Travels"

BRITISH EDITIONS

Although Bruce returned to England in 1774, the publication of his *Travels* was inexplicably delayed until 1790. This caused much concern to his contemporaries who were anxious to read the full account of his journey and the remarkable discoveries he made. Perhaps he was expecting the British government to finance its publication, especially the engravings which would have been too expensive to be commercially viable. The delay did nothing to allay the rumours that were current at the time that Bruce might not have even reached Ethiopia after all, that he murdered Balugani and that his account of certain Ethiopian practices, such as carving living beef, were figments of his imagination. An essay in the *London Magazine* of August and September 1774 had merely whetted the appetite of armchair travellers, many of whom must have lost interest or even died during the intervening years.

The first edition was published under the title *Travels to discover the source of the Nile in the years 1768, 1769, 1770, 1771, 1772, and 1773*, 5 volumes (Vol. 5 "Select specimens of Natural History collected in travels to discover the course of the Nile" etc. including 24 plant engravings of 18 species). J. Ruthen: Edinburgh; G.G.J. & J. Robinson: London. 1790. Quarto.

Many versions of *Travels* appeared in editions abridged to one volume and without the section on natural history. These popular editions started appearing in the same year as the first edition and new versions were issued as late at 1860. Recently (1972) a reprint of the first edition has been published in England by Gregg International Publishers Ltd. of Farnborough.

Second and third editions were prepared by Alexander Murray, the oriental scholar, who had access to the MSS and drawings then in Scotland which he used for the addition of certain information on the natural history and several more plates which were engraved by Heath and dated as "published 10 October 1804".

"The second edition corrected and enlarged. To which is prefixed a life of the Author". 8 volumes (7 vols. text, octavo, and one vol. plate, quarto). A Constable & Co.: Manners & Miller: Edinburgh. 1805.

The new material in the second edition was also issued separately in Quarto for the benefit of the owners of the first edition: Murray, A.: *Account of the life and writings of James Bruce*. 1 volume Quarto. Edinburgh 1808. (Contains also the additions to the "Select specimens of Natural History" in the 2. ed.).

The third edition of 1813 was similar but with different pagination. In both editions the plates themselves are marked "Vol. VII" which is a reference to the text in vol. 7 that they illustrate.

FRENCH EDITIONS

Bruce's *Travels* were translated immediately into French. Two versions, differing only in title and format, were published: *Voyage en Nubie et en Abyssinie entrepris pour decouvrir les sources du Nil, pendant les années 1768, 1769, 1770, 1771, 1772 & 1773*. Traduit de l'anglais par M. Castéra. Quatre voyages dans le pays des Hottentots et la Cafrérie, en 1777, 1778 & 1779, par le lieutenant William Paterson, traduit de l'anglais sur la seconde édition par M. Castéra.

6 volumes, Quarto. (vol. 5 contains the section on natural history and the travels of Lt. Paterson. vol. 6 contains the plates and maps of both travellers). 1790–1792. Paris.

This is a confusing work to use since the plates of Bruce's Abyssinian plants are followed by South African plants discovered by Paterson.

Voyages aux sources du Nil, en Nubie et en Abyssinie, pendant les années 1768.... 1772. Traduit de l'anglais par M. Castéra. Quatre voyages... par le lieutenant William Paterson, faisant le tome 14ᵉ du voyage de Bruce, traduit... par M. Castéra. 14 vols. in Octavo, 1 volume in Quarto ("Cartes et Figures de voyage en Nubie et en Abyssinie") Paris & Londres. 1790–1792.

The natural history in these two versions are unannotated translations of the first English edition, but the plates of the plants are provided with generic and sometimes also with specific names in Latin and French.

GERMAN EDITIONS

A German translation of the first English edition was also immediately prepared.

Reisen zur Entdeckung der Quellen des Nils in den Jahren 1768, 1769, 1770, 1771, 1772 und 1773. In fünf Bändern von James Bruce von Kinnaird, Esq., F.R.S. In Teutsche übersetst von J.J. Volkmann, mit einer Vorrede und Anmerkung versehen von Joh.Fr. Blumenbach und Th.Chr. Tychson. 5 vols. Octavo. 1790–1791. Leipzich. The section on natural history is almost identical with that of the English edition. Blumenbach's comments on natural history are in a separate part of vol. 5. It contains critical notes but no new names are proposed.

An abbreviated version appeared in E.W. Culin's *Sammlung merkwürdiger Reisen in das Innere von Afrika* 2: 305–444 (1790) and 3: 1–432 (1791). Leipzich.

This translation formed the basis of a separate edition:

Reisen in das Innere von Afrika, nach Abyssinien an die Quellen des Nils. Aus dem Englischen mit nötiger Abkürzungen in das Deutsche übersetst von E.W. Cuhn. Mit zur Naturgeschichte gehörigen Berichtigungen und Zusässen versehen von J.F. Gmelin. 2 volumes in Octavo. 1791. Leipzich.

This is one of the most interesting foreign editions of the *Travels* because of J.F. Gmelin's detailed notes on the natural history, which appears in a separately paginated part "Anhang zu James Bruce Reisen..." (I–XXV, 1–176) which was also published separately. It includes the diagnoses of *Mimosa sanguinea*, *Mimosa cornuta*, *Euphorbia abyssinica*, and of the genera *Racka* and *Hagenia*. These diagnoses may possibly antedate those published by J.F. Gmelin in *Systema Naturae*, ed. 13, 2, 1 (late September–November 1791) & 2, 2 (April–October 1792), where more new species based on Bruce's drawings and descriptions are proposed.

· Appendix IV ·

General Bibliography and Abbreviated References

PRIMARY SOURCES

Bruce archives. MSS and printed documents collected by James Bruce of Kinnaird, consisting of drawings, watercolours, journals, letters and other records relating to his travels. Acquired from Lord Elgin of Broomhall in 1968 and in 1977 by Mr. Paul Mellon and donated by him to the Yale Center for British Art in 1977.

Lord Elgin's collection, Broomhall. Drawings and other documents, belonging to James Bruce of Kinnaird, remaining in the Earl of Elgin's possession at Broomhall, Dunfermline.

Oretti MS. Marcello Oretti, "Notizie de Professori del Disegno cioè Pittori, Scultori et Architetti Bolognesi e de Forestieri di sua scuola." Largely unpublished MS (c. 1760–1780) including an account of Luigi Balugani (vol. 10, pp. 123–135) and other artists who worked on his or Bruce's architectural drawings. Biblioteca Communale Dell' Archeginnasio, Bologna.

SECONDARY SOURCES

Beckingham (1964). *Travels to discover the source of the Nile by James Bruce.* Selections and introduction by C.F. Beckingham. Edinburgh, 1964.

——— (1984). *The Itinerário of Jerónimo Lobo.* Transl. D.M. Lockhart, Portuguese text ed. M.G. da Costa. Introduction and notes by C.F. Beckingham. Hakluyt Soc., London, 1984.

Bruce *Travels* (1790). James Bruce of Kinnaird, *Travels to discover the source of the Nile in the years 1768, 1769, 1770, 1771, 1772 & 1773.* 5 vols. Edinburgh, 1790.

Bruce *Travels* (1805). Second edition ed. Alexander Murray, with amendments and additions, including a life of the author. 8 vols. Edinburgh, 1804–1805.

Bruce *Travels* (1813). Third edition ed. Alexander Murray. 8 vols. Edinburgh, 1813.

Boswell (1960). James Boswell's private papers. *Boswell for the defence 1769–1774,* ed. W.K. Wimsatt, jr, and F.A. Pottle. London, 1960.

Boswell (1934–1950). James Boswell, *Life of Johnson,* ed. G.B. Hill, rev. L.F. Powell. 6 vols. Oxford, 1934–1950.

Buffon (1775). Georges Leclerc, Comte de Buffon, *Histoire naturelle...* vol. 18, *Histoire naturelle des oiseaux,* vol. 3. Paris, 1775.

Burney (1907). *The early diary of Frances Burney 1768–1778,* ed. Annie Raine Ellis. 2 vols. London, 1907.

Catalogue (1974). *Catalogo generale... della Pinacoteca Nazionale di Bologna, Gabinetto delle Stampe.* Sezione III. *Incisori bolognesi ed emiliani del sec. XVIII.* Catalogue ed. Giovanna Gaeta Bertalà and Stefano Ferrara. Bologna, 1974.

Catalogue (1979). Biennale d'Arte Antiqua. *L'Arte del Settecento Emiliano. La pittura. L'Accademia Clementina.* Catalogue ed. Andrea Emiliani *et al.* Bologna, 1979.

Cheesman (1936). R.E. Cheesman, *Lake Tana and the Blue Nile.* London, 1936.

Chiovenda (1940). Emilio Chiovenda, "Documenti relativi a James Bruce e Luigi Balugani che visitarono l'Etiopia nel 1769–72" in *Atti R. Accademia d'Italia, Rendiconti Scienze Fisiche* etc., December 1940, ser. 7, vol. 2 (Rome), pp. 439–496.

Cuppini (1974). Giampiero Cuppini, *I palazzi senatorii a Bologna. Architettura come immagine del potere.* Bologna, 1974.

Degli Azzi (1908). G. Degli Azzi in *Allgemeine Lexicon der bildender Kunst,* ed. U. Thieme & F. Becker, vol. 2 (Leipzig, 1908), pp. 427–428.

Dennistoun (1855). James Dennistoun, *Memoirs of Sir Robert Strange and Andrew Lumisden.* 2 vols. London, 1855.

Emiliani (1971). Andrea Emiliani. *L'opera dell 'Accademia Clementina per il patrimonio artistico e la formazione della Pinacoteca Nazionale di Bologna.* Bologna, 1971.

Farrer (1904). R. Farrer, "Filippo Balugani," in *Biographical dictionary of medallists,* vol. 1 (London, 1904), pp. 119–120.

Friis (1981). Ib Friis, "James Bruce—en pioner i udforskningen af Afrikas planteverden" in *Naturens verden,* vol. 12. (Copenhagen, 1981), pp. 404–422.

Garnett (1921–1922). R. Garnett, "James Bruce" in *Dictionary of national biography,* vol. 3 (London 1921–1922), pp. 98–102.

George (1938). Mary Dorothy George, *Catalogue of political and personal satires*, vol. 5. London, 1938.
Graves (1905). Algernon Graves, *The Royal Academy of Arts. A complete dictionary of contributors*, vol. 1. London, 1905.
Hallett (1965). R Hallett, *The penetration of Africa*, vol. 1. London, 1965.
Head (1836). F.B. Head, *The life of Bruce, the African traveller*. Second edition. London, 1836.
Hepper (1980). F. Nigel Hepper, "On the botany of James Bruce's expedition to the source of the Nile 1768–1773" in *Journal of the Society for the Bibliography of Natural History* (London, 1980), vol. 9 (4), p. 527–537.
——— (1988). "Taxonomic aspects of James Bruce's travels of the source of the Blue Nile", *Monog. Syst. Bot.*, Vol. 25 *Missouri Botanical Garden*.
Lewis (1961). Leslie Lewis, *Connoisseurs and secret agents*. London, 1961.
Lobo (1735). *A voyage to Abyssinia, by Father Jerome Lobo a Portuguese Jesuit*. Transl. Samuel Johnson from the French of Le Grand. London, 1735.
Lorenzini (1894). D. Lorenzini, *Guida dei bagni della Poretta e dintorni*. Bologna, 1894.
Moorehead (1962). Alan M. Moorehead, *The Blue Nile*. London, 1962.
Murray (1808). Alexander Murray, *Account of the life and writings of James Bruce of Kinnaird, Esq. F.R.S., author of the travels to discover the source of the Nile*. Reprinted from the second edition of the *Travels* with additions and emendations. Edinburgh, 1808.
Oppé (1950). A.P. Oppé, *English drawings at Windsor Castle*. London, 1950.
Panzacchi (1897). Enrico Panzacchi, "Un architetto bolognese in Abissinia nel secolo passato" in *La Vita Italiana*, new ser. 3, vol. 1, pp. 295–298.
Playfair (1877). R.L. Playfair, *Travels in the footsteps of Bruce in Algeria and Tunis*. Illustrated by facsimiles of his original drawings. London, 1877.
Reid (1968). J.M. Reid, *Traveller extraordinary. The life of James Bruce of Kinnaird*. London, 1968.
Scott Elliot (1950). A. Scott Elliot, "The Early Publications of the Temples of Paestum" in *Journal of the Courtauld and Warburg Institutes*, vol. 13 (London, 1950), pp. 48–64.
Skjoldebrand (1777). Eric Skjoldebrand, "På et litet och rart Djur infrån Africa, hörande til Räfslägtet" in *Kongliga Vetenskaps-Academiens Handlingar*, vol. 38 (Stockholm, 1777), pp. 265–267 and pl. 6.
Sparrmann (1783). Andreas Sparrmann, *Resa till Goda Hoppa-Udden...*, vol. 1. Stockholm, 1783.
Tarchiani (1930). N. Tarchiani, "Luigi Balugani" in *Enciclopedia italiana di scienze, lettere ed arti*, ed. G. Trecciani, vol. 6 (Milan, 1930), p. 10.
Ullendorff (1953). Edward Ullendorff, "James Bruce of Kinnaird" in *Scottish Historical Review*, vol. 32, no. 114 (Edinburgh, Oct. 1953), pp. 128–143.
——— (1960). Edward Ullendorff, *The Ethiopians. An introduction to country and people*. London, 1960.
Zaghi (1963). C. Zaghi, "Balugani, Luigi" in *Dizionario biografico degli Italiani*, ed. A.M. Ghisalbert *et al.*, vol. 5 (1963), pp. 632–634.

1. *Acanthus sennii* (c) p. 70

2. *Acanthus sennii* (e) p. 70

3. *Ruttya speciosa* (c) p. 70

4. *Acokanthera schimperi* p. 71/72

5. *Saba comorensis* (c) p. 72

6. *Commiphora gileadensis* (b) p. 75

7. *Carthamus tinctorius* (b) p. 77

8. *Senecio gigas* (c) p. 78

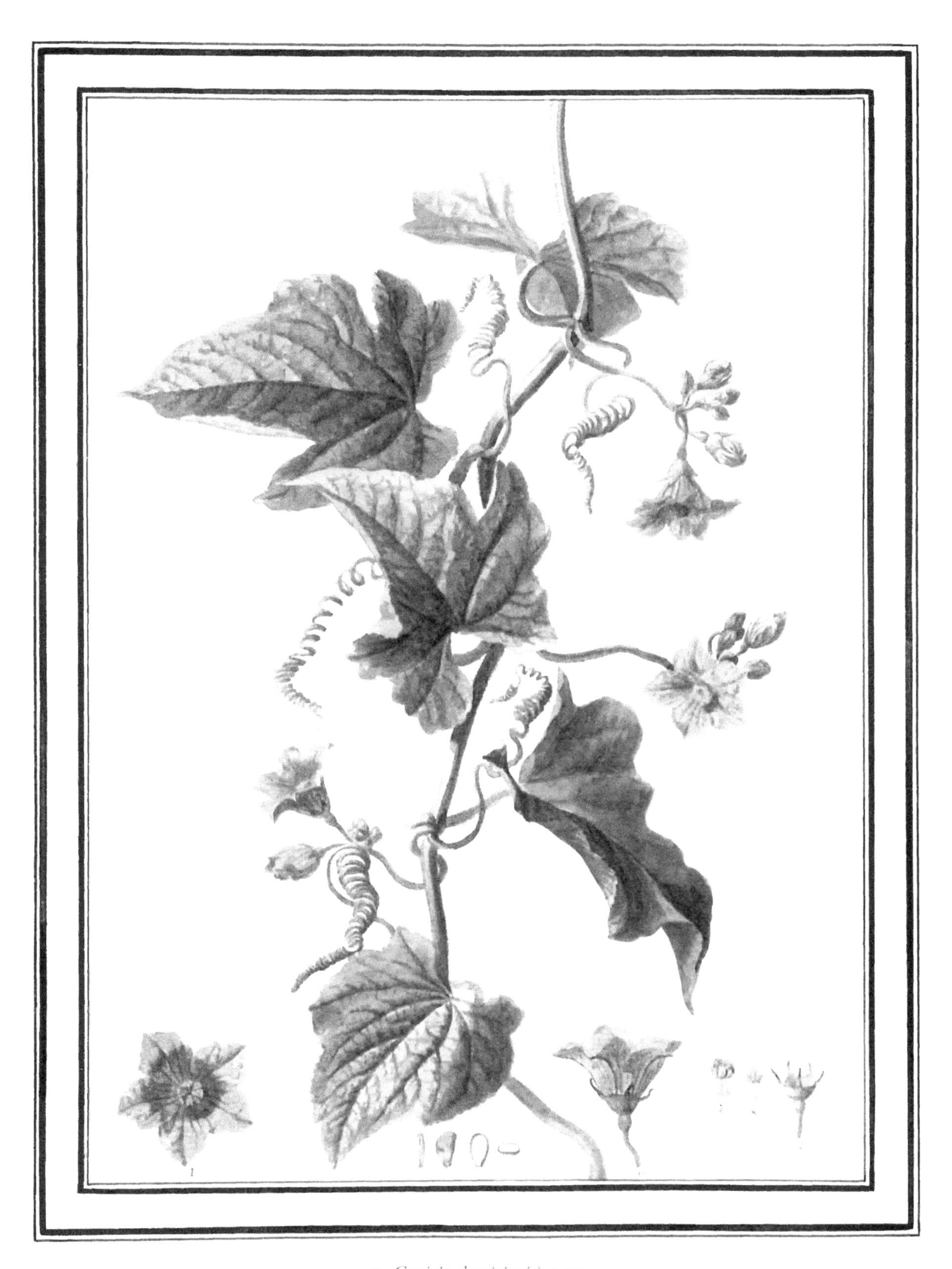

9. *Coccinia abyssinica* (c) p. 79

10. *Coccinia abyssinica* (f) p. 79

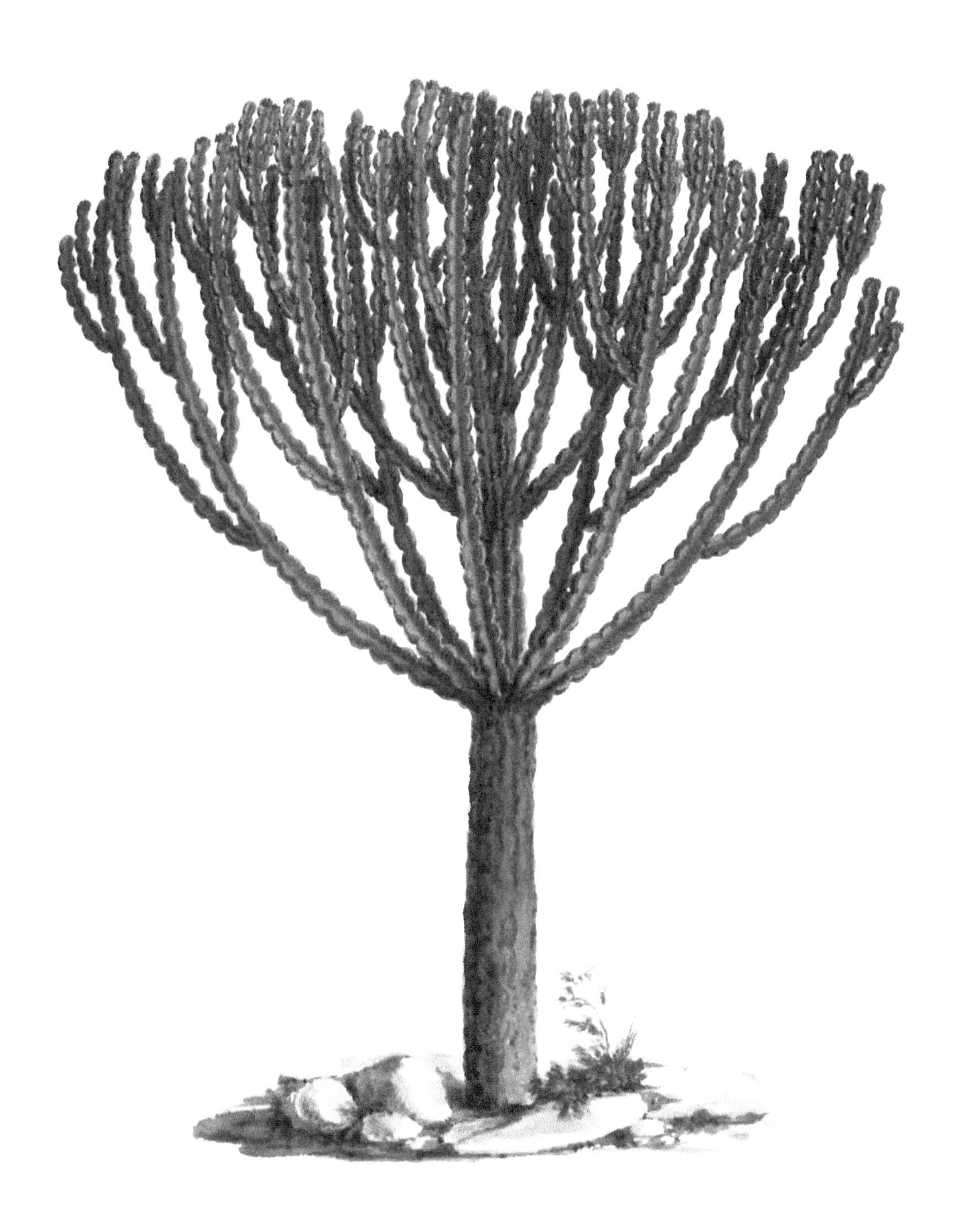

11. *Euphorbia abyssinica* (c) p. 81

12. *Euphorbia abyssinica* (d) p. 81

13. *Hypericum quartinianum* (b) p. 82

14. *Coleus edulis* (b) p. 82/83

15. *Phlomis herba-venti* (b) p. 83

16. *Salvia nilotica* (b) p. 84

17. *Cassia fistula* (c) p. 86

18. *Millettia ferruginea* (d) p. 88

19. *Pterolobium stellatum* (c) p. 89

20. *Buddleia polystachya* (b) p. 90

21. *Bersama abyssinica* (b) p. 92

22. *Jasminum dichotomum* (c) p. 93

23. *Pittosporum viridiflorum* (c) p. 94

24. *Protea gaguedi* (f) p. 94/95

25. *Delphinium wellbyi* (c) p. 95/96

26. *Hagenia abyssinica* (c) p. 96/97

27. *Hagenia abyssinica* (f) p. 96/97

28. *Gardenia ternifolia* subsp. *jovis-tonantis* (f) p. 97/98

29. *Clausena anisata* (c) p. 98

30. *Verbascum sinaiticum* (d) p. 100

31. *Brucea antidysenterica* (f) p. 100/101

32. *Discopodium penninervium* (b) p. 101

33. *Solanum adoense* (c) p. 102

34. *Solanum incanum* (c) p. 102

35. *Solanum marginatum* (b) p. 103

36. *Dombeya torrida* (d) p. 103

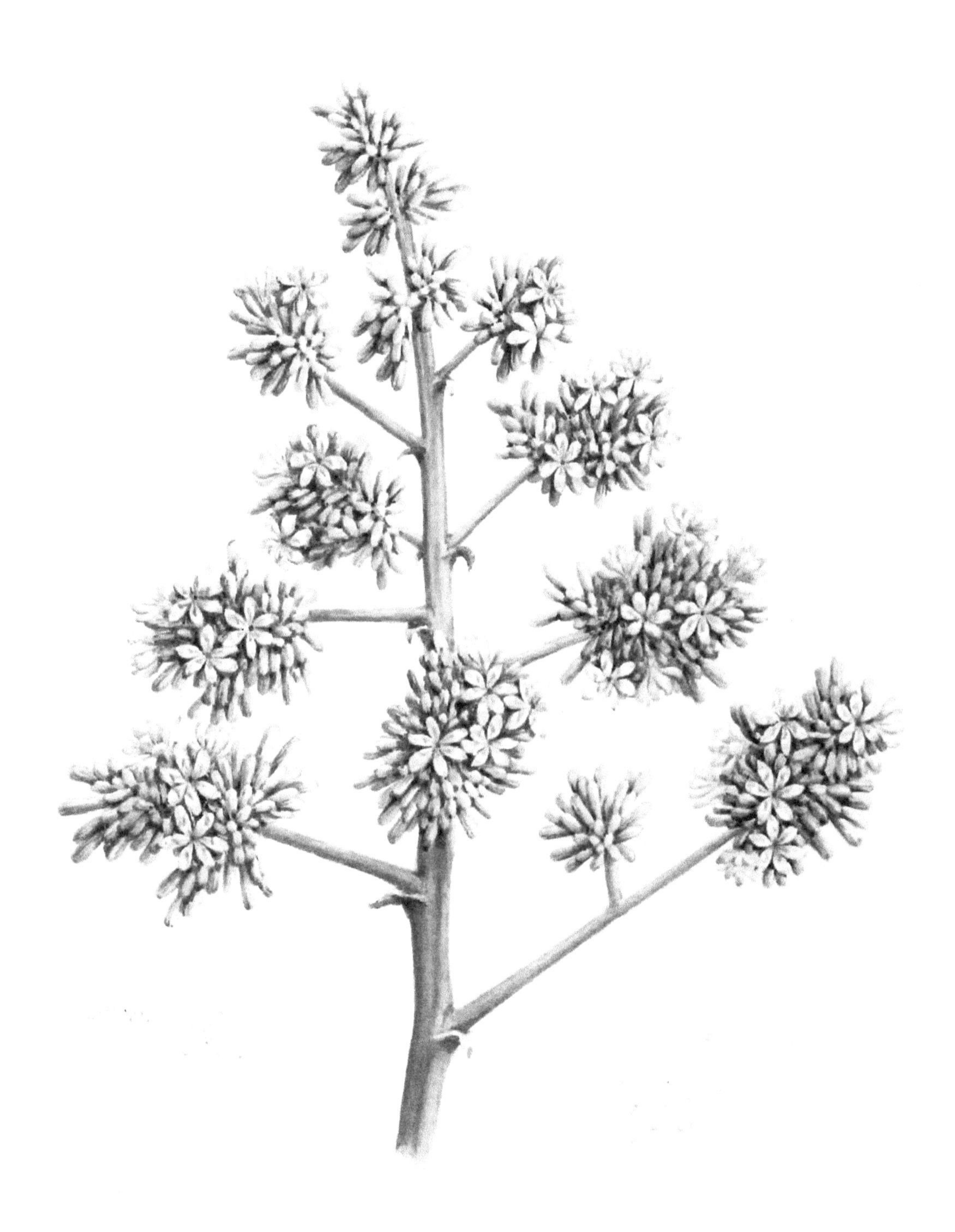

37. *Dracaena steudneri* (d) p. 105

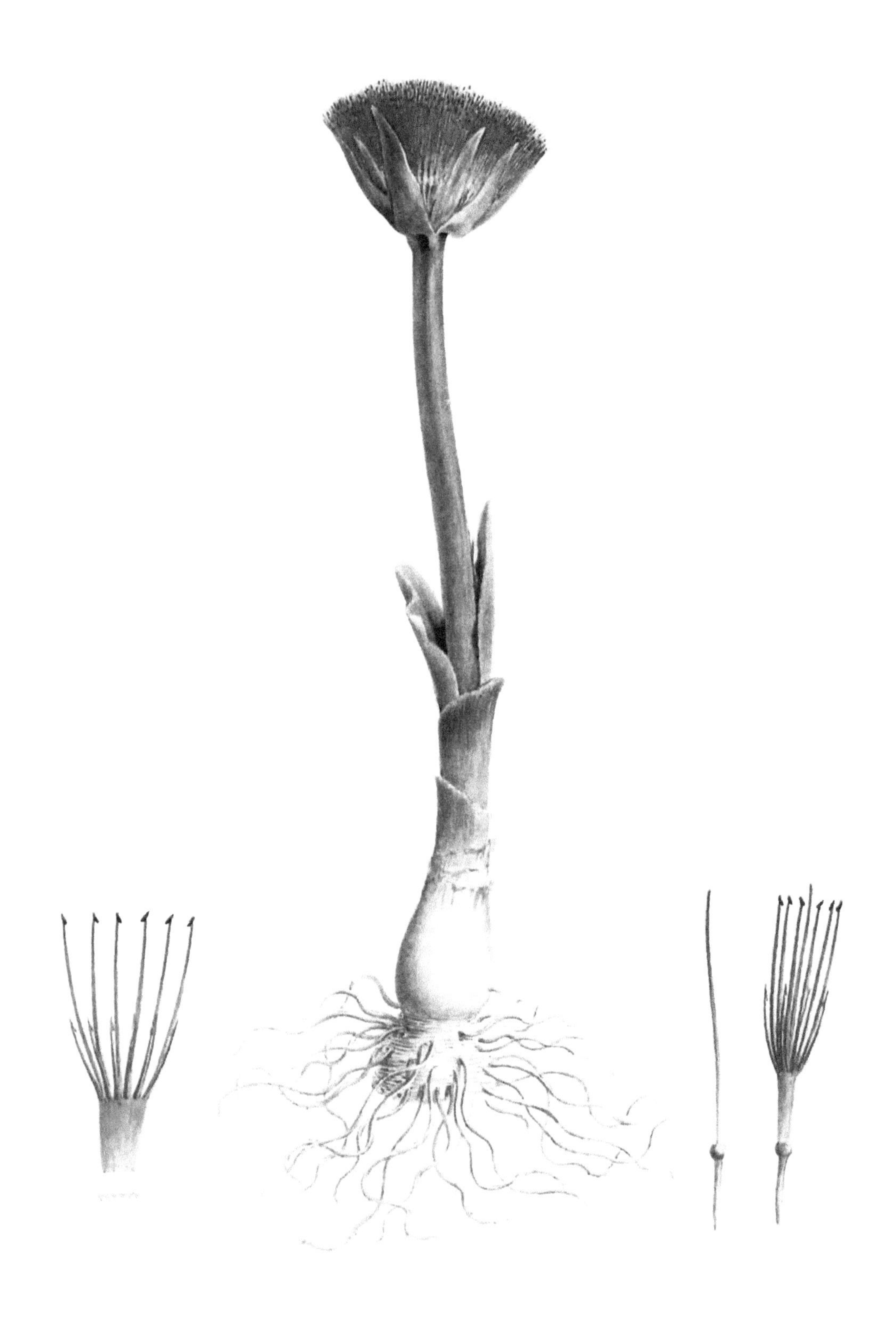

38. *Scadoxus puniceus* (d) p. 106

39. *Canna bidentata* (d) p. 106

40. *Cyperus papyrus* (a) p. 107

41. *Cyperus papyrus* (b) p. 107

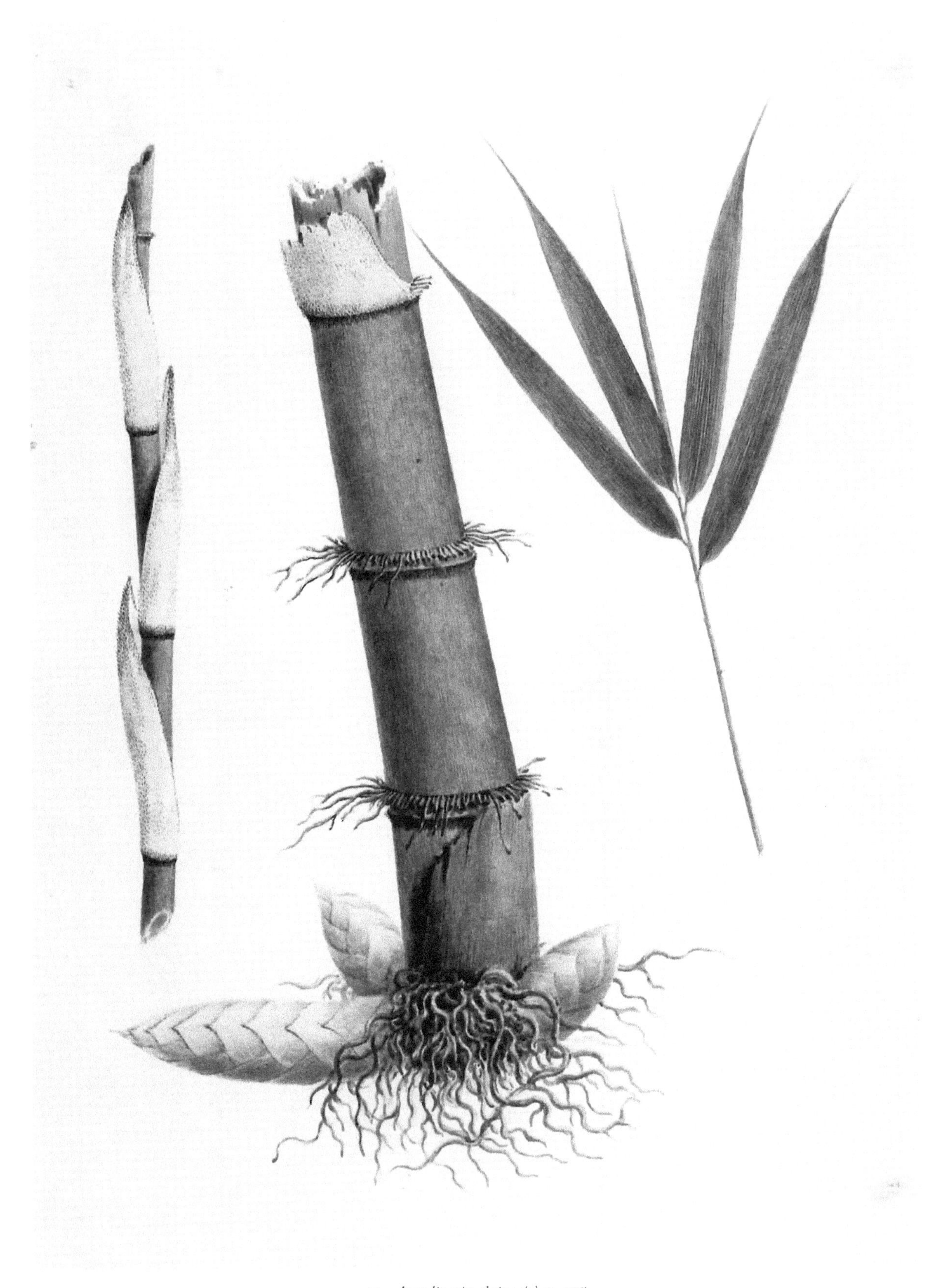

42. *Arundinaria alpina* (c) p. 108

43. *Ischaemum afrum* (b) p. 109

44. *Oenostachys abyssinica* (c) p. 110

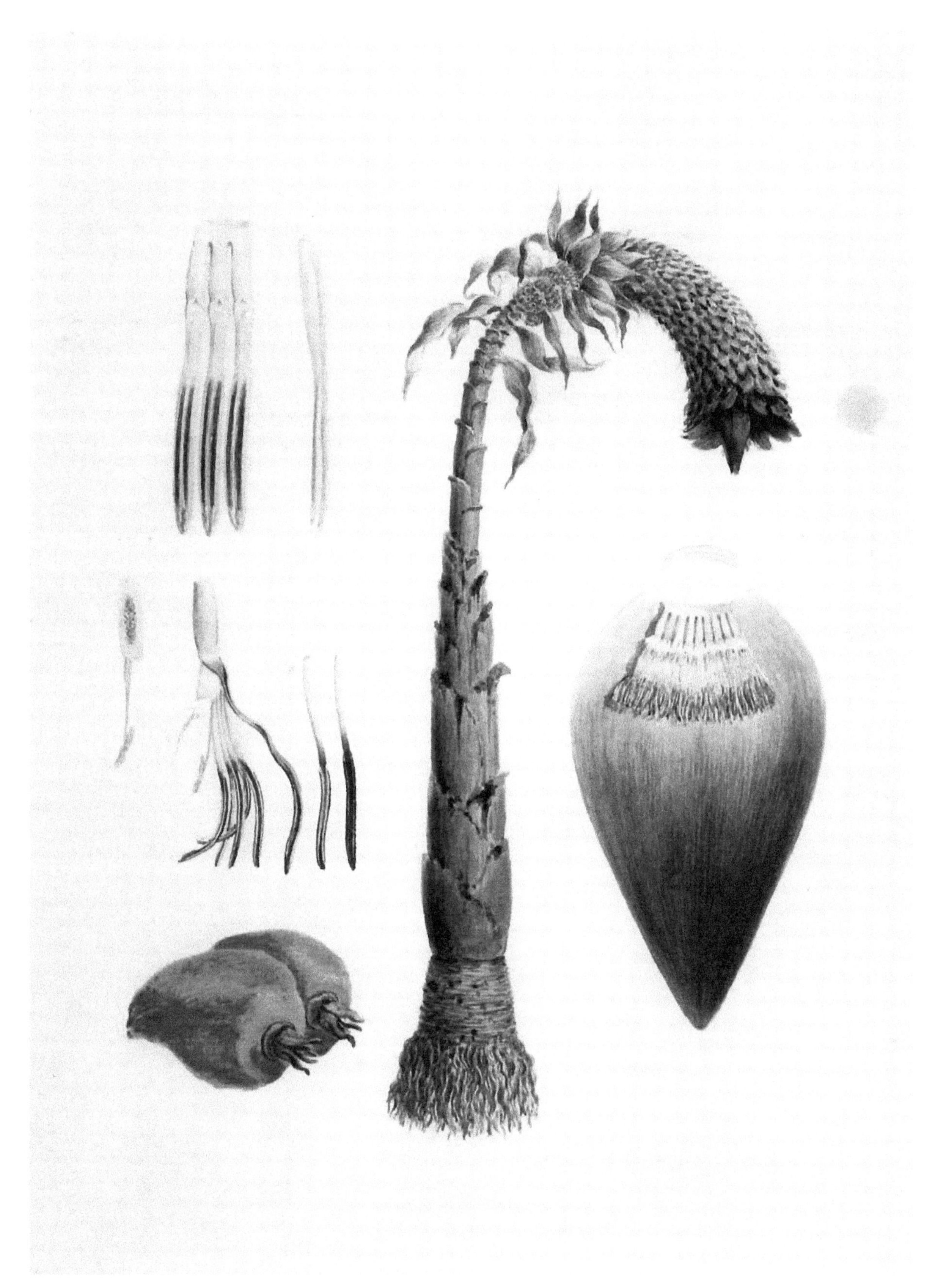

45. *Ensete ventricosum* (j) p. 113

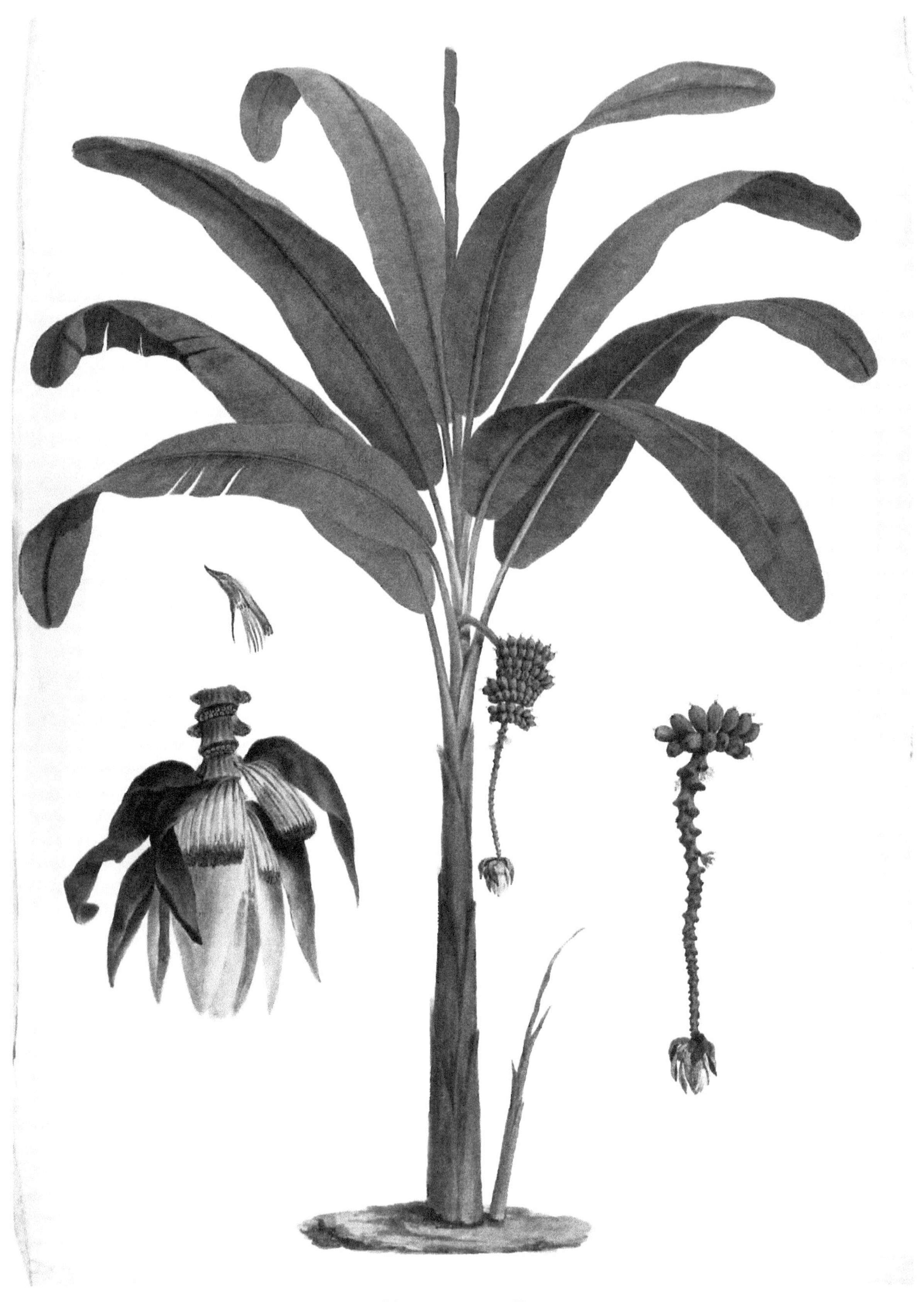

46. *Musa* × *sapientum* (b) p. 113

47. *Pandanus* cf. *odoratissimus* p. 114

48. *Juniperus procera* (b) p. 114

49. (left) *Acanthus sennii* (a) p. 70
(right) *Coriandrum sativum* p. 104

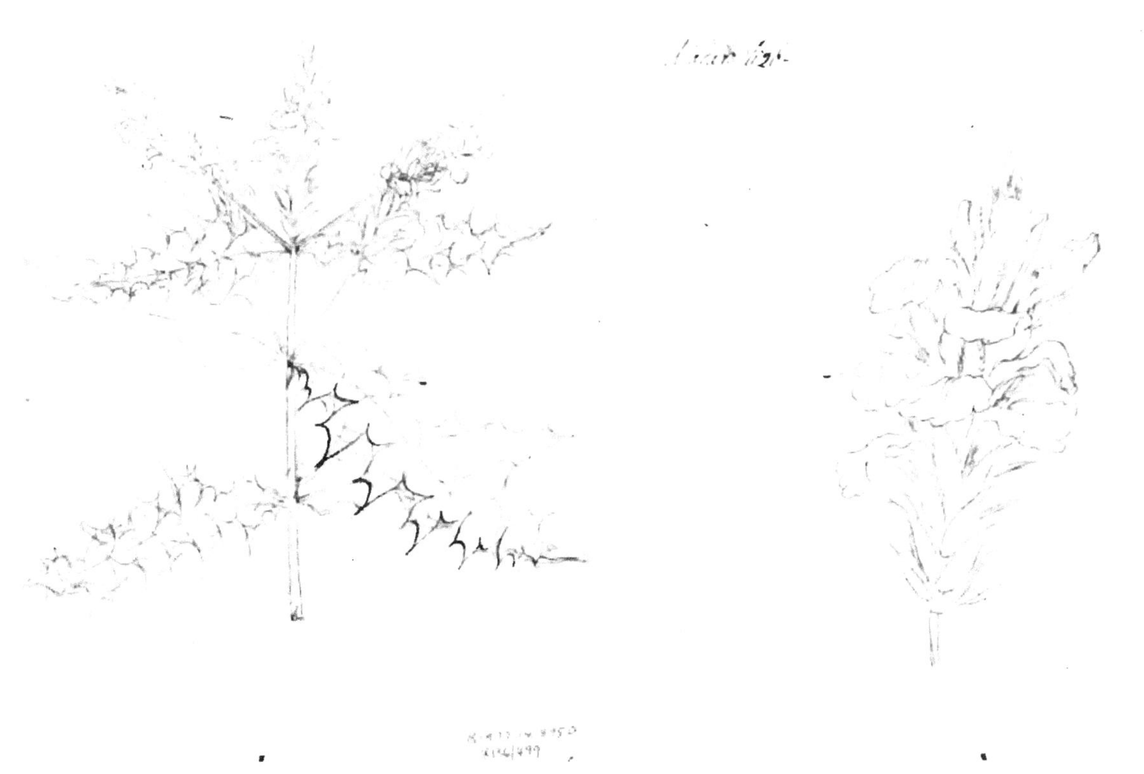

50. *Acanthus sennii* (b) p. 70

51. *Acanthus sennii* (d) p. 70

52. *Acanthus sennii* (c) p. 70

53. (left) *Hygrophila auriculata* p. 70
(middle) *Euphorbia abyssinica* (a) p. 81
(right) *Ficus vasta* p. 92

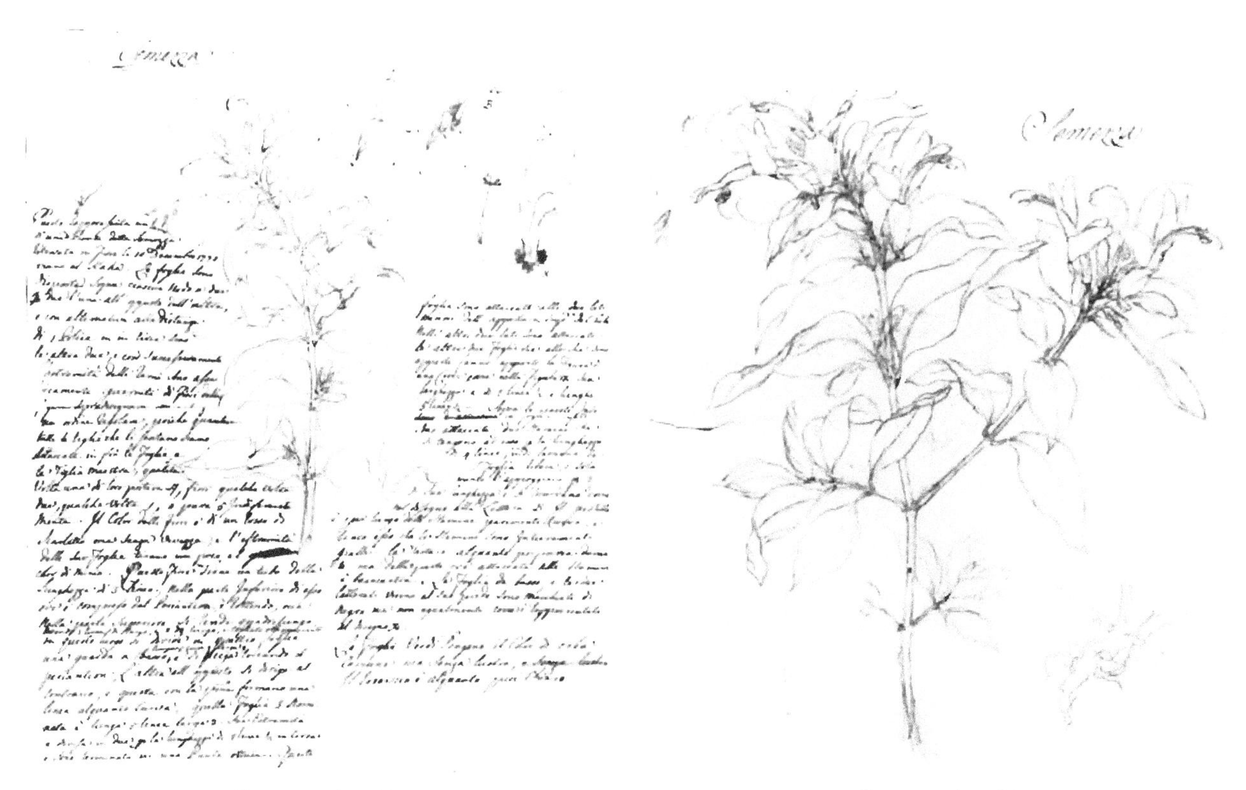

54. *Ruttya speciosa* (a) p. 70

55. *Ruttya speciosa* (b) p. 70

56. *Rhus glutinosa* (b) p. 71

57. *Rhus glutinosa* (c) p. 71

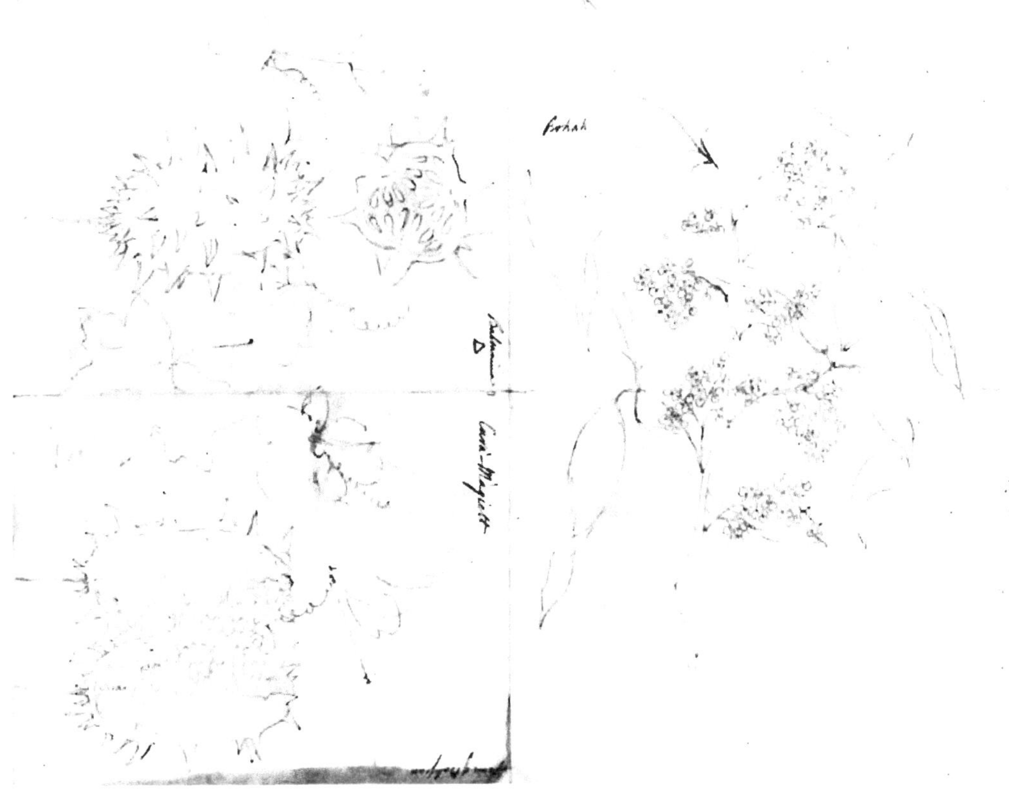

58. (left) *Cucumis metuliferus* (a) p. 79
(right) *Rhus? retinorrhoea* (b) p. 71

59. *Rhus glutinosa* (d) p. 71

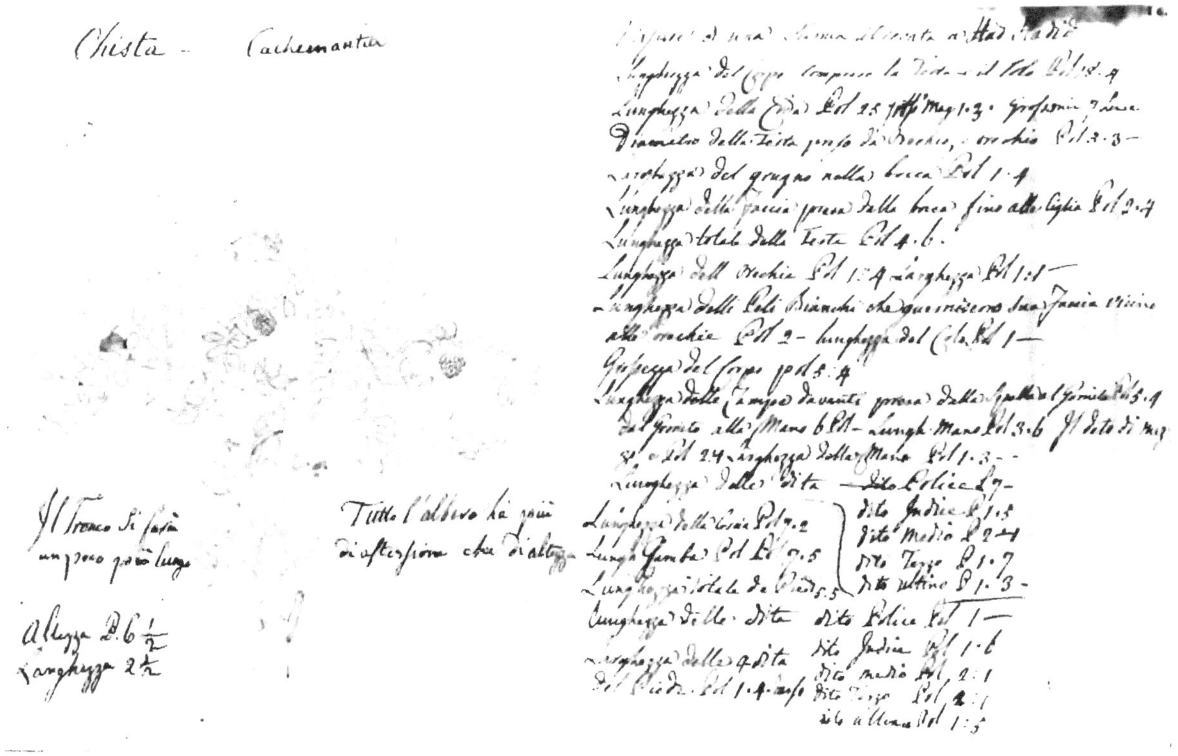

Chista – Cachemantia

[illegible] di una Scimia allevata a [illegible]

Lunghezza del Corpo compreso la Testa e il Collo Pol. 8. 4

Lunghezza della Coda Pol. 25 [illegible] 1. 3. grossezza [illegible]

Diametro della Testa preso da Orecchio, e orecchio Pol. 3. 3 –

Larghezza del grugno nella bocca Pol. 1. 4

Lunghezza della faccia presa dalla bocca fino alle ciglia Pol. 2. 4

Lunghezza totale della Testa Pol. 4. 6.

Lunghezza dell Orecchia Pol. 1. 4 Larghezza Pol. 1. 1 –

Lunghezza delle Peli Bianchi che guarniscono la Faccia vicino alle orecchie Pol. 2 – lunghezza del Collo Pol. 1 –

Grossezza del Corpo Pol. 5. 4

Lunghezza delle Zampe davanti presa dalla Spalla al Gomito Pol. 5. 1 dal Gomito alla Mano 6 Pol. – Lungh. Mano Pol. 3. 6 Il dito di mezzo [illegible] Pol. 2. 4 Larghezza della Mano Pol. 1. 3. –

Lunghezza delle dita – dito Police P. 7 –

dito Indice P. 1. 5

dito Medio P. 2 –

dito Terzo P. 1. 7

dito ultimo P. 1. 3 –

Lunghezza della Coscia Pol. 7. 2

Lungh. Gamba Pol. 7. 5

Lunghezza totale de Piedi 5. 5

Lunghezza delle dita dito Police Pol. 1 –

Larghezza delle 4 dita del Piede Pol. 1. 4. [illegible]

dito Indice Pol. 1. 6

dito medio Pol. 2. 1

dito Terzo Pol. 2. 1

dito ultimo Pol. 1. 5

Il Tronco si farà un poco più lungo

Tutto l'albero ha più di estensione che di altezza

Altezza P. 6 ½
Larghezza 2 ½

60. *Annona squamosa* (a) p. 71

61. *Annona squamosa* (c) p. 71

62. *Rhus? retinorrhoea* (c) p. 71

63. *Annona squamosa* (b) p. 71

64. *Annona squamosa* (d) p. 71

65. *Saba comorensis* (a) p. 72

66. *Saba comorensis* (b) p. 72

67. *Carissa edulis* (b) p. 72

68. *Calotropis procera* p. 72

69. *Carissa edulis* (a) p. 72

70. *Dregea abyssinica* (a) p. 73

71. (left) *Cynanchum altiscandens* (a) p. 72/73
(right) *Tacazzea venosa* p. 73/74

72. *Dregea abyssinica* (b) p. 73

73. *Gomphocarpus fruticosus* p. 73

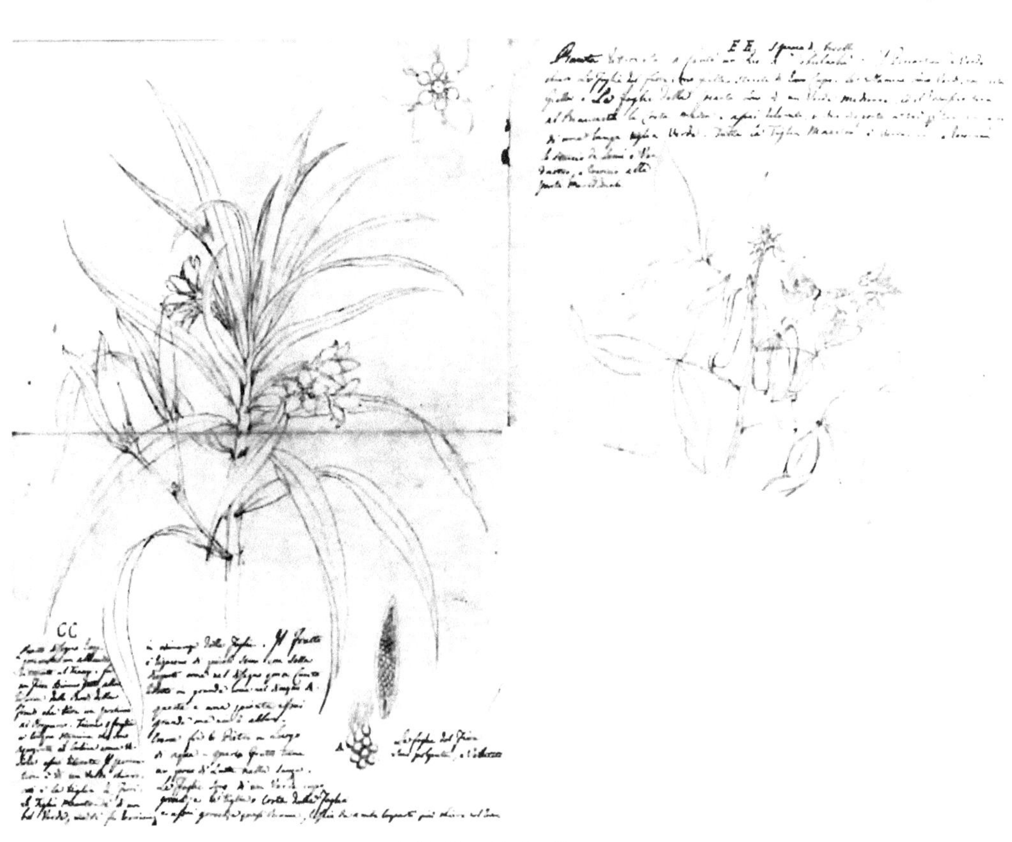

74. (left) *Kanahia laniflora* p. 73
(right) *Crotalaria pallida* (a) p. 86

75. *Dregea abyssinica* (c) p. 73

76. *Avicennia marina* p. 74

77. *Tacazzea galactagoga* (a) p. 73

78. *Impatiens rothii* (b) p. 74

79. (left) *Tacazzea galactagoga* (b) p. 73
(right) *Lagenaria siceraria* p. 79/80

80. *Impatiens rothii* (c) p. 74

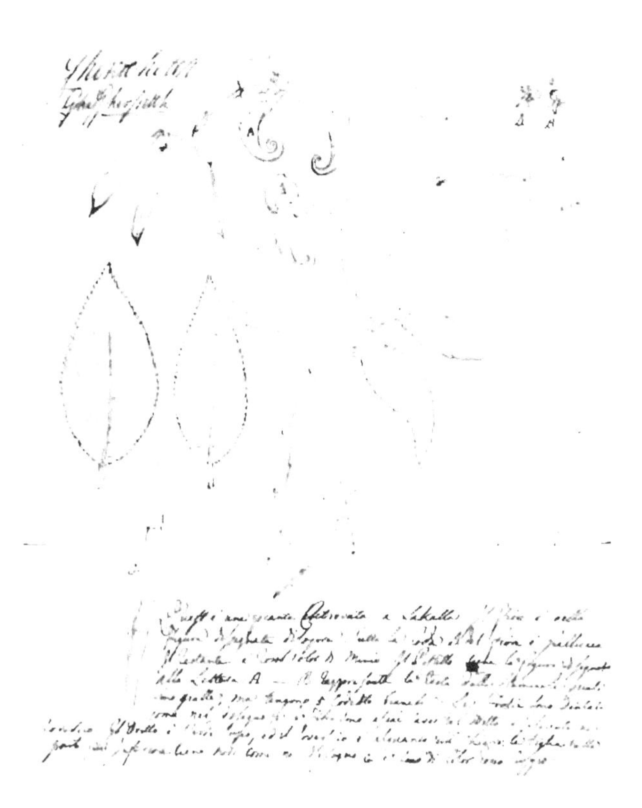

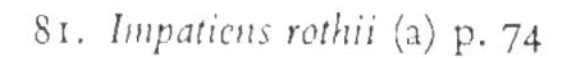

81. *Impatiens rothii* (a) p. 74

82. *Cordia abyssinica* (a) p. 74

83. *Boswellia papyrifera* (a) p. 75

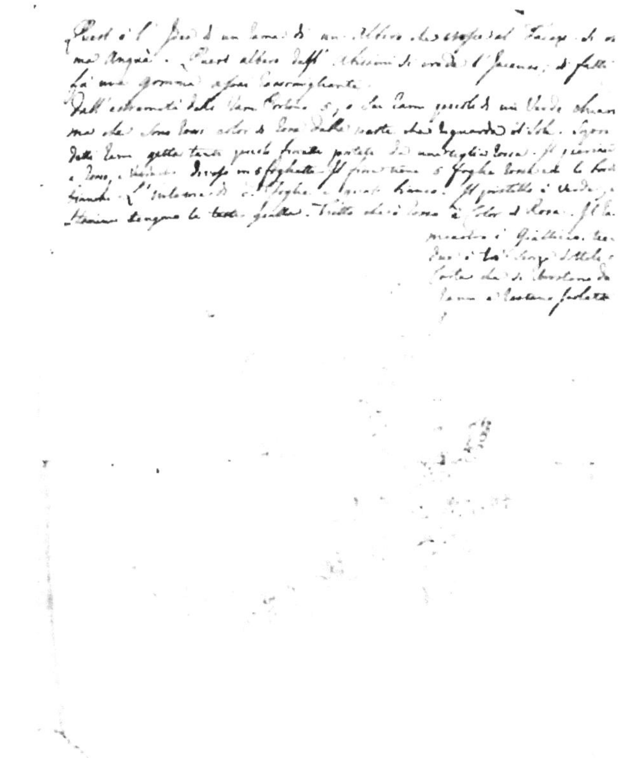

Quest'è il fiore d'un ramo di un Albero che [illegible] dal Tacazè [illegible] Anguà. Quest'albero dagli Abissini si crede l'Incenso; di fatti fa una gomma assai somigliante.
Dall'estremità delle rami [illegible] 5, o sei rami piccoli di un verde oscuro ma che sono [illegible] color di rosa dalla parte che riguarda il sole. Ognuno delle rami getta tanti piccoli [illegible] portate da un [illegible] rosa. Il [illegible] è rosso, e [illegible] diviso in 5 foglietti. Il fiore tiene 5 foglie [illegible] le [illegible] bianche. L'interno di [illegible] foglie è [illegible] bianco. Il pistillo è verde, e [illegible] stamine tengono le teste gialle. Tutto che è rosso è color di rosa. Il [illegible] è giallino. [illegible]
[illegible]
[illegible]
[illegible]

84. *Boswellia papyrifera* (b) p. 75

85. *Boswellia papyrifera* (d) p. 75

86. *Cordia abyssinica* (b) p. 74

87. *Ehretia cymosa* p. 75

88. *Commiphora gileadensis* (a) p. 75

89. *Commiphora gileadensis* (c) p. 75/76

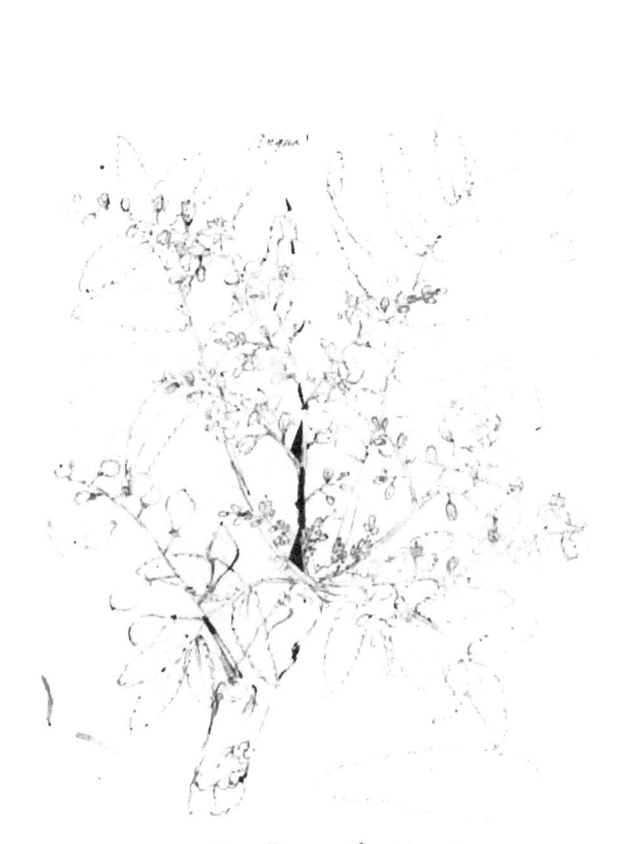

90. *Boswellia papyrifera* (c) p. 75

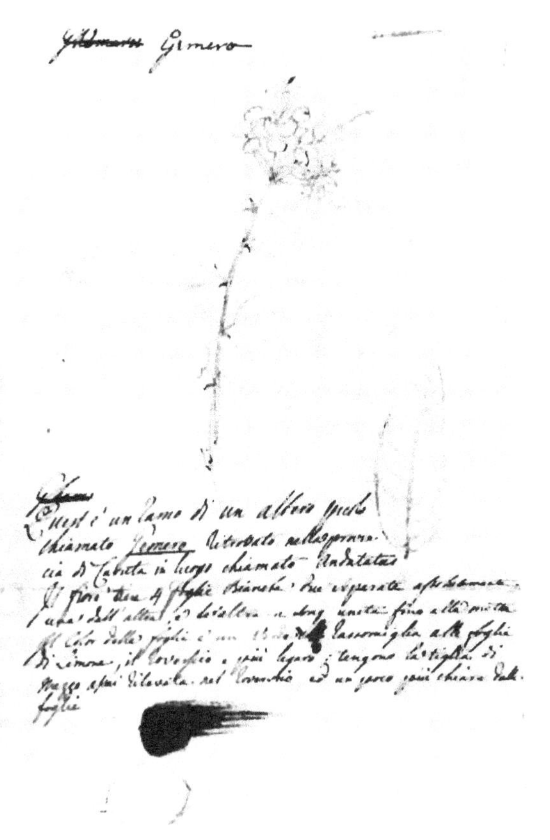

91. *Capparis tomentosa* (b) p. 76

92. *Capparis tomentosa* (c) p. 76

1.A

Disegno di un [illegible] Albero [illegible] confronto
[illegible] il frutto [illegible] con [illegible]
[illegible] il frutto [illegible]
grandezza di un Arboscello [illegible] grande [illegible] foglie [illegible]
[illegible] alternativamente

1.B

Disegno di un Albero [illegible] come una grande [illegible]
Le foglie sono di un bel verde [illegible]
[illegible]

93. *Capparis tomentosa* (a) p. 76

94. (left) *Michauxia campanuloides* p. 76
(right) *Hagenia abyssinica* (a) p. 96

95. *Capparis tomentosa* (d) p. 76

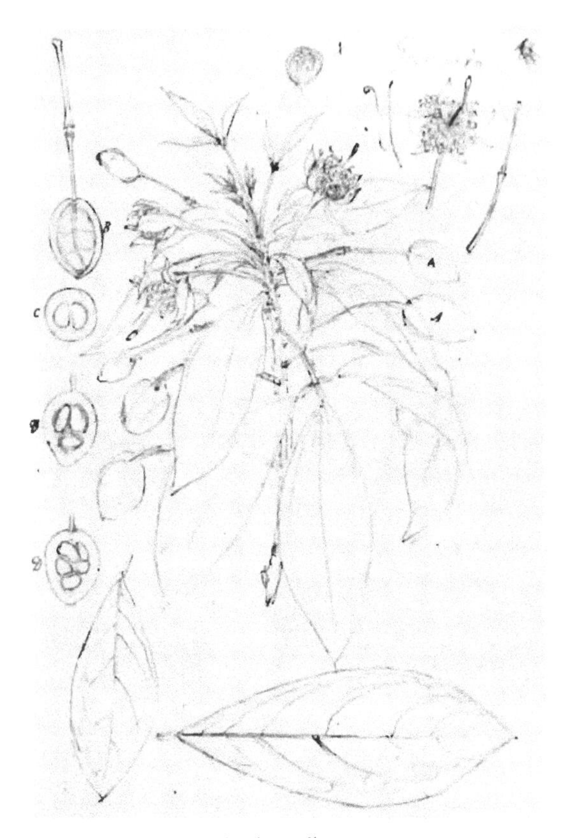

96. *Ritchiea albersii* p. 76

97. *Catha edulis* (a) p. 76/77

98. *Catha edulis* (b) p. 76/77

99. *Echinops macrochaetus* p. 77

100. *Maytenus arguta* (a) p. 77

101. *Maytenus arguta* (b) p. 77

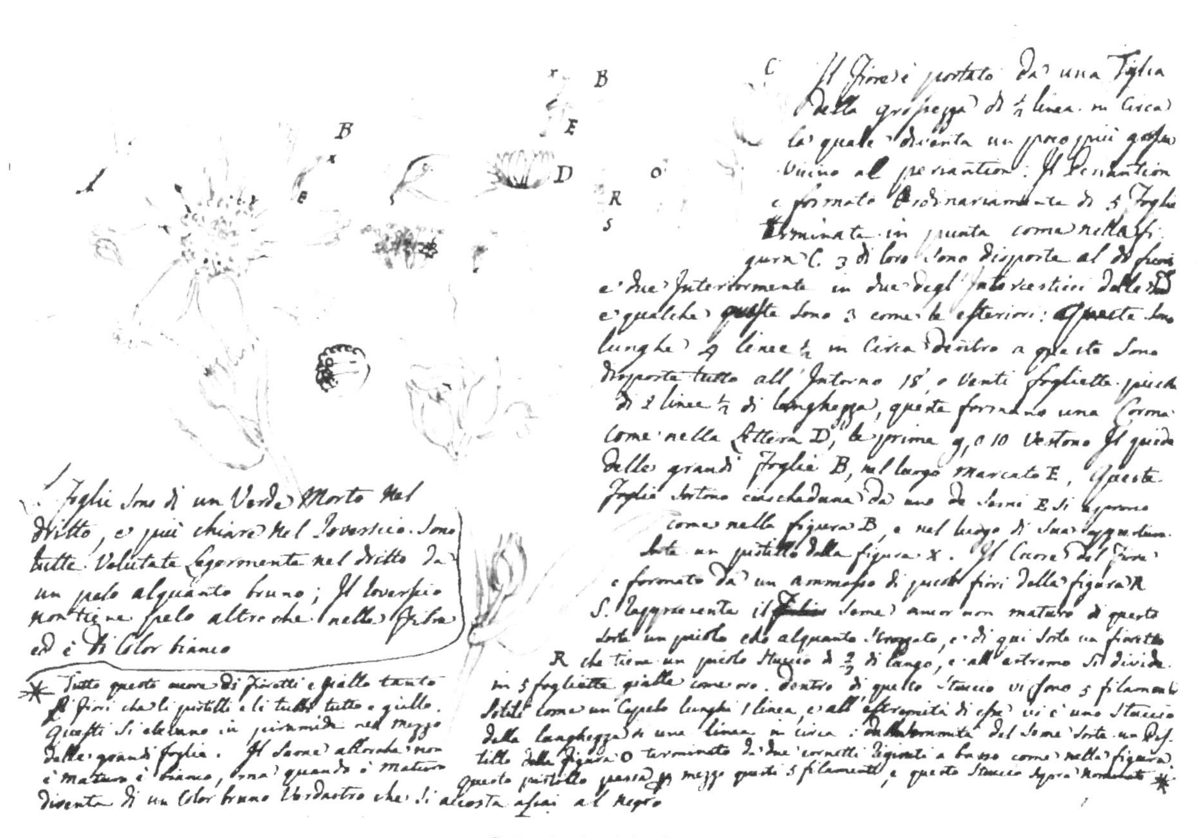

Il Fiore è portato da una Foglia della grossezza di ½ linea in circa la quale diventa un poco più grossa vicino al perianteo. Il Perianteo è formato ordinariamente di 5 Foglie terminate in punta come nella figura C. 3 di loro sono disposte al di fuori e due interiormente in due degl'intersticii delle [illegible] e qualche [illegible] sono 3 come le esteriori: queste sono lunghe 4 linee ½ in circa dentro a queste sono disposte tutte all'intorno 18, o venti foglietta [illegible] di 2 linee ½ di lunghezza, queste formano una Corona come nella Lettera D, le prime 9, o 10 vestono il [illegible] delle grandi Foglie B, nel luogo marcato E, queste Foglie sortono ciascheduna da uno de Semi E si [illegible] come nella figura B, e nel luogo di sua [illegible] sorte un pistillo della figura X. Il Cuore del Fiore è formato da un ammasso di questi fiori della figura R S. Rappresenta il Seme ancor non maturo di questo sorte un pistillo che alquanto [illegible], e di qui sorte un fioretto R che tiene un picolo stuccio di ⅔ di lungo, e all'estremo si divide in 5 foglietta gialle come oro. Dentro di questo stuccio vi sono 5 filamenti sottili come un capello lunghi 1 linea, e all'estremità di essi vi è uno stuccio della lunghezza di una linea in circa: dall'estremità del Seme sorte un pistillo della figura O terminato di due cornetti piegati a basso come nella figura, questo pistillo passa per mezzo questi 5 filamenti, e questo stuccio sopra nominato *

Le Foglie sono di un Verde Morto nel Ditto, e più chiaro nel Roverscio. Sono tutte Vellutate leggermente nel Ditto da un pelo alquanto bruno; il Roverscio non tiene pelo altro che nelle Fibre ed è di Color bianco

* Tutto questo mucchio di Fioretti è giallo tanto li Fiori che li pistilli e li tubbi tutto è giallo. Questi si elevano in piramide nel mezzo delle grandi Foglie. Il Seme allorchè non è maturo è bianco, ma quando è maturo diventa di un Color bruno Verdastro che si accosta assai al Negro

102. *Guizotia abyssinica* (a) p. 77/78

103. *Guizotia abyssinica* (d) p. 77/78

104. *Guizotia abyssinica* (b) p. 77/78

105. *Guizotia abyssinica* (c) p. 77/78

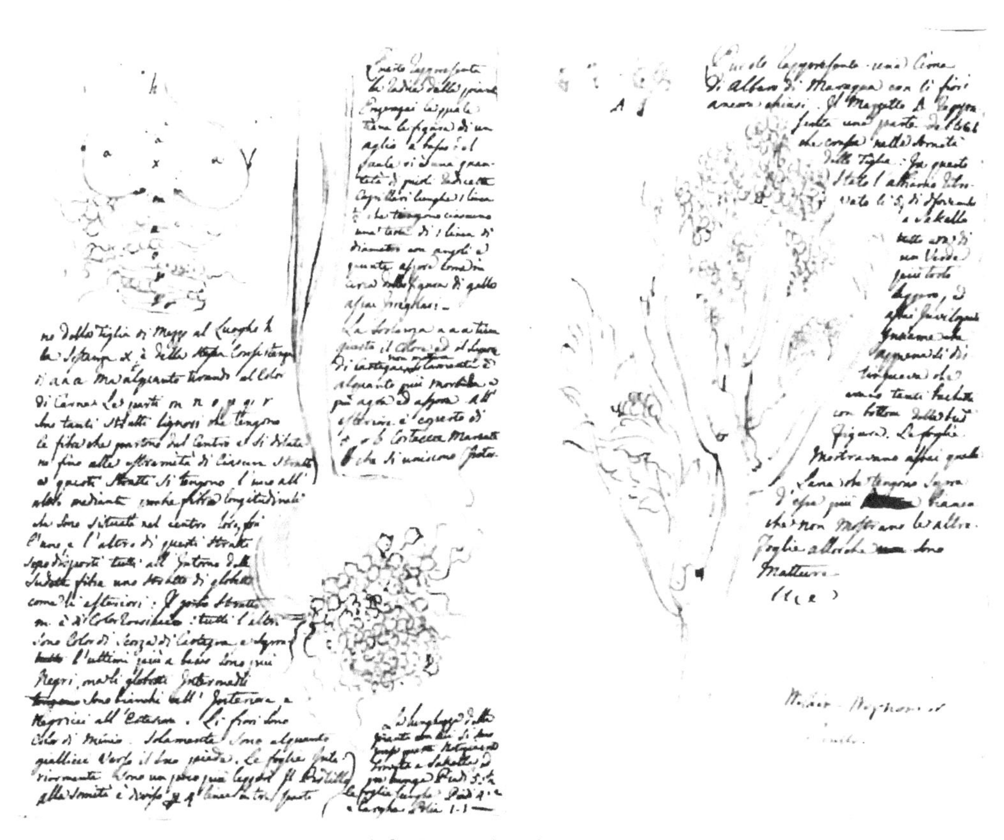

106. (left) *Oenostachys abyssinica* (a) p. 110
(right) *Senecio gigas* (a) p. 78

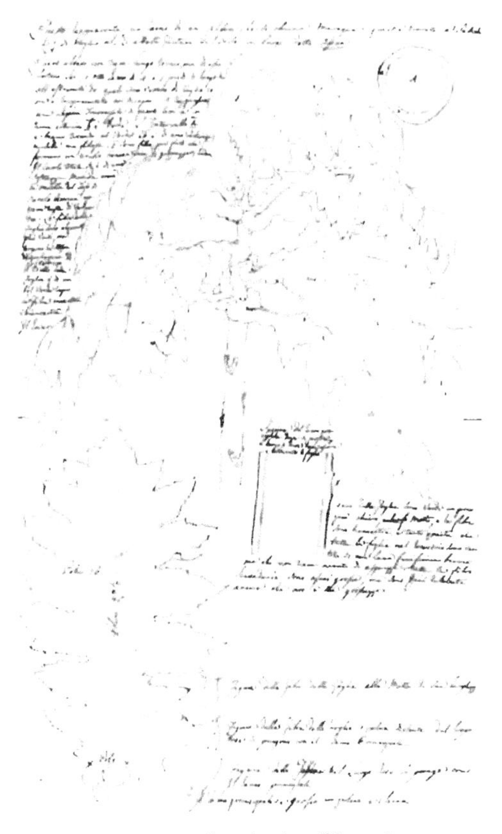

107. *Senecio gigas* (b) p. 78

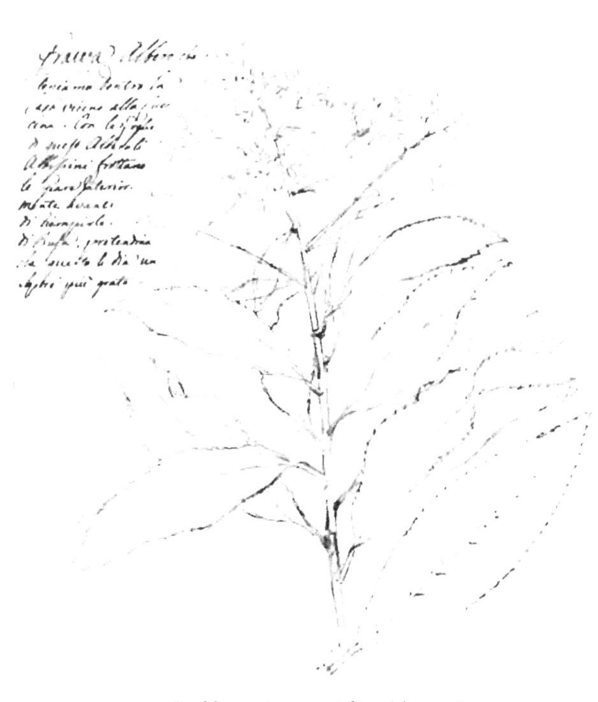

108. *Vernonia amygdalina* (a) p. 78

109. *Coccinia abyssinica* (a) p. 79

110. *Coccinia abyssinica* (b) p. 79

111. *Vernonia amygdalina* (b) p. 78

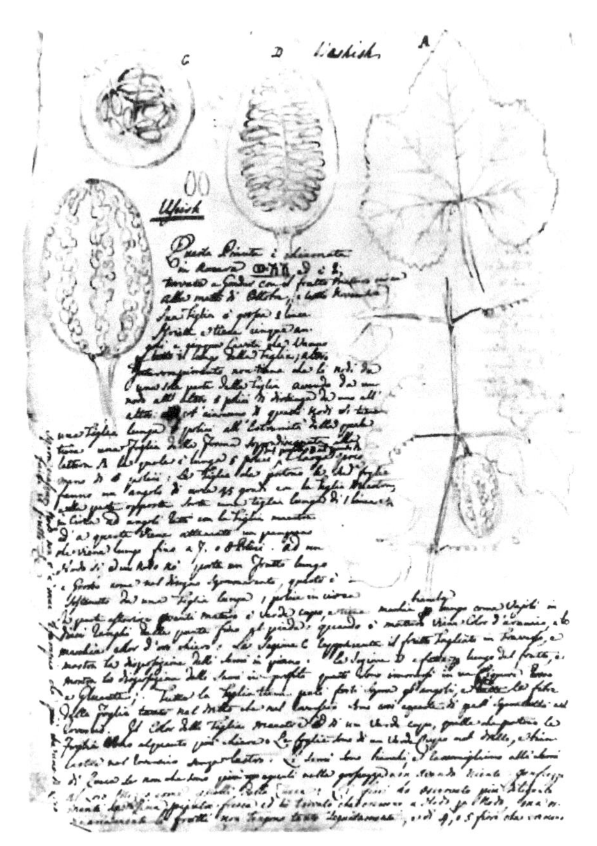

112. *Coccinia abyssinica* (c) p. 79

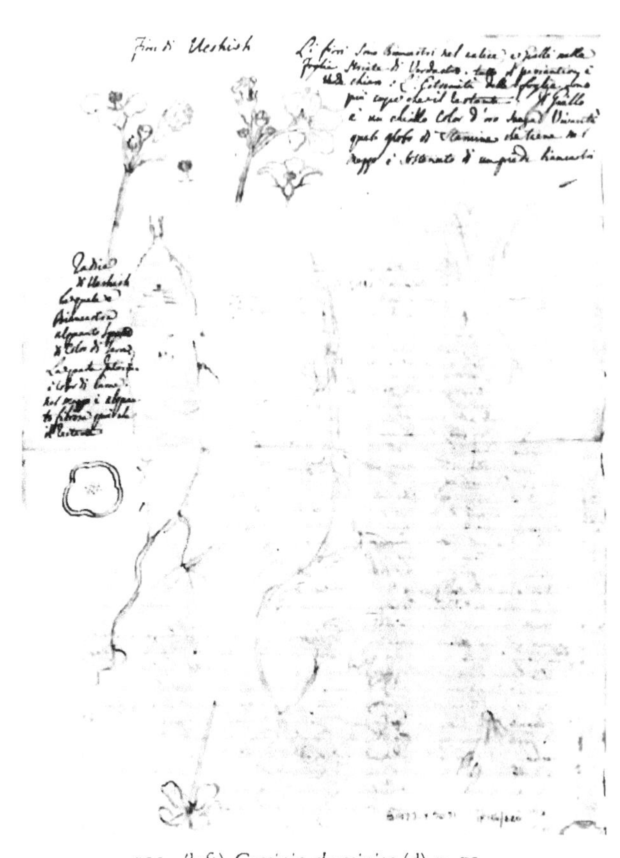

113. (left) *Coccinia abyssinica* (d) p. 79
(right) *Pelargonium alchemilloides* subsp. *multibracteatum* p. 82

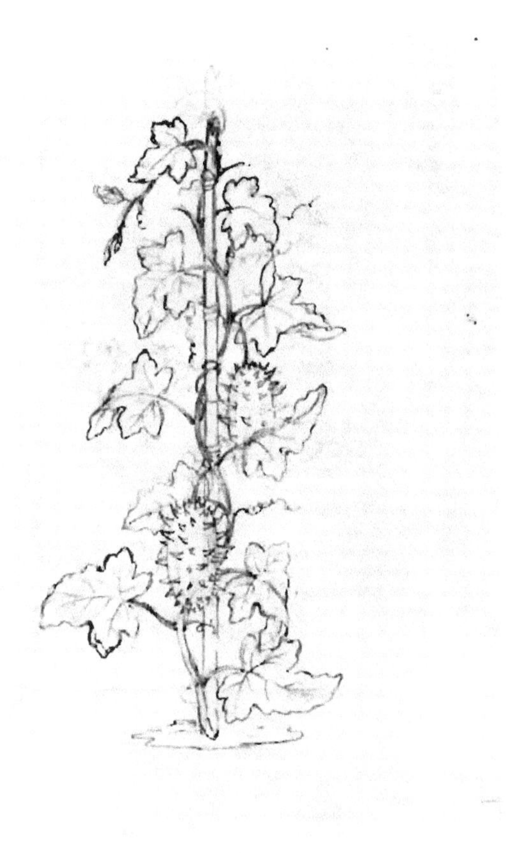

114. *Cucumis metuliferus* (b) p. 79

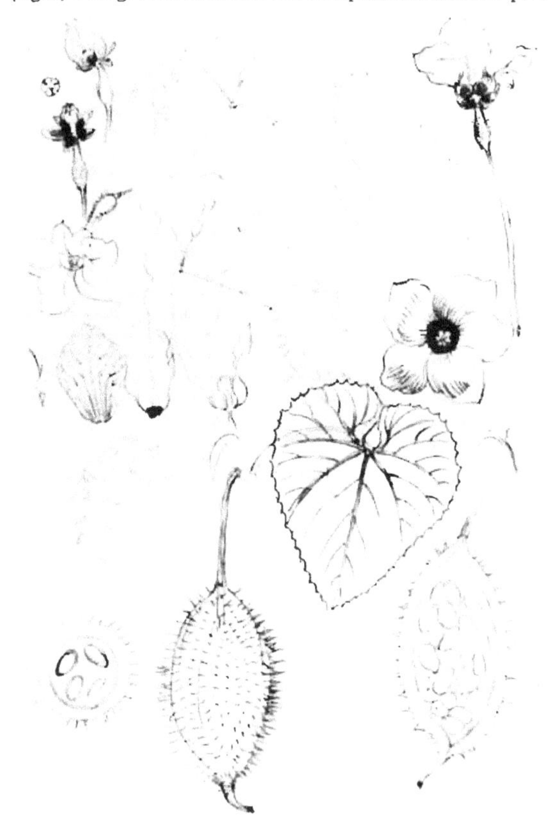

115. *Momordica foetida* p. 80

116. *Coccinia abyssinica* (g) p. 79

117. *Cucumis metuliferus* (c) p. 80

118. *Peponium vogelii* (a) p. 80

119. *Peponium vogelii* (b) p. 80

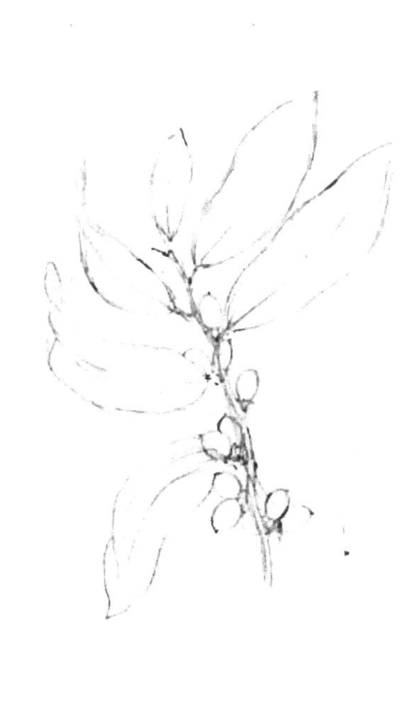

120. *Diospyros abyssinica* (a) p. 80

[illegible] Albero grande [illegible] Le foglie [illegible] Il frutto [illegible]

121. *Diospyros abyssinica* (b) p. 80

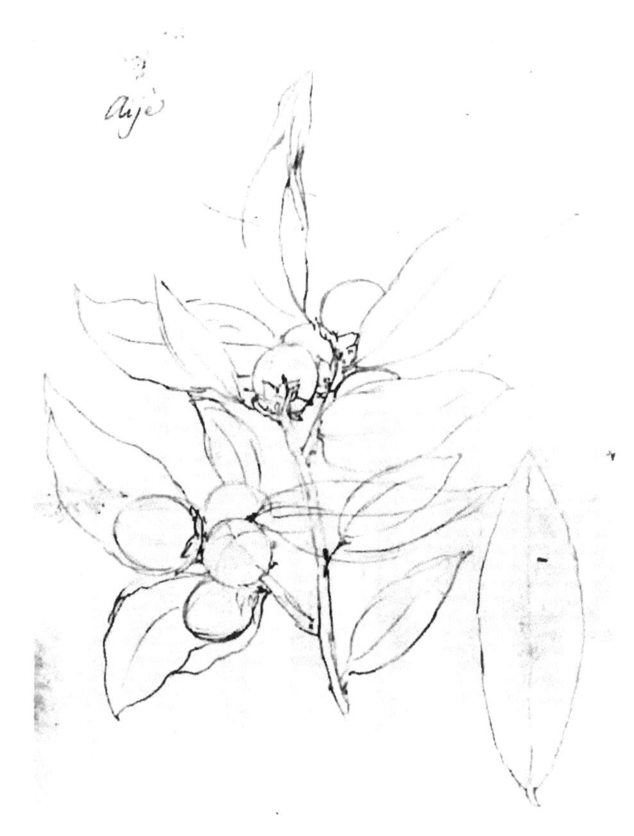

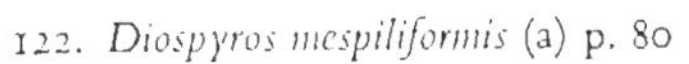

122. *Diospyros mespiliformis* (a) p. 80

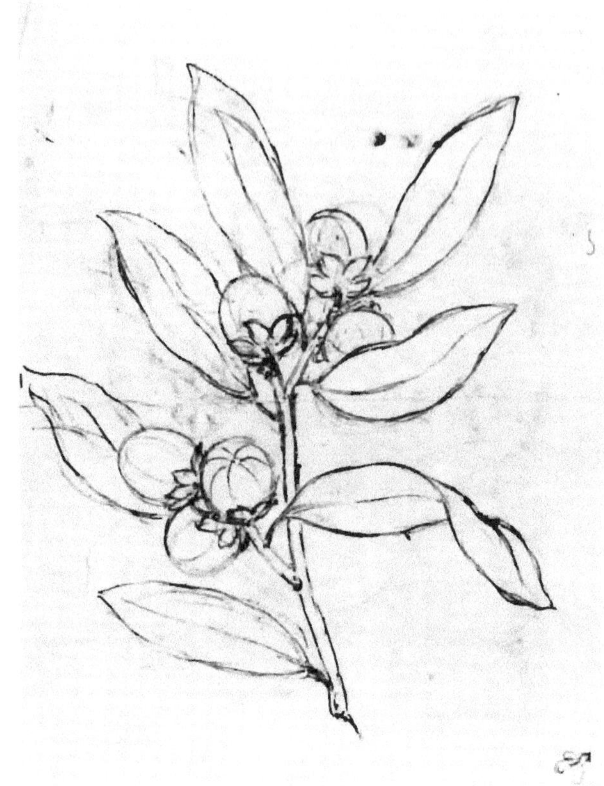

123. *Diospyros mespiliformis* (c) p. 80

Aijè Albero che porta frutti rottondi un poco angolari in sei luoghi, e che si mangiono. Le foglie di quest Albero sono di un bel Verde più tosto cupo e lustro alla dritta, sono di un Verde più chiaro, e alquanto nel rovescio meno palido senza lustro. La costa principale è gialla chiara in rovescio ed assai rilevata; nel dritto tira al Verde. Le altre fibre sono tutte sottili assai fine, nel rovescio non appariscono niente. Quando il frutto viene maturo diventa negro e morbido

Sezione Transversale — Sezione per lungo.

Nel profilo A le due luoghi marcati a sono di una sostanza umida, e gluante; li semi sono come a basso lo descritto, ed il restante tiene assai di ligneo

Li Semi sono come Amandole. Il vano tiene una sostanza assai umida, e gluante. Avanti lor maturità rassomigliano ancora assai all' amandola se non che sua scorza è verdastra ma sottilissima

[illegible] Albero Aquariti

Quest albero porta frutti come una noce, ma un poco più lunghi terminando quasi in punta da ambe le parti.

Le fronde sono di un Verde bello nel dritto e lustre come [illegible]. Nel rovescio sono di un Verde assai leggero tirando al giallo. Le sue fibre nel dritto tirano al bianco; nel rovescio quella di mezzo è bianca, e le altre pure sono tirando al verde ma tutte assai rilevate. Il frutto di quest Albero usano per tingere le vela con il suo [illegible] che è ripieno di semi come il disegno qui a basso. Le foglie sono disposte sopra li rami a tre per tre ed formano nel mezzo come una rosa perché sono disposte alternativamente una verso dell' altra ed in poca distanza.

Qualche volta le foglie ancora sono disposte ad quattro per quattro alternativamente come le [illegible] —

Il fiore di quest Albero è tutto bianco; tiene il [illegible] che è tutto verde, ed di sei foglie disposte come nel disegno A, e B. C rappresenta una sezione del fiore ove si mostra l'ovario, il pistillo, il calice, e le stamine le quali poco dopo pongo [illegible] perché non sopravanzano niente l'altezza del calice, ed il pistillo è assai più lungo. Le dette stamini sono disposte nella sponda del calice, e sono attaccate da una piccola cordella che è nel mezzo di loro alla [illegible] di ciascuna foglia

124. *Diospyros mespiliformis* (b) p. 80

125. *Diospyros mespiliformis* (d) p. 80/81

126. *Croton macrostachyus* p. 81

127. *Euphorbia abyssinica* (b) p. 81

128. (above) *Euphorbia nubica* (a) p. 82
(below) *Cassia fistula* (a) p. 86

129. *Euphorbia nubica* (b) p. 82

130. *Euphorbia helioscopia* p. 81

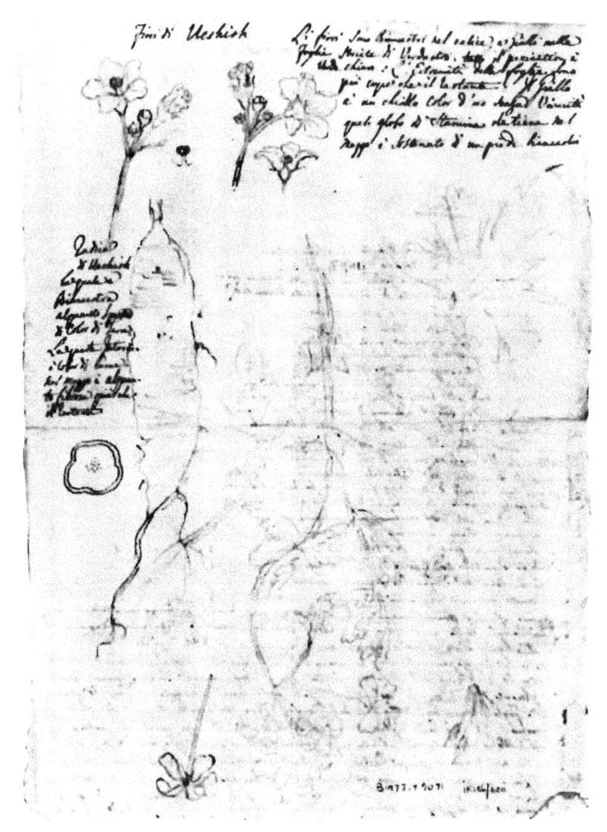

131. (left) *Coccinea abyssinica* (d) p. 79
(right) *Pelargonium alchemilloides* subsp. *multibracteatum* p. 82

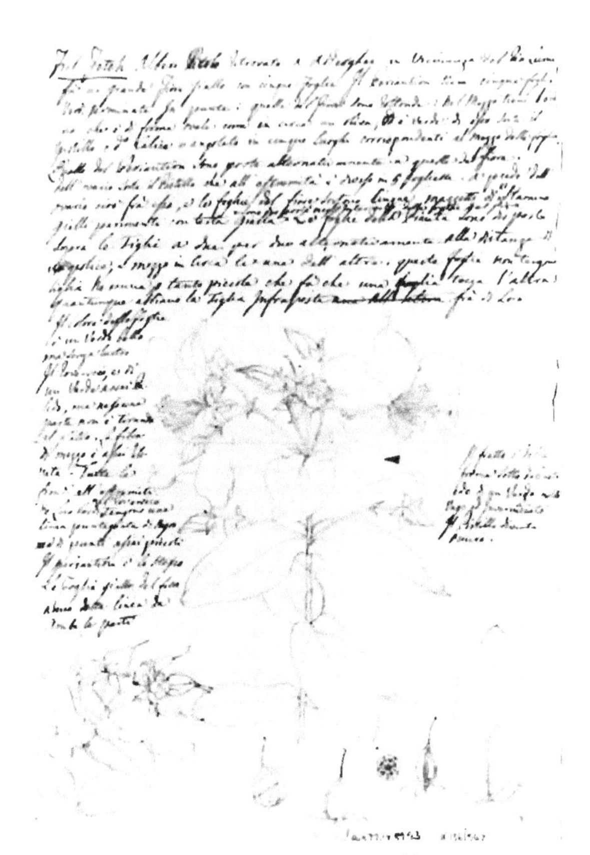

132. *Hypericum quartinianum* (a) p. 82

133. *Coleus edulis* (a) p. 82/83

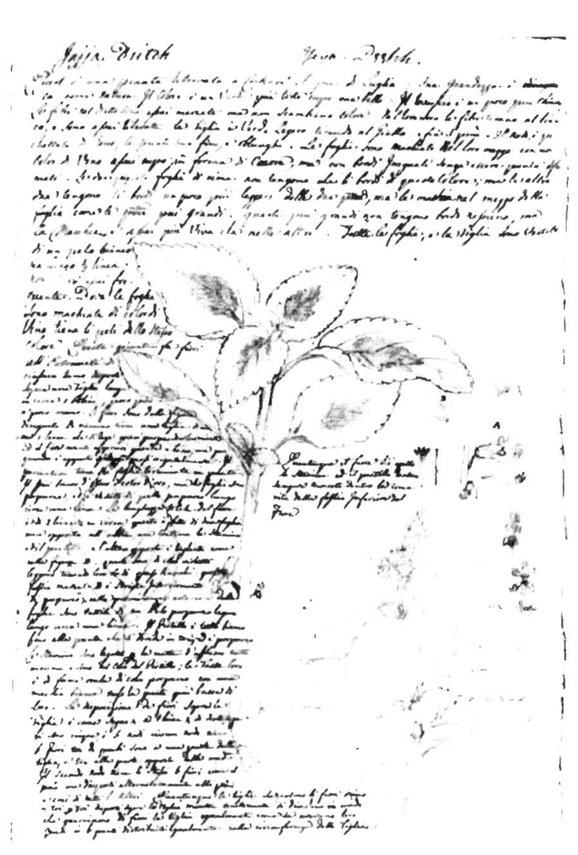

134. *Coleus* species 2 p. 83

135. *Coleus* species I p. 83

136. *Isodon ramosissimus* p. 83

137. *Otostegia tomentosa* subsp. *ambigens* p. 83

138. *Phlomis herba-venti* (a) p. 83

139. *Pycnostachys abyssinica* p. 84

140. *Salvia nilotica* (a) p. 84

141. *Labiatae* indeterminate p. 84

142. *Acacia* species (a) p. 84

143. *Acacia* species (b) p. 84

144. *Acacia* species (c) p. 84

145. *Albizia gummifera* (a) p. 84/85

146. *Albizia gummifera* (b) p. 84/85

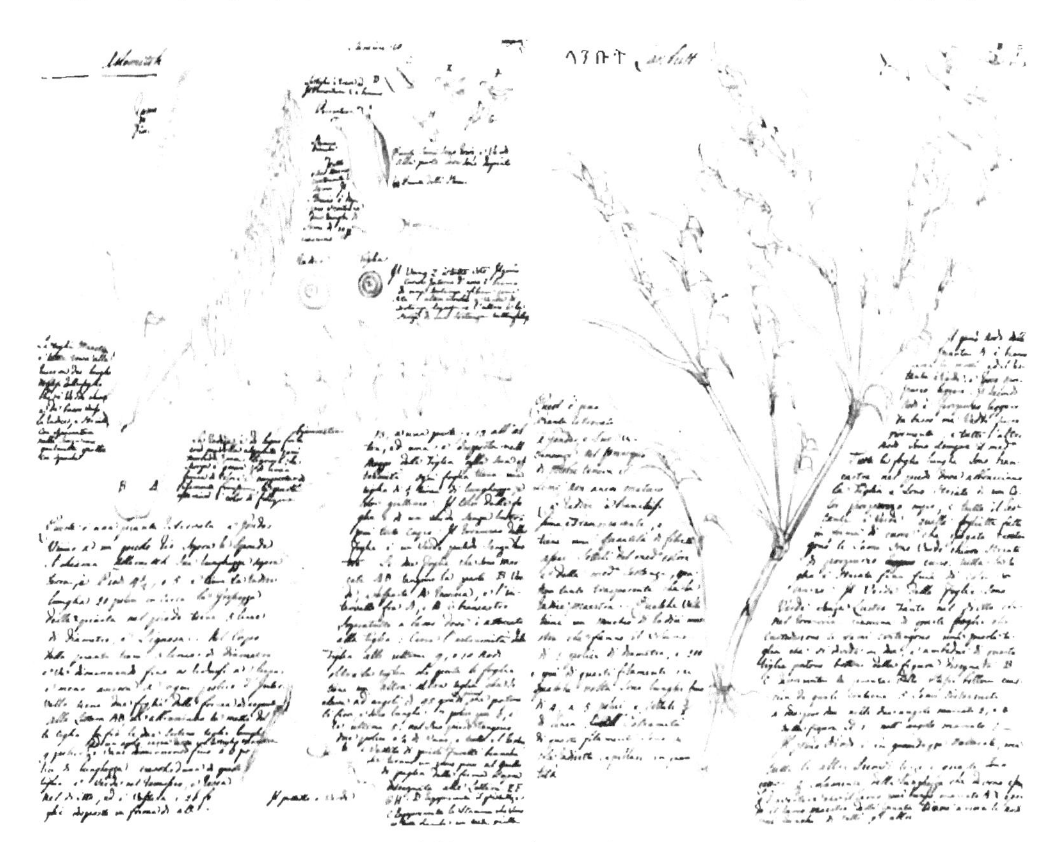

147. (left) *Astragalus atropilosus* p. 85
(right) *Commelina africana* p. 107

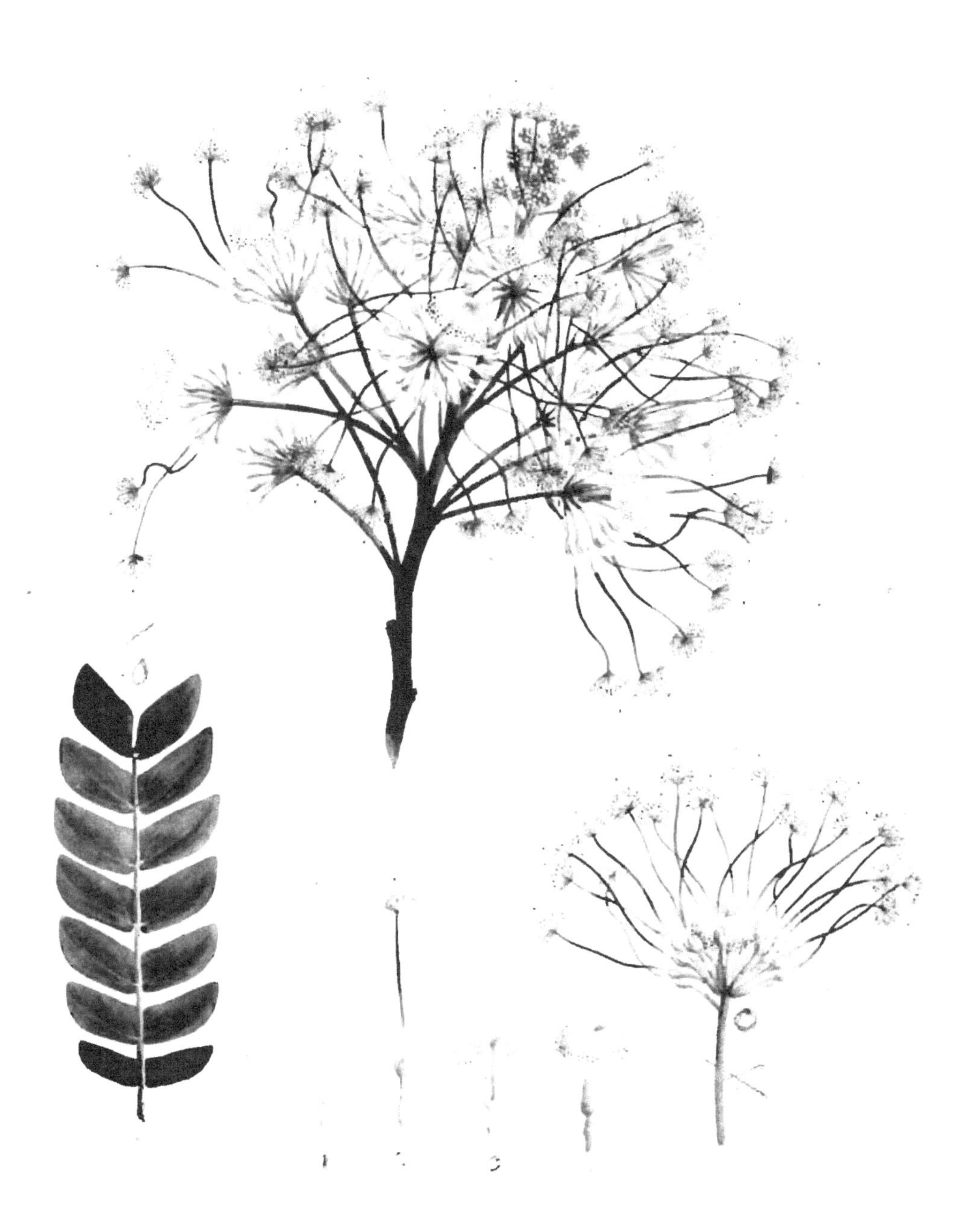

148. *Albizia gummifera* (d) p. 85

149. *Albizia malacophylla* (a) p. 85

150. *Albizia malacophylla* (b) p. 85

151. *Bauhinia farec* p. 85

152. *Albizia gummifera* (c) p. 84/85

153. *Cassia fistula* (b) p. 86

154. *Cassia fistula* (c) p. 86

155. *Crotalaria pallida* (b) p. 86

156. *Dichrostachys cinerea* (a) p. 86/87

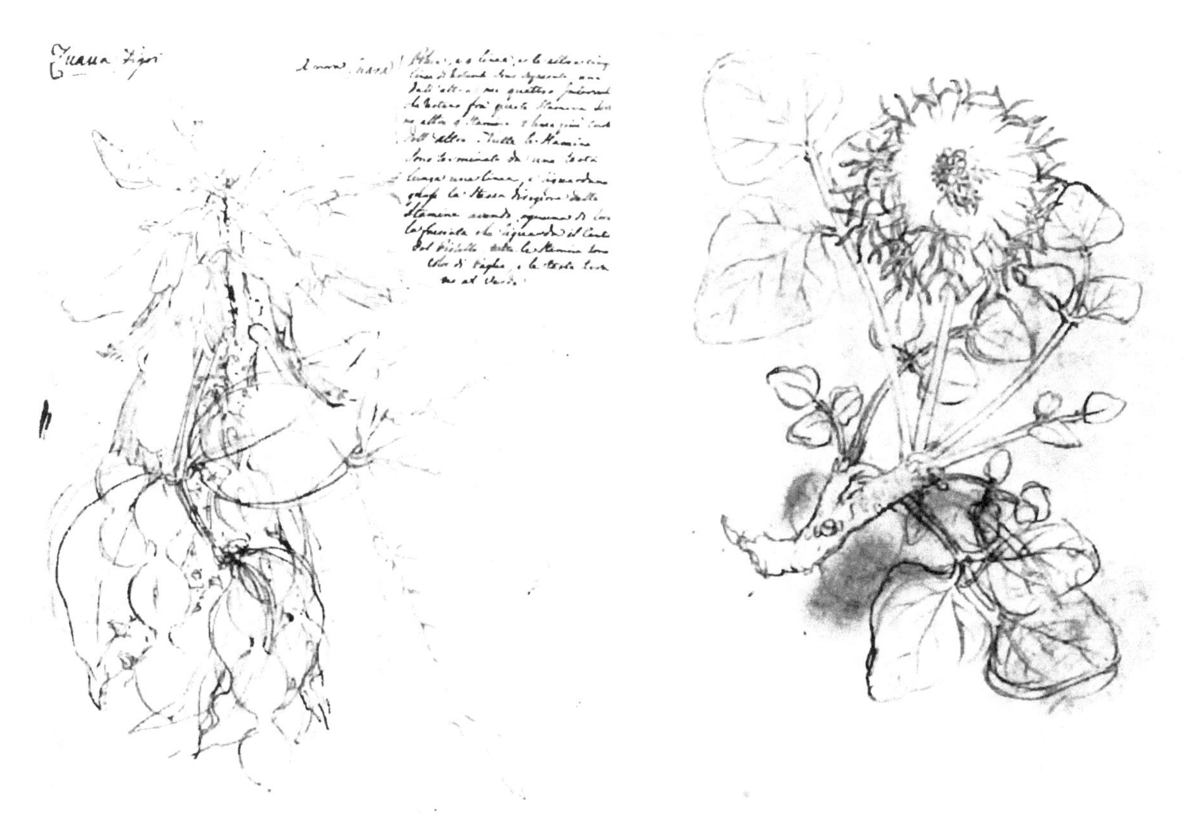

157. *Erythrina abyssinica*, fruits (a) p. 87
Erythrina brucei, flowers (a) p. 87/88

158. *Erythrina abyssinica* (c) p. 87

159. *Dichrostachys cinerea* (b) p. 86/87

160. *Erythrina abyssinica*, fruits; *Erythrina brucei*, flowers (b) p. 87

161. *Erythrina abyssinica* (d) p. 87

Qui a basso è rappresentato un fiore dell' Albero di Ciarà. Con tutte
le sue parti in grandezza naturale, perchè per se stesso sono assai visibili
senza alterazione di misura. Il [illegible] di questo fiore è [illegible] in una
[illegible] circa 4 linee di lunghezza. Il [illegible] [illegible] che tiene
[illegible] circonferenza, e di 8 linee in [illegible] [illegible]
una foglia che termina quasi in punta della lunghezza di 7 linee [illegible]
[illegible] due parti della foglia. Questo è tutto vellutato e color di fuligine [illegible]
[illegible] un poco con porporino. Nella parte [illegible] [illegible]
[illegible] foglia disposta nel modo che si vede. La prima [illegible] marcata N° 1 [illegible]
la parte di [illegible] che si divide e questa [illegible] le pistillo e le stamine
che sono nel seno del fiore.
Oltre di questa sono due foglie della forma di quella marcata N° 2, disposte lateral-
mente alla [illegible] ricoprendone una gran porzione. Sopra queste vi è una
grande foglia che [illegible] [illegible] in forma del fiore, ed
è disposta in fronte della [illegible] questa è tutta color di [illegible]
[illegible]

162. *Erythrina brucei* (b) p. 87/88

163. *Desmodium* species (a) p. 88

164. *Millettia ferruginea* (a) p. 88

165. *Millettia ferruginea* (b) p. 88

166. *Desmodium* species (b) p. 88

167. *Millettia ferruginea* (c) p. 88

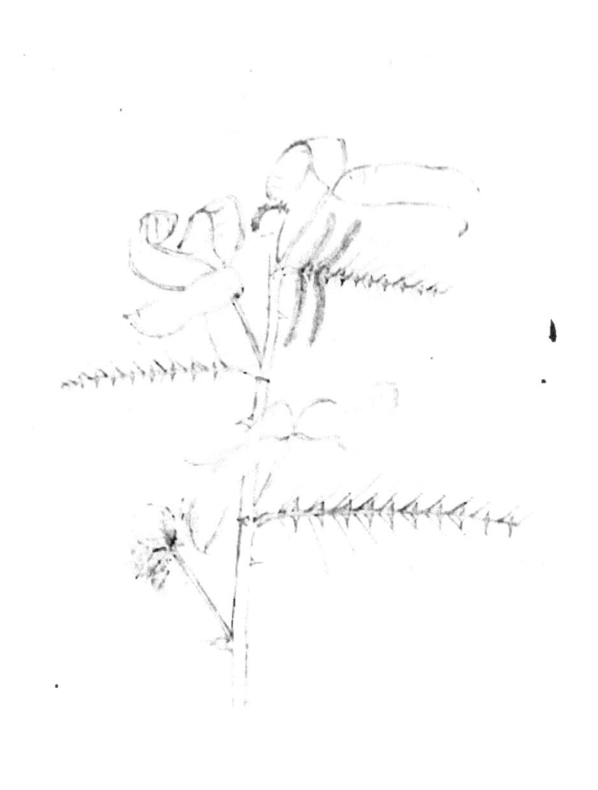

168. *Mimosa pigra* (a) p. 88

169. *Pterolobium stellatum* (a) p. 89

170. *Pterolobium stellatum* (b) p. 89

171. *Mimosa pigra* (b) p. 88

[illegible]

172. (left) *Trifolium rueppellianum* (a) p. 89
(right) *Trifolium polystachyum* p. 89

173. (left) *Trifolium schimperi* p. 89-90
(right) *Trifolium usambarense* p. 90

174. *Trifolium rueppellianum* (b) p. 89

175. *Nuxia oppositifolia* (c) p. 90

176. *Woodfordia uniflora* (b) p. 91

[illegible]

A

177. *Buddleia polystachya* (a) p. 90

178. *Nuxia oppositifolia* (b) p. 90

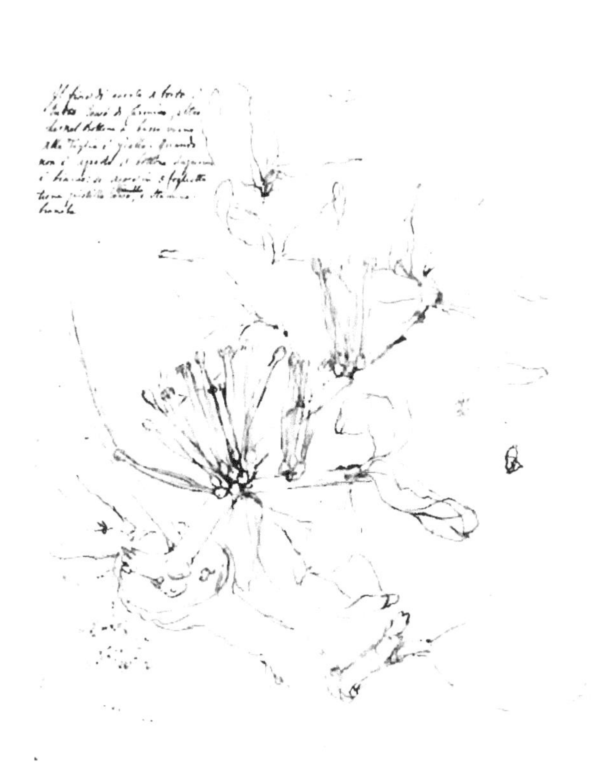

179. *Loranthus globiflorus* p. 90, 91

180. *Loranthus heteromorphus* p. 91

181. *Lawsonia inermis* p. 91

182. *Woodfordia uniflora* (a) p. 91

183. *Abutilon longicuspe* (a) p. 91

[illegible] [illegible] [illegible] Pianta i un
albero [illegible] in [illegible] alli 21 di
Novembre. Le foglie [illegible] disposte sopra
l'estremità delle rami alla distanza di
[illegible] una dall'altra disposte in triangolo, ogni foglia tiene una
lunga [illegible] [illegible] in circa. La larghezza della foglia [illegible] [illegible] [illegible]
[illegible] [illegible] [illegible] di lungo; [illegible] [illegible]
[illegible] [illegible] [illegible] all'estremità delle rami
le foglie vengono più [illegible] e [illegible]
[illegible] [illegible] più corta dell'altra. [illegible]
delle foglie è di un verde [illegible], ed il [illegible]
[illegible] è di un verde chiaro tirando al [illegible]
[illegible]. [illegible] [illegible] [illegible] [illegible] e
tirano al giallo chiaro. [illegible] [illegible]
le [illegible] sono di un verde chiaro
tirando al giallo. [illegible] alle estremità delle rami oltre le foglie [illegible]
una altra [illegible] che [illegible] fiori
come nel disegno A. [illegible] [illegible]
B rappresenta un fiore in grandezza
naturale. CCC rappresentano
[illegible] [illegible] [illegible] [illegible]
[illegible] [illegible] [illegible] [illegible]

184. *Abutilon longicuspe* (b) p. 91

185. *Hibiscus cannabinus* p. 92

186. *Abutilon longicuspe* (c) p. 91

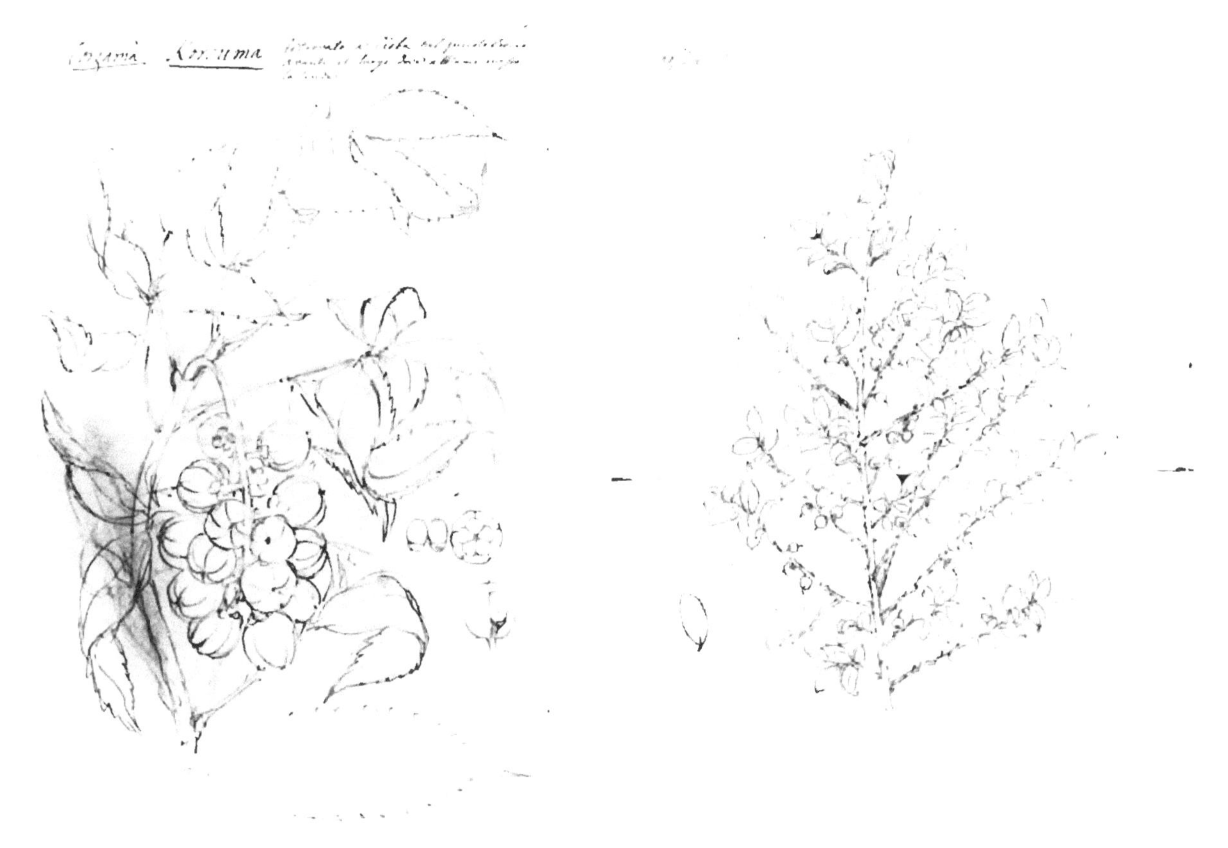

187. *Bersama abyssinica* (a) p. 92

188. *Myrsine africana* (a) p. 92

189. *Myrsine africana* (b) p. 92

190. *Jasminum abyssinicum* (a) p. 93

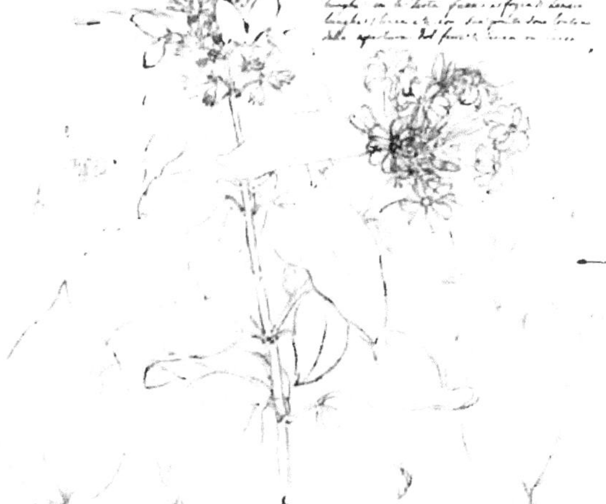

191. *Jasminum dichotomum* (a) p. 93

192. *Olea europaea* subsp. *africana* p. 93

193. *Oxalis corniculata* p. 93

194. *Jasminum abyssinicum* (b) p. 93

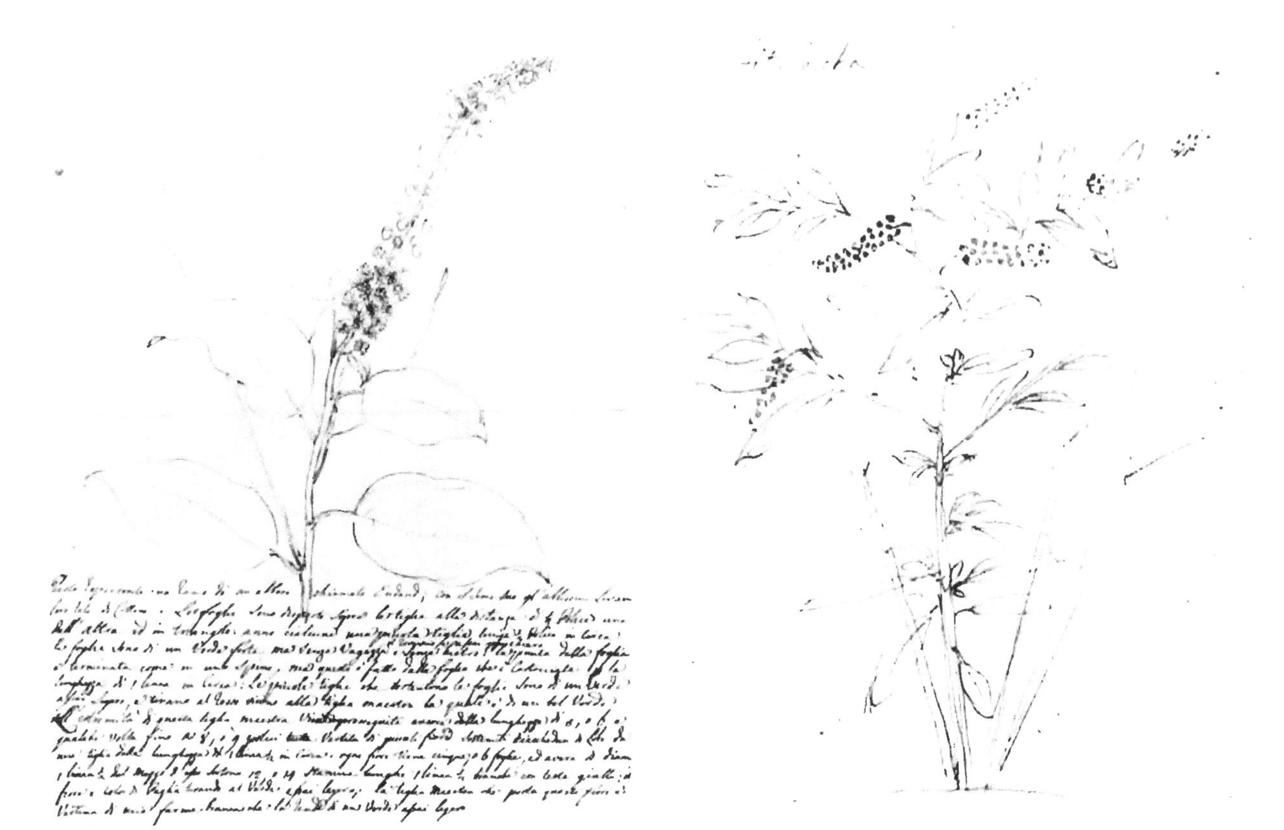

195. *Phytolacca dodecandra* (a) p. 93/94

196. *Phytolacca dodecandra* (b) p. 93/94

197. *Phytolacca dodecandra* (c) p. 93/94

198. *Pittosporum viridiflorum* (a) p. 94

199. *Pittosporum viridiflorum* (b) p. 94

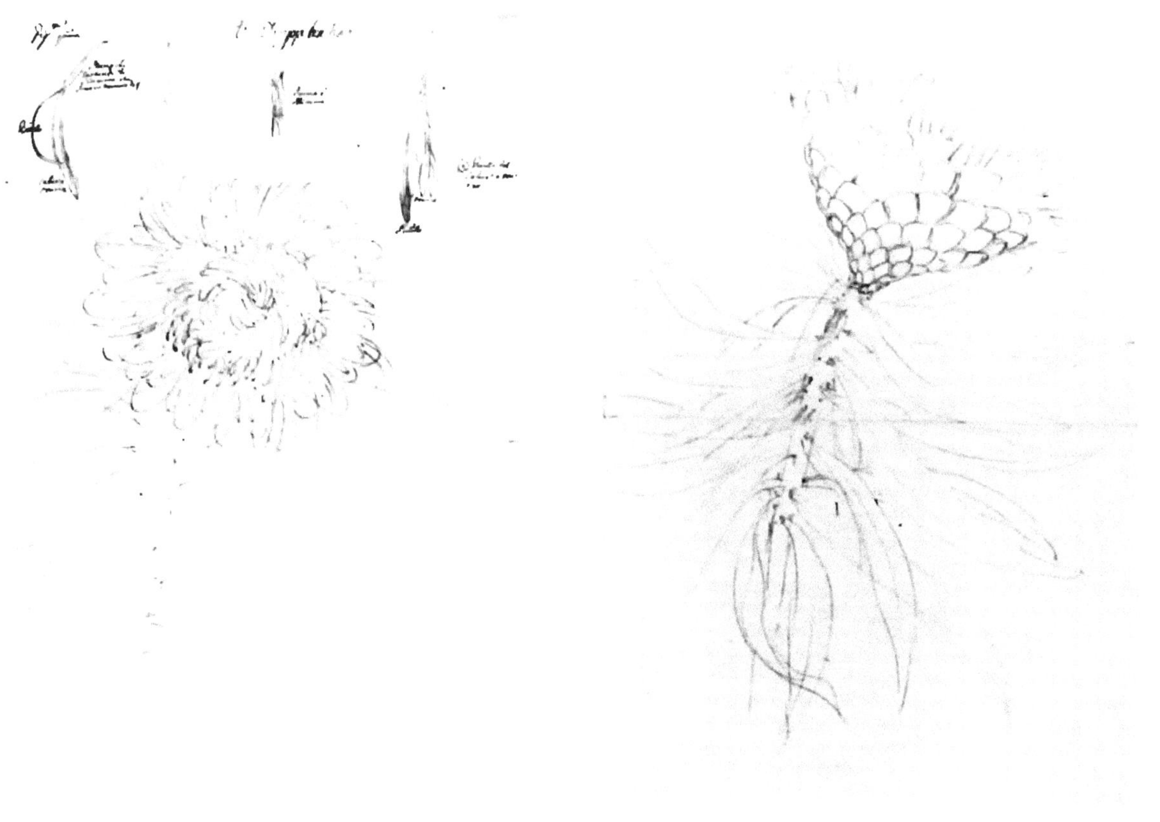

200. *Protea gaguedi* (a) p. 94

201. *Protea gaguedi* (b) p. 94

202. *Protea gaguedi* (c) p. 94

[illegible] albero ritrovato a Dyppabahar, [illegible]

[illegible]

203. *Protea gaguedi* (d) p. 94

Gaguedi. Albero ritrovato a [illegible] miglia di distanza [illegible]

[illegible] dell' Albero [illegible] l'estremità [illegible] rami [illegible]

[illegible]

204. *Protea gaguedi* (e) p. 94/95

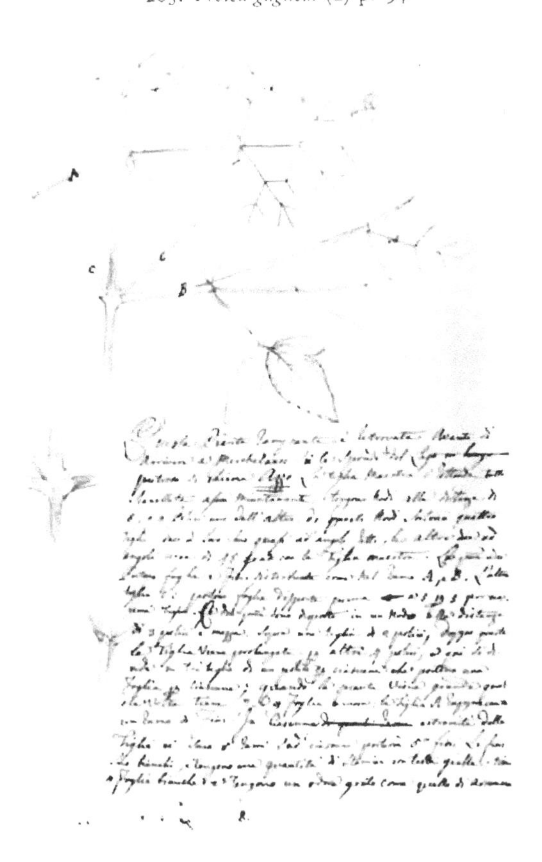

205. *Clematis hirsuta* (a) p. 95

206. *Protea gaguedi* (g) p. 94/95

207. *Punica granatum* (b) p. 95

208. *Punica granatum* (c) p. 95

209. *Delphinium wellbyi* (a) p. 95

210. *Delphinium wellbyi* (b) p. 95/96

211. *Rhamnus prinoides* (a) p. 96

212. *Ziziphus abyssinica* (a) p. 96

213. *Rhamnus prinoides* (b) p. 96

214. *Ziziphus abyssinica* (b) p. 96

215. *Hagenia abyssinica* (b) p. 96/97

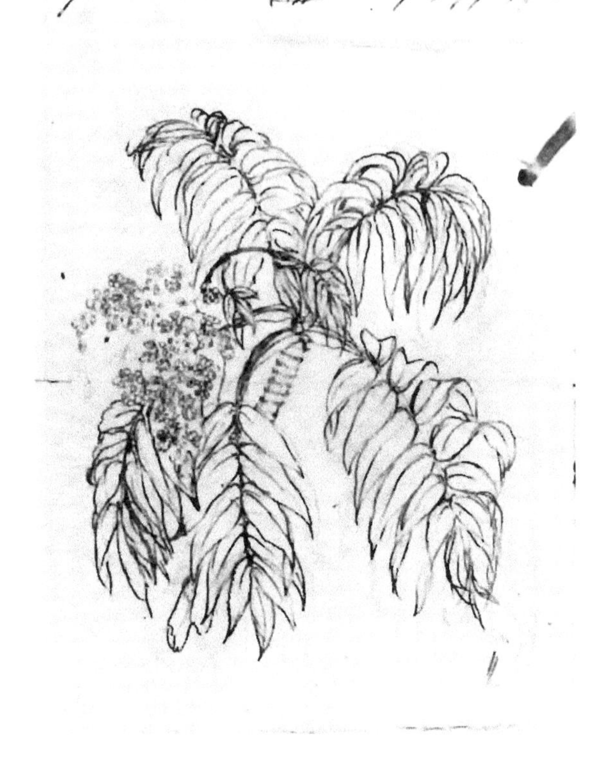

216. *Hagenia abyssinica* (c) p. 96/97

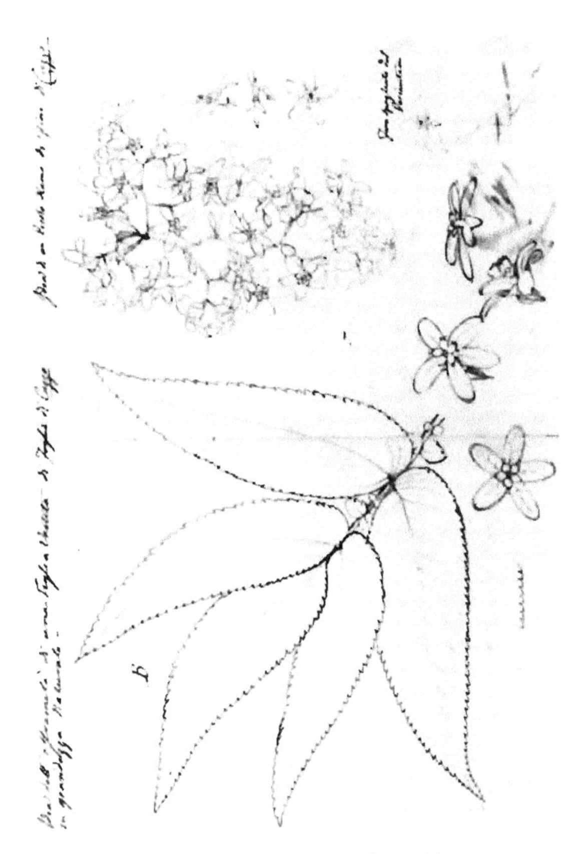

217. *Hagenia abyssinica* (d) p. 96/97

218. *Coffea arabica* (a) p. 97

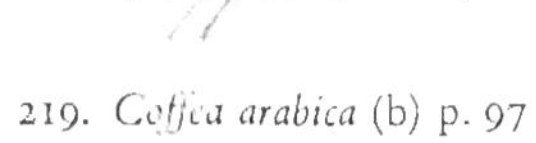

219. *Coffea arabica* (b) p. 97

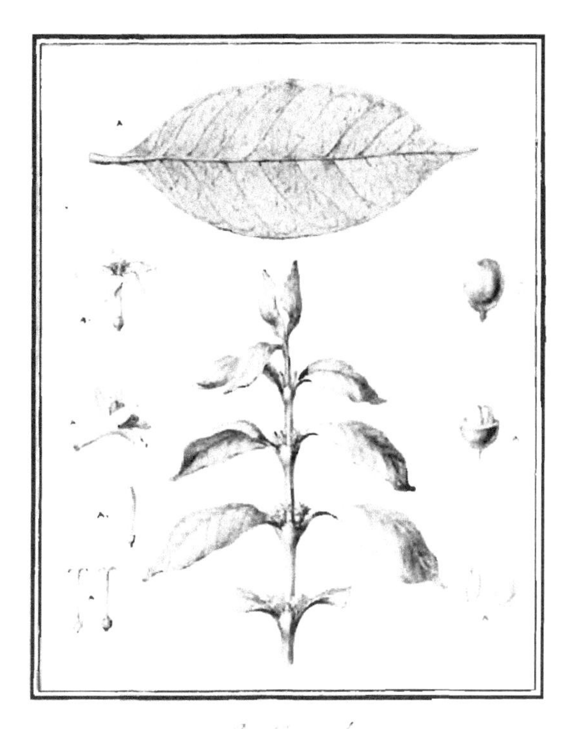

220. *Coffea arabica* (c) p. 97

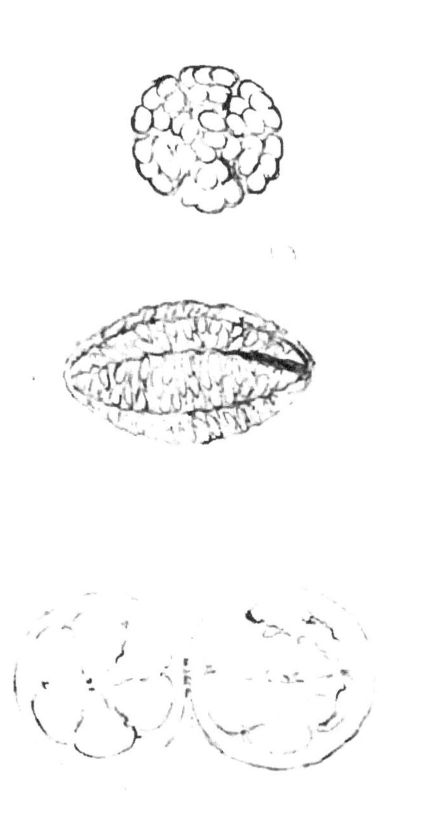

221. *Gardenia ternifolia* subsp. *jovis-tonantis* (a) p. 97

222. *Gardenia ternifolia* subsp. *jovis-tonantis* (b) p. 97

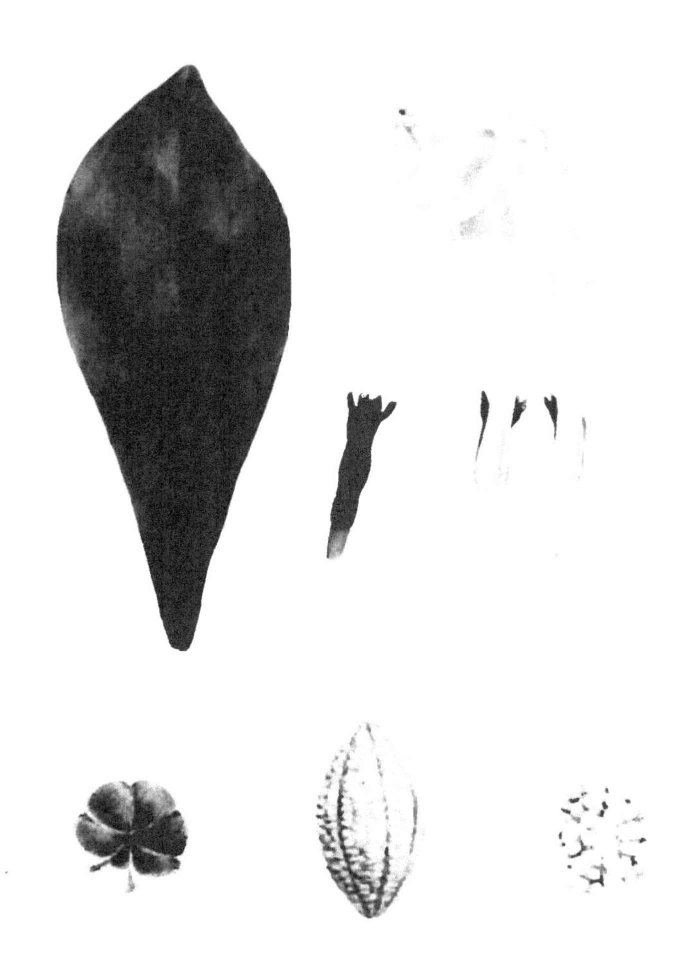

223. *Gardenia ternifolia* subsp. *jovis-tonantis* (g) p. 97/98

224. *Gardenia ternifolia* subsp. *jovis-tonantis* (c) p. 97

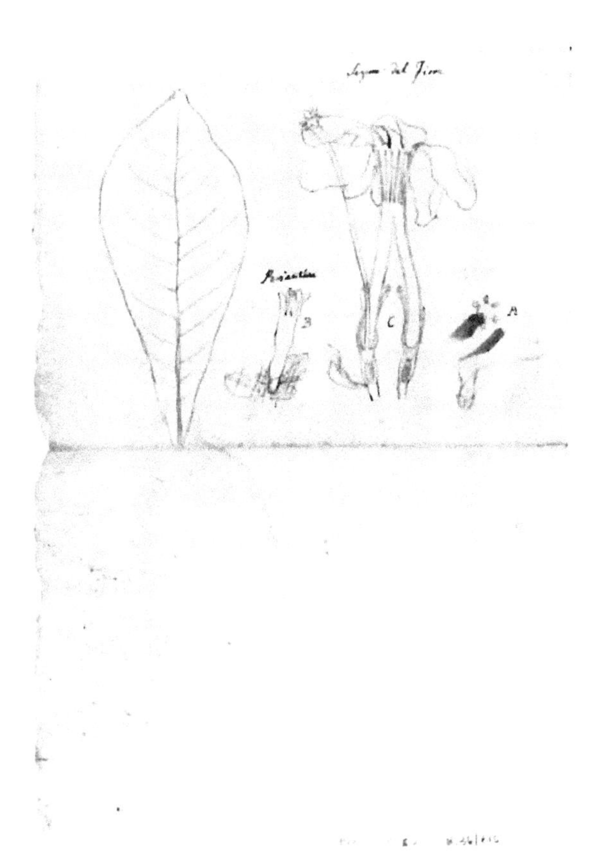

225. *Gardenia ternifolia* subsp. *jovis-tonantis* (d) p. 97

226. *Mussaenda arcuata* (a) p. 98

227. *Clausena anisata* (a) p. 98

228. *Mussaenda arcuata* (c) p. 98

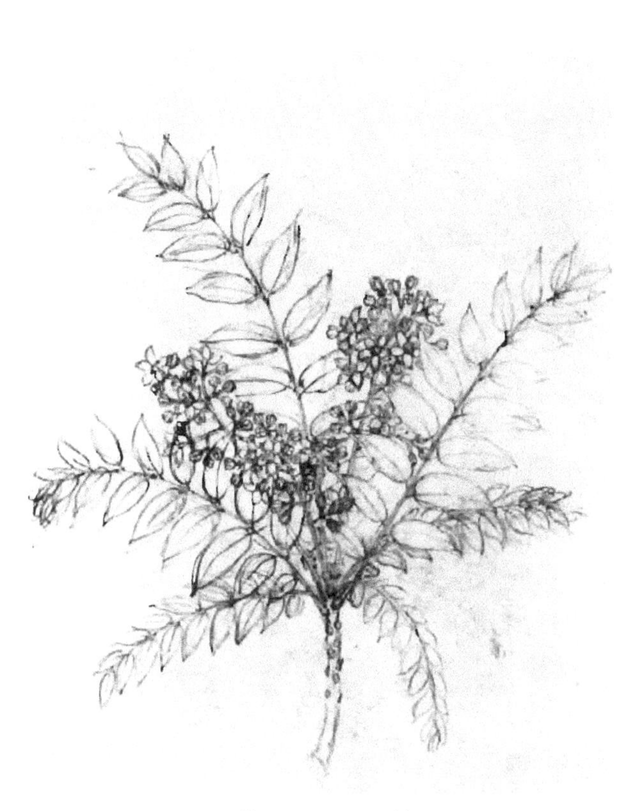

229. *Clausena anisata* (b) p. 98

230. *Salix subserrata* (a) p. 98

231. *Osyris lanceolata* (b) p. 99

232. *Osyris lanceolata* (a) p. 99

233. *Salix subserrata* (b) p. 98/99

234. *Cardiospermum halicacabum* (a) p. 99

235. *Cardiospermum halicacabum* (b) p. 99

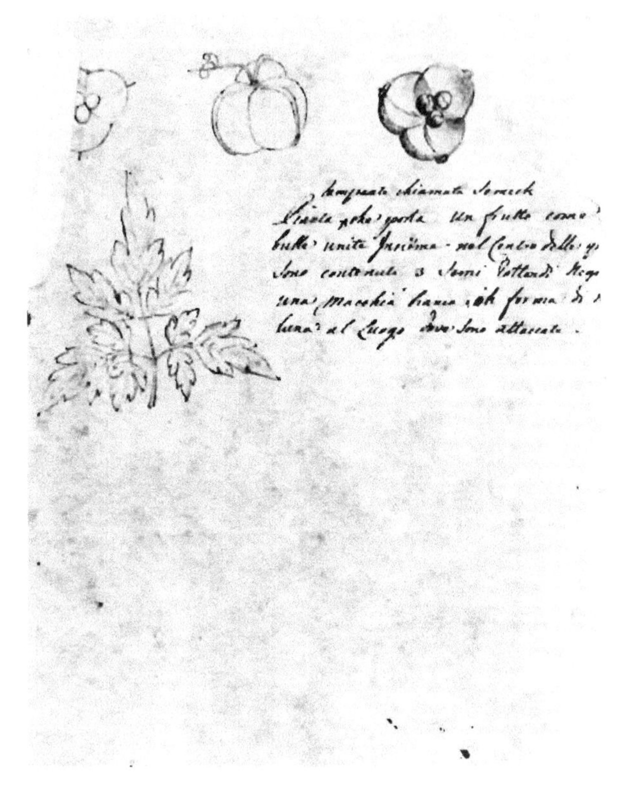

236. *Cardiospermum halicacabum* (c) p. 99

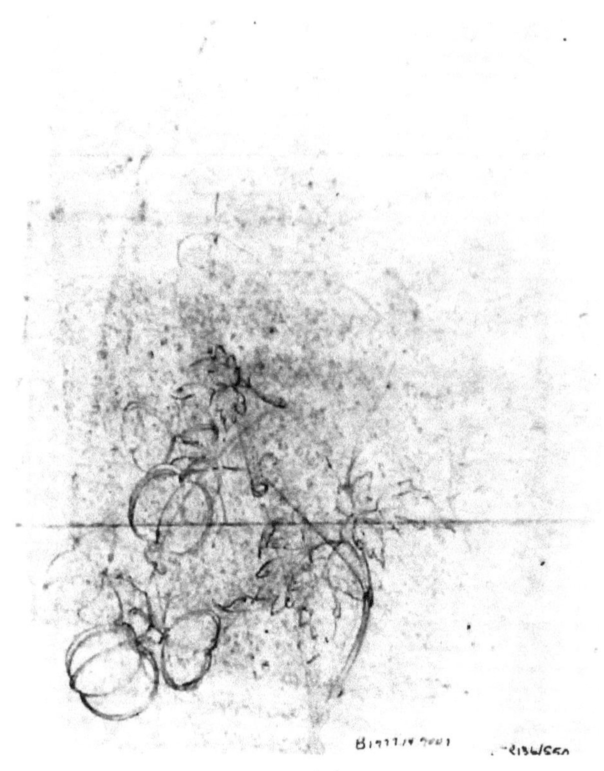

237. *Cardiospermum halicacabum* (d) p. 99

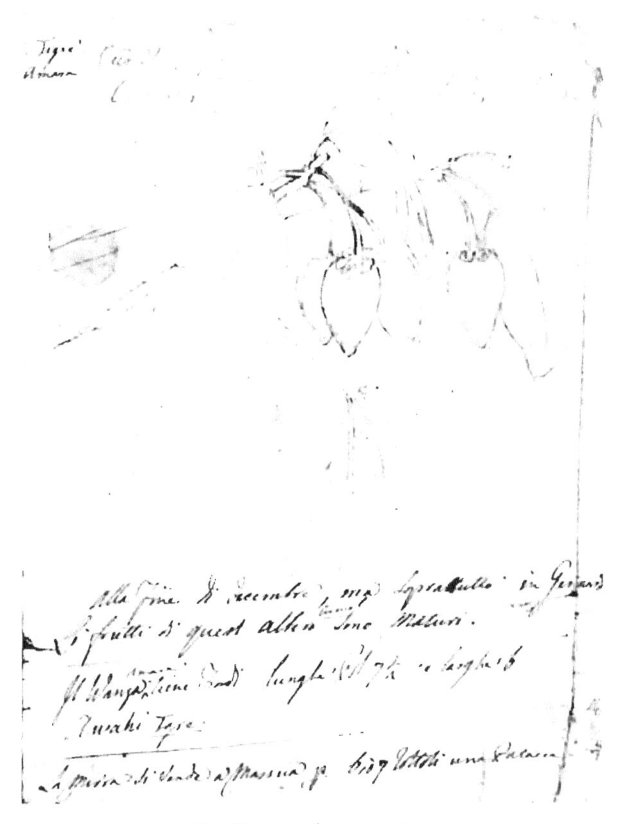

238. *Mimusops kummel* (a) p. 99

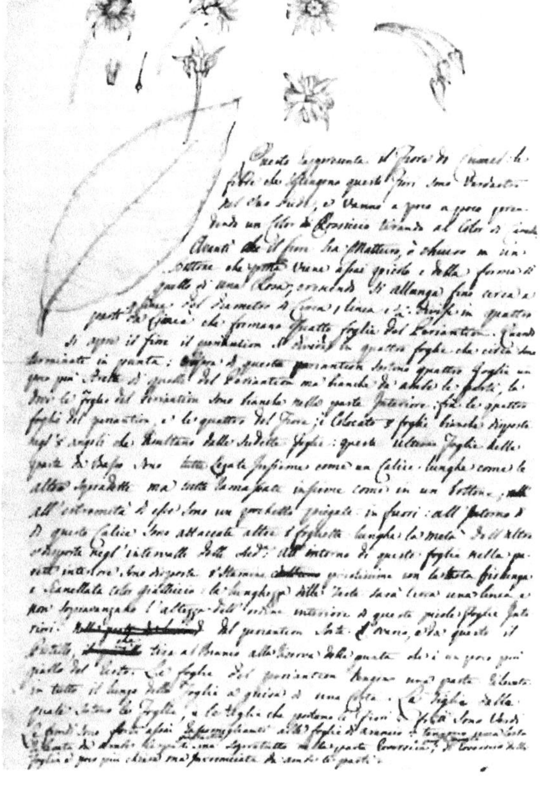

239. *Mimusops kummel* (c) p. 99

240. *Mimusops kummel* (d) p. 99/100

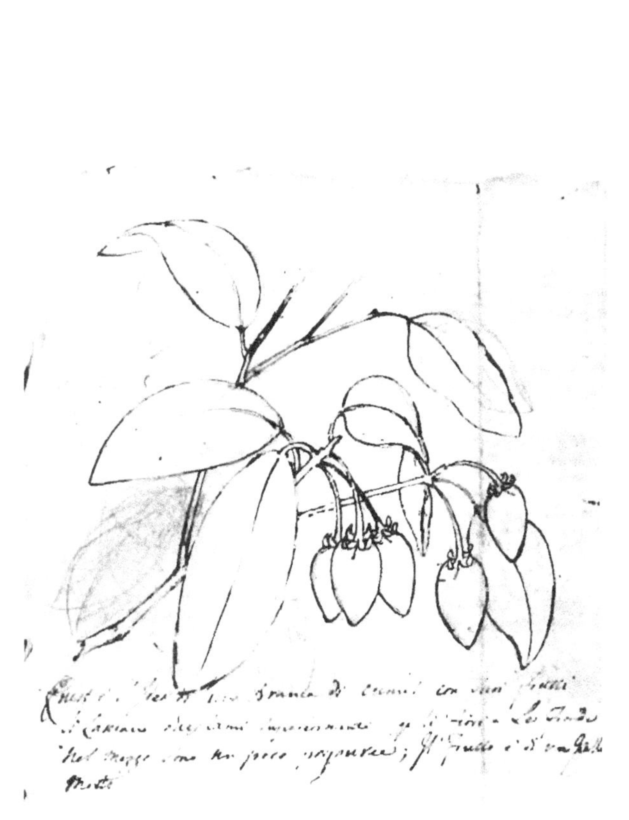

241. *Mimusops kummel* (b) p. 99

242. *Mimusops kummel* (c) p. 99/100

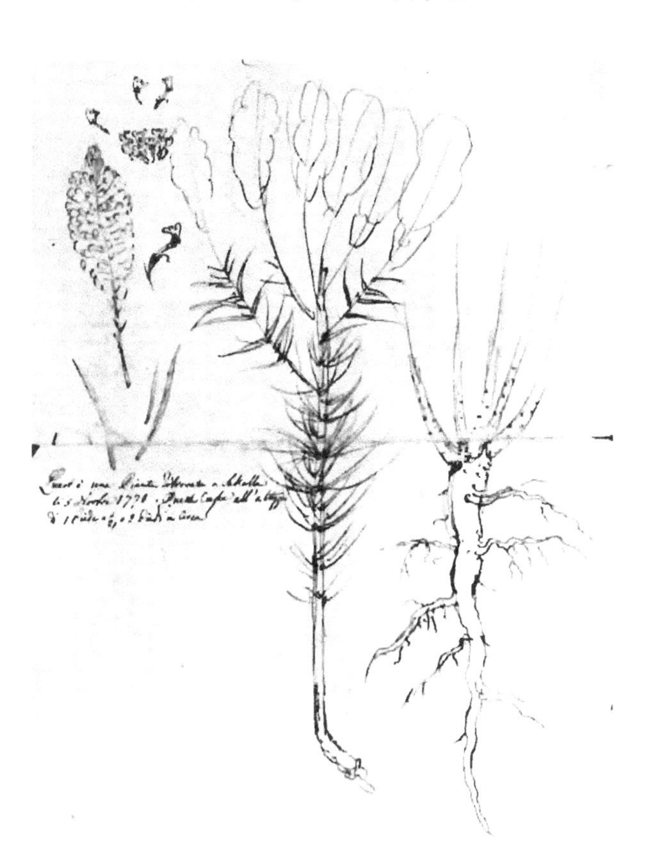

243. *Hebenstreitia dentata* p. 100

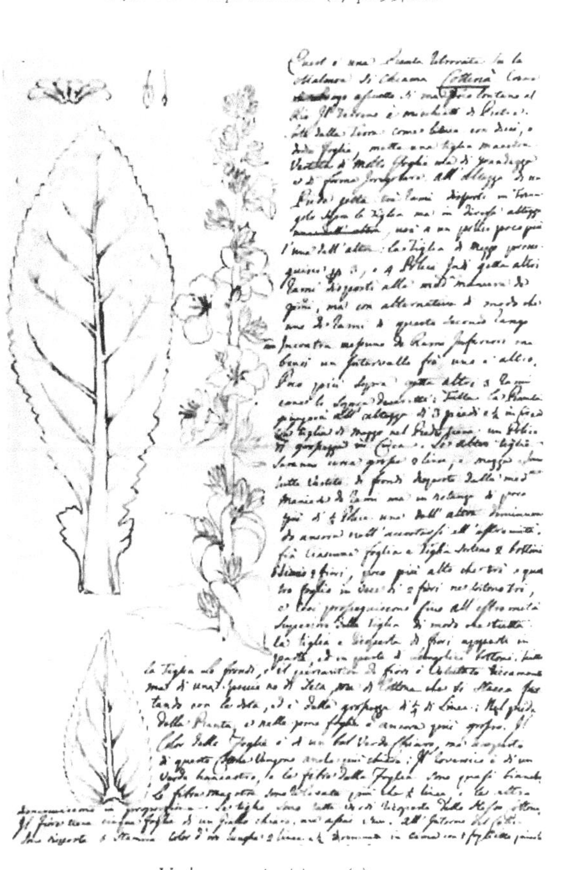

244. *Verbascum sinaiticum* (a) p. 100

245. *Verbascum sinuaticum* (?) p. 100

246. *Verbascum sinaiticum* (b) p. 100

247. *Brucea antidysenterica* (a) p. 100/101

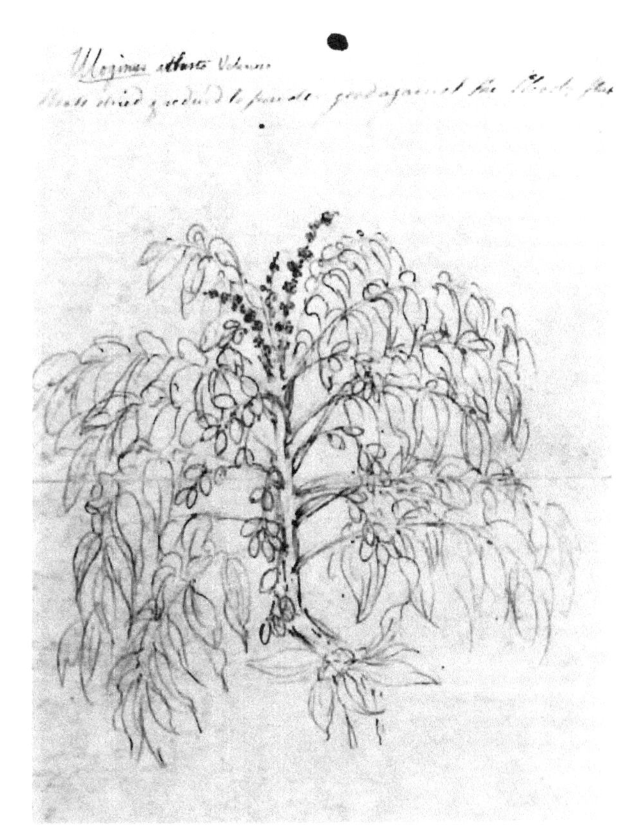

248. *Brucea antidysenterica* (b) p. 100/101

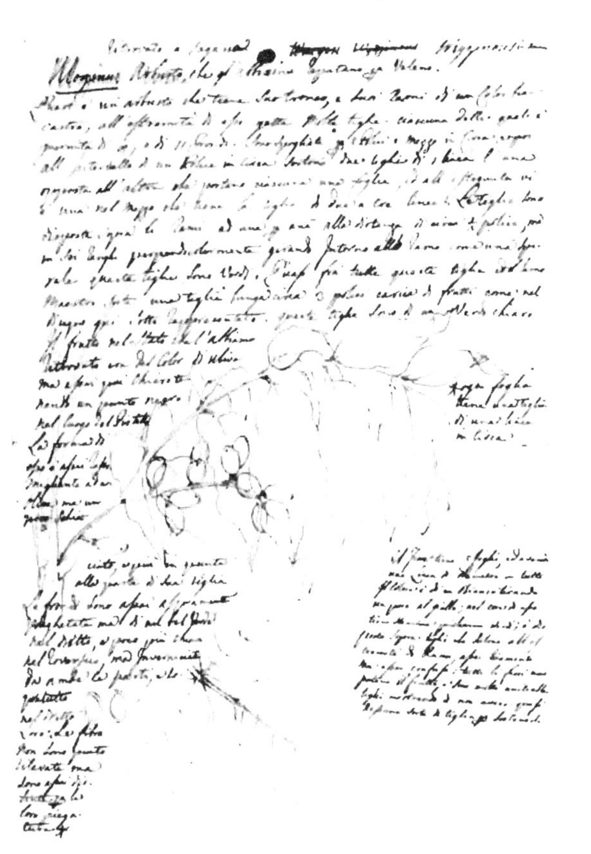

249. *Brucea antidysenterica* (c) p. 100/101

250. *Brucea antidysenterica* (c) p. 100 101

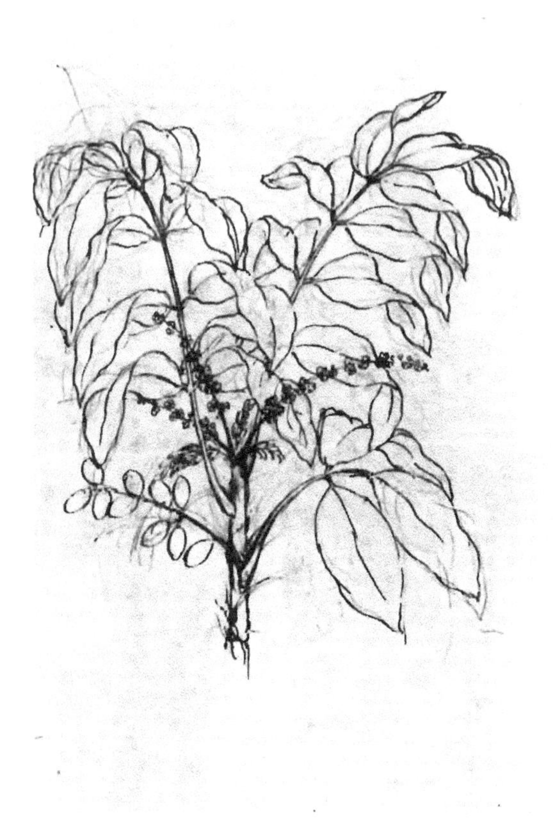

251. *Brucea antidysenterica* (d) p. 100/101

252. *Discopodium penninervium* (a) p. 101

253. *Solanum adoense* (a) p. 102

254. *Solanum adoense* (b) p. 102

255. *Solanum capsicoides* p. 102

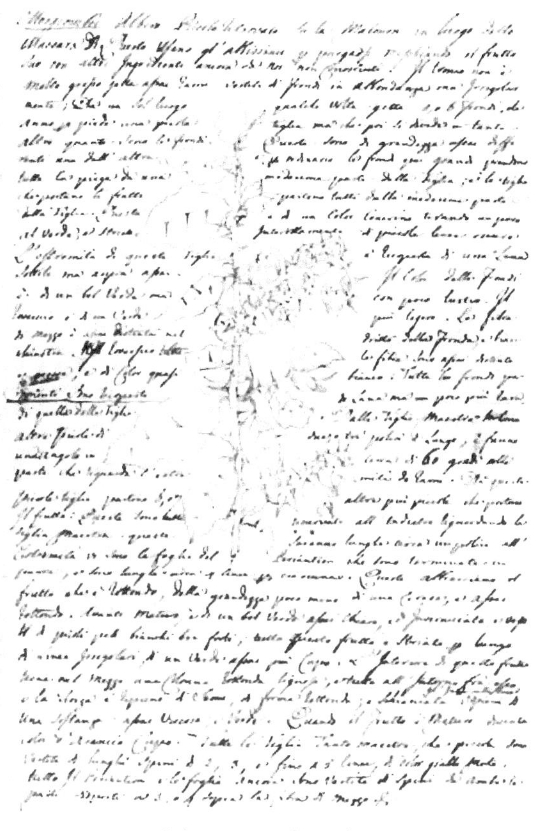

256. *Solanum piperiferum* (a) p. 103

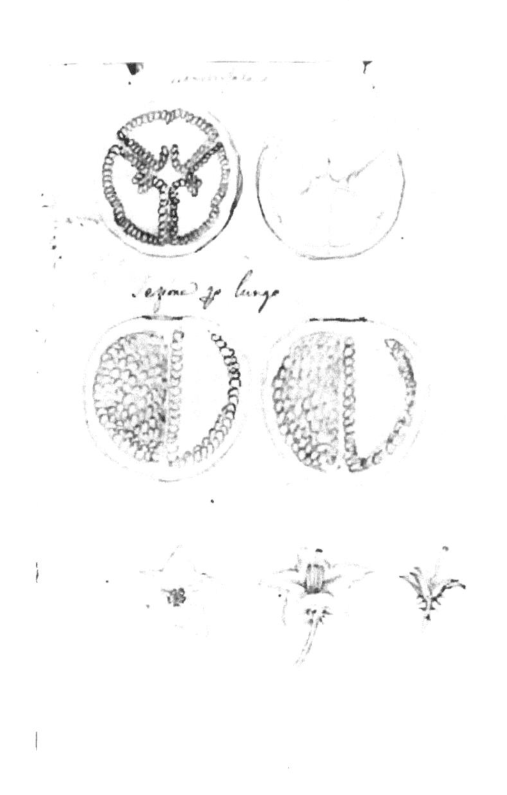

Menghi – Albero che fa piccoli frutti assai (copiosamente) e che si mangiano – li frutti sono rotondi come ciriegie e grandi come un Pisello

Azazo – Albero (grande) che fa un seme ovale dentro di uno Stuccio lungo 10, o 12 pollici. Con il Legno di questo albero si profumano le Vestite delle Donne

Hatcaru – Albero che tiene piccoli semi come finocchio

Mesgombee Arbusto spinoso [illegible] a [illegible]

Tothalo – Albero che fa fiori Bianchi, e dopo un piccolo frutto che non si Mangia

Bohak – Albero che fa piccoli frutti come [illegible] e che si Mangiano. Le foglie sono lunghe come quelle di Amandola

Zestata Albero che porta Gomma

Avalé Albero con semi quadrilatteri

Cursà Albero che porta frutti rotondi come aranci che sono ripieni di semi in una sostanza umida e gluanta – si chiama Gammari in Amara

257. (left) *Solanum incanum* (b) p. 102

258. *Solanum marginatum* (a) p. 102

259. *Solanum piperiferum* (b) p. 103

260. *Dombeya torrida* (a) p. 103

261. *Dombeya torrida* (b) p. 103

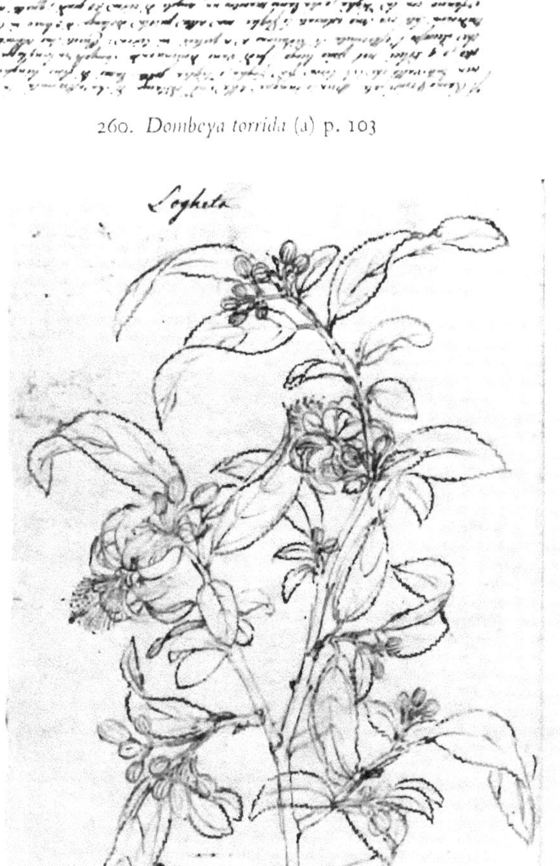

Il disegno qui a Canto rappresenta un Ramo di un Arbusto
chiamato Logheta, e ritrovato a [illegible]. Il luogo dove
cresceva era [illegible], un luogo vicino all'Aqua. Questo
Luogo pittoreso è tutto di Marmo Rosso tirante al porpora.
Con la Scorza di quest albero, tirano una [illegible] e
gl'abitanti lavano Loro Capele —
La Foglia è di un Verde allegro puntoto tirando al Giallo di
sovverso Loro è più Chiaro. tiene tutte le bordi dentati, questo
contatura Rivolta in alto. Le fibre delle foglie sono gialle
Il perianthion di questo fiore è Verde esteriormente
è velutato. le foglie sono 5 e assai grosse
nel Mezzo di esso tiene un Calice pentagono dai
lati del quale escono 5 Foglie Bianche, sottili, e [illegible]
più meno di quelle del Perianthion, le quali Interior
mente sono bianche; nel Mezzo del Calice Sorte l'
ovario che è [illegible] ovale e
tiene un pistillo con cima verde.
tutto all'Intorno di esso escono quantità di Sta
mine bianche con una testa gialla. tutte le Foglie
che portano li Fiori sono Verdi, e Velutate, la Fo
glia Vechia è Rossiccia tirando al Color di Fuligine
chiaro; Il Fiore è in assai abbondanza. [illegible] Ramo [illegible]
di Fiori e le buona piccolissima foglietta [illegible]

262. *Grewia ferruginea* (a) p. 104

263. *Sterculia africana* p. 104

264. *Grewia ferruginea* (b) p. 104

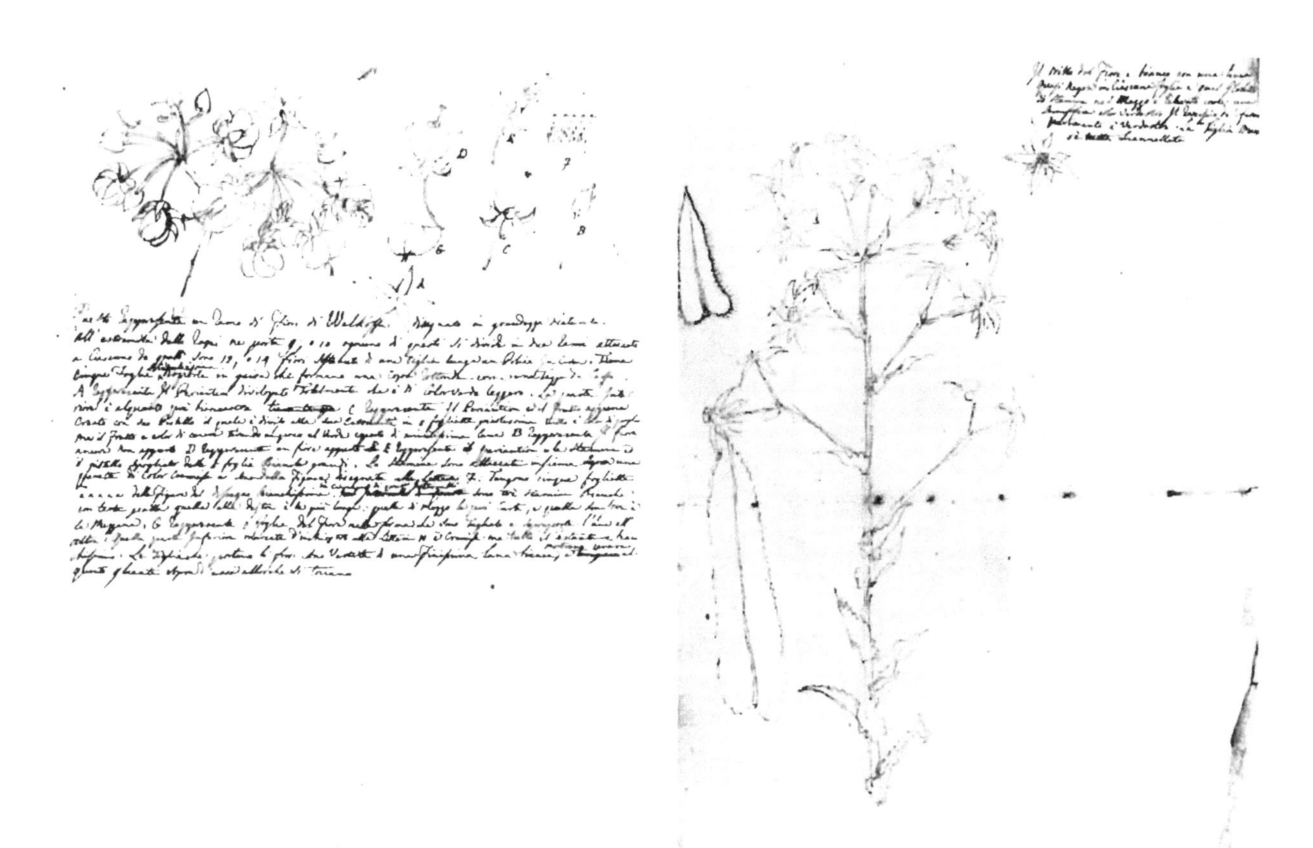

265. *Dombeya torrida* (c) p. 103

266. *Alepidea peduncularis* p. 104

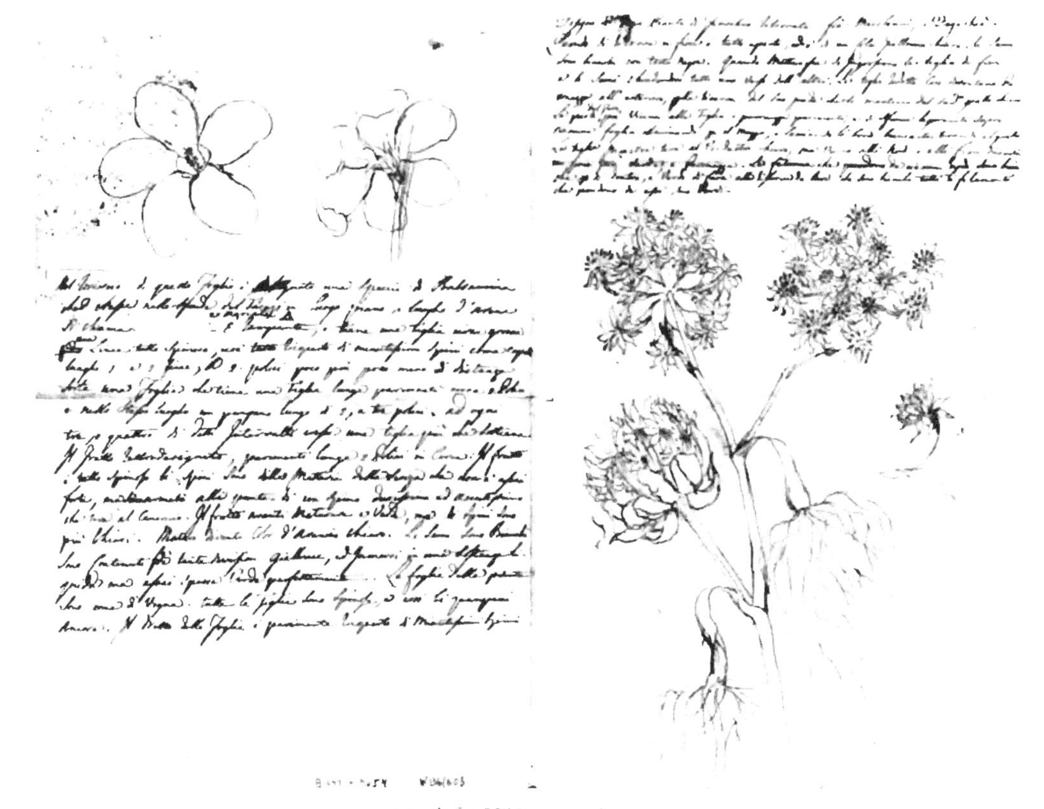

267. (left) *Hibiscus* species p. 92
(right) *Diplolophium africanum* p. 104/105

268. *Steganotaenia araliacea* p. 105

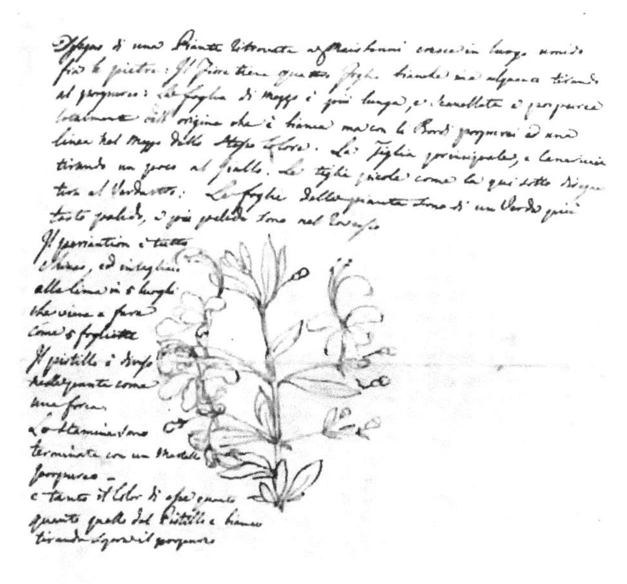

269. *Clerodendrum myricoides* p. 105

270. *Dracaena steudneri* (b) p. 105

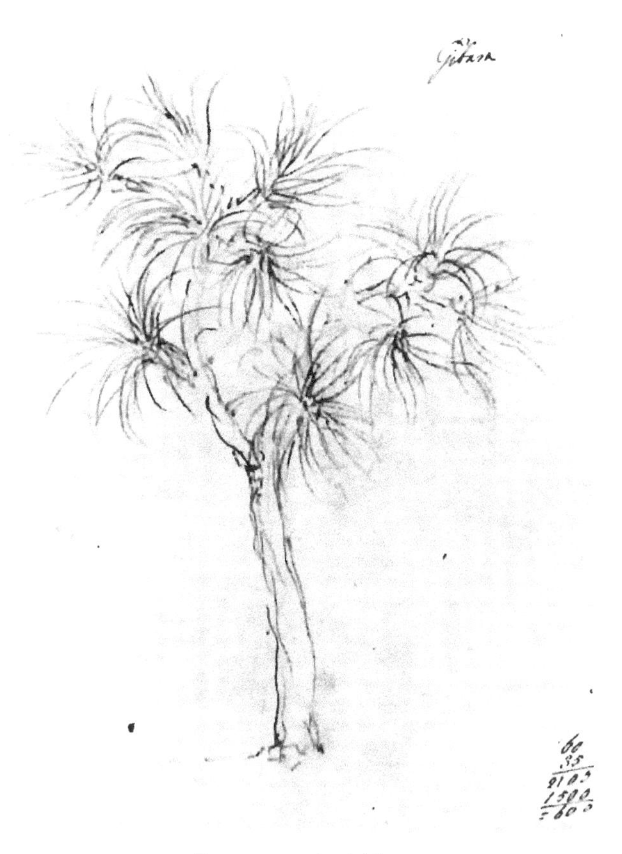

271. *Dracaena steudneri* (a) p. 105

272. (upper) *Punica granatum* p. 95
(lower) *Vitis vinifera* p. 105

273. *Dracaena steudneri* (c) p. 105

274. *Crinum schimperi* p. 105

275. *Dracaena steudneri* (c) p. 105

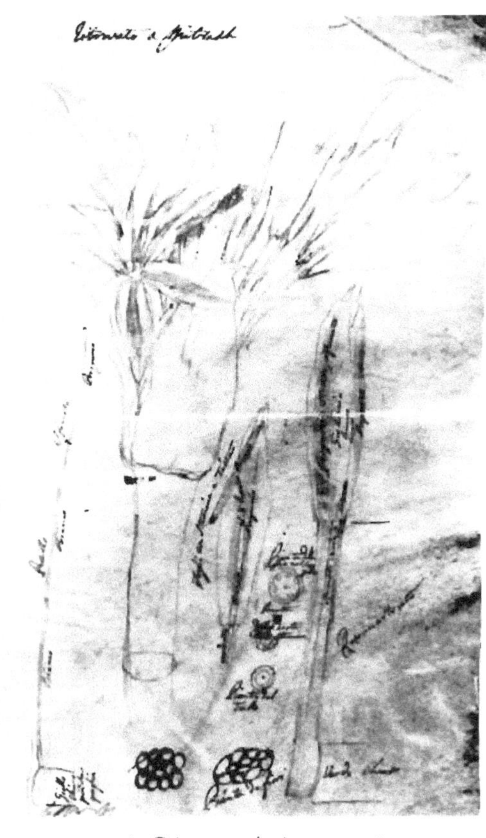

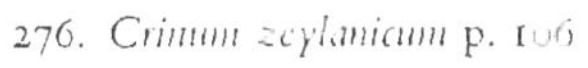

276. *Crinum zeylanicum* p. 106

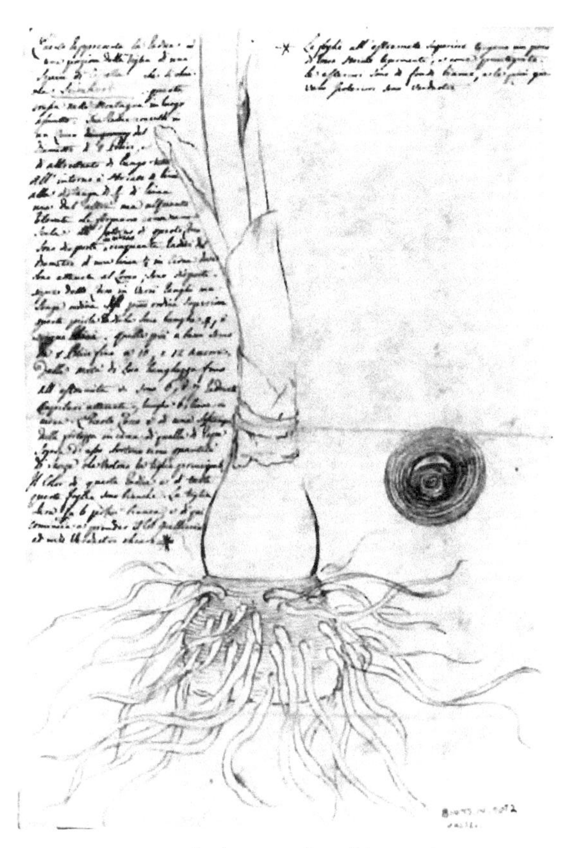

277. *Scadoxus puniceus* (b) p. 106

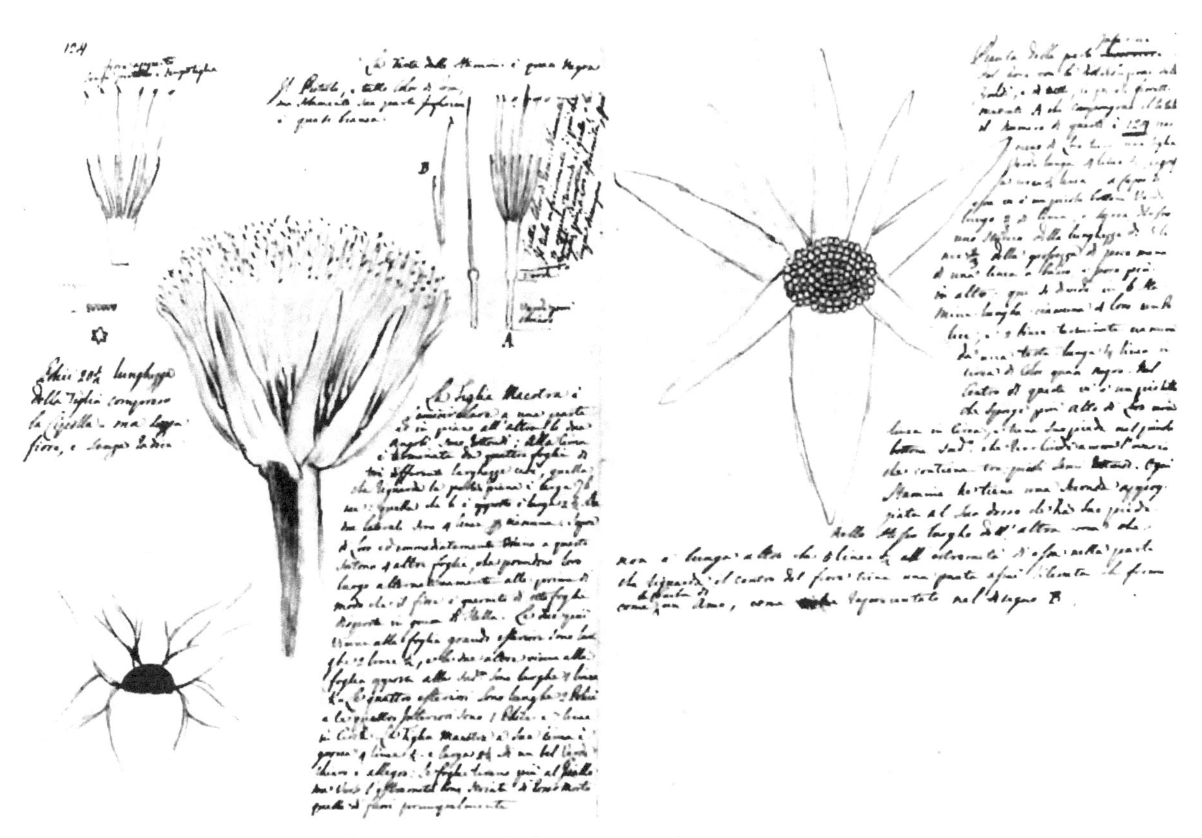

278. *Scadoxus puniceus* (a) p. 106

279. *Scadoxus puniceus* (c) p. 106

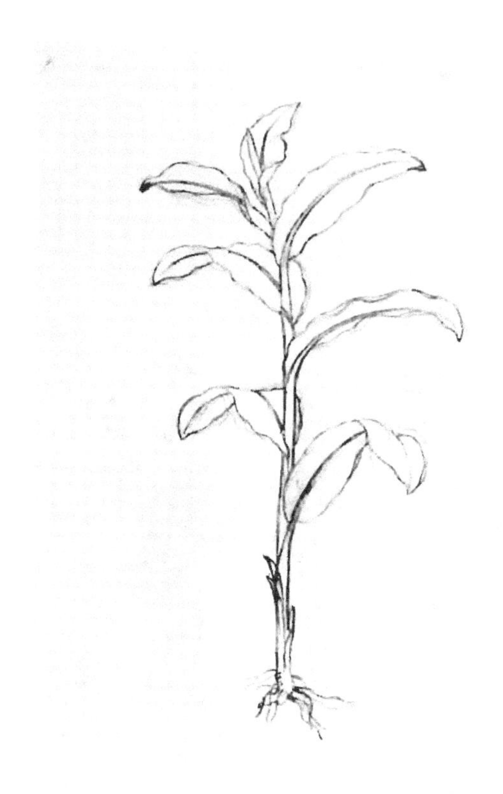

280. *Canna bidentata* (a) p. 106

281. *Canna bidentata* (b) p. 106

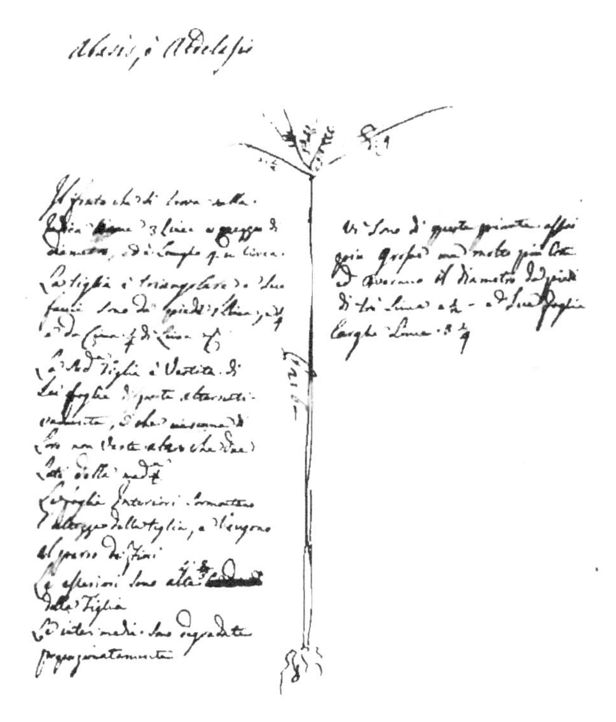

282. *Cyperus esculentus* (b) p. 107

283. *Canna bidentata* (c) p. 106

284. *Canna bidentata* (d) p. 106

285. *Cyperus esculentus* (c) p. 107

286. *Cyperus* species 1 (b) p. 107/108

287. *Cyperus* species 2 p. 108

288. *Mariscus cyperoides* p. 108

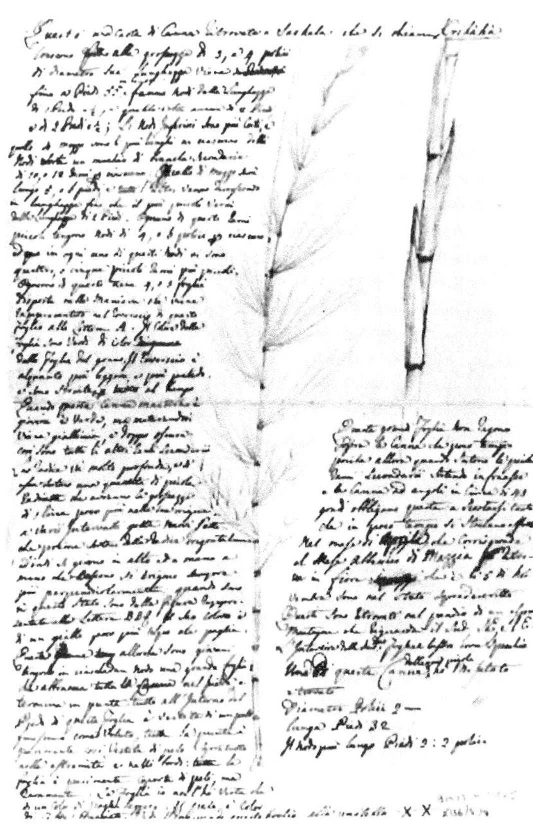

289. *Arundinaria alpina* (a) p. 108

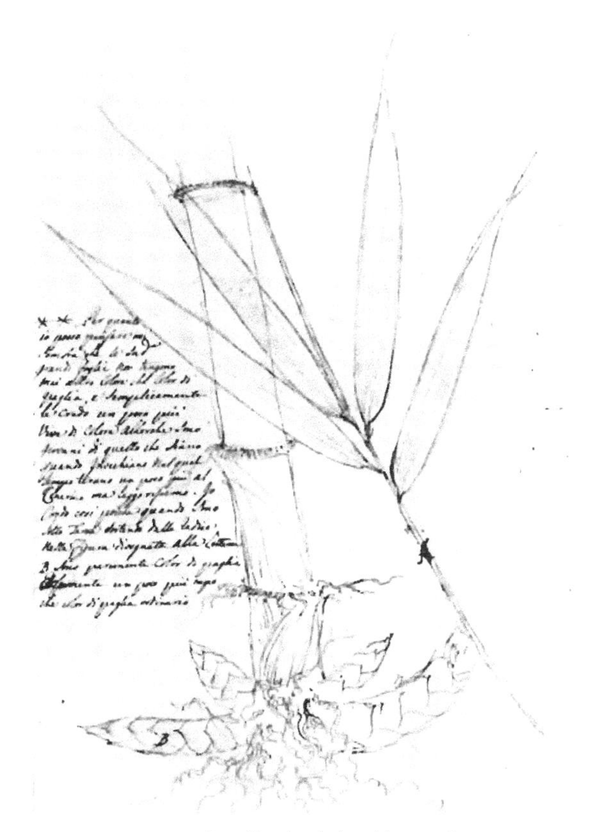

290. *Arundinaria alpina* (b) p. 108

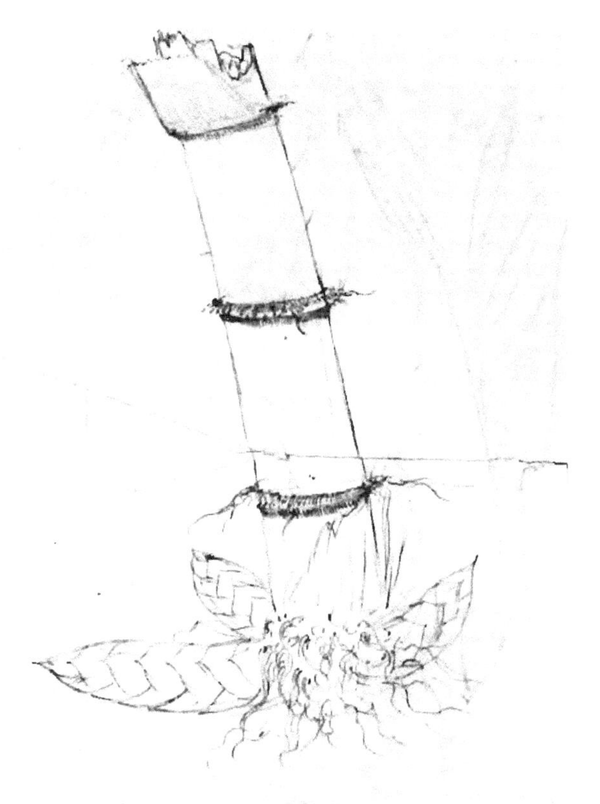

291. *Arundinaria alpina* (c) p. 108

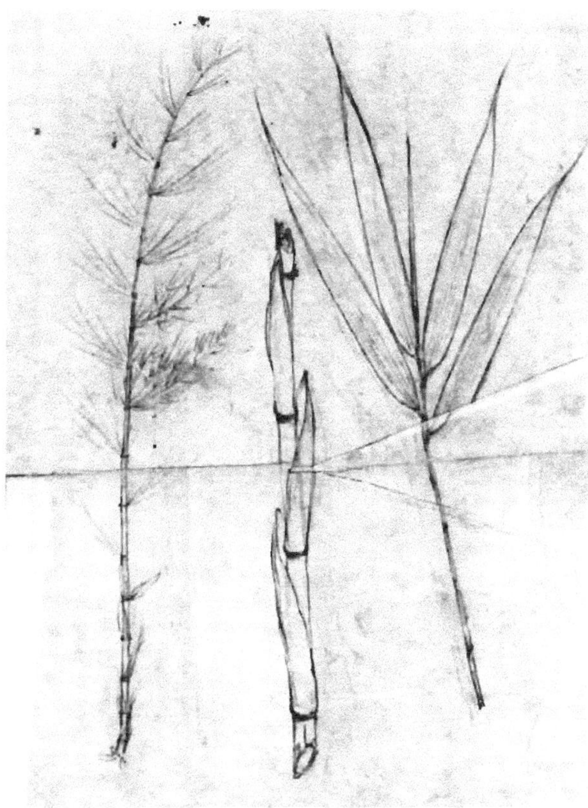

292. *Arundinaria alpina* (d) p. 108

293. *Arundinaria alpina* (f) p. 108

294. *?Chloris* species p. 108/109

295. *Eragrostis tef* (a) p. 109

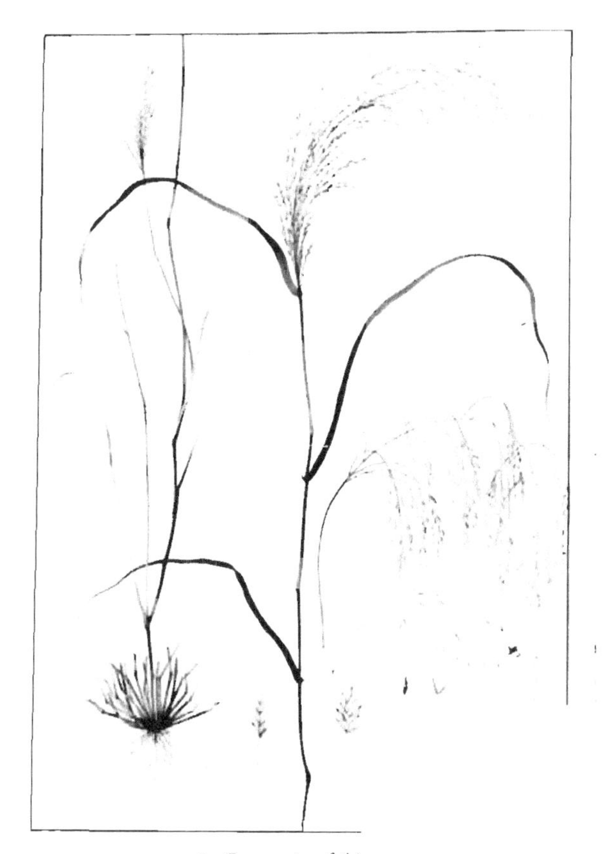

296. *Eragrostis tef* (b) p. 109

297. *Ischaemum afrum* (a) p. 109

298. *Eragrostis tef* (c) p. 109

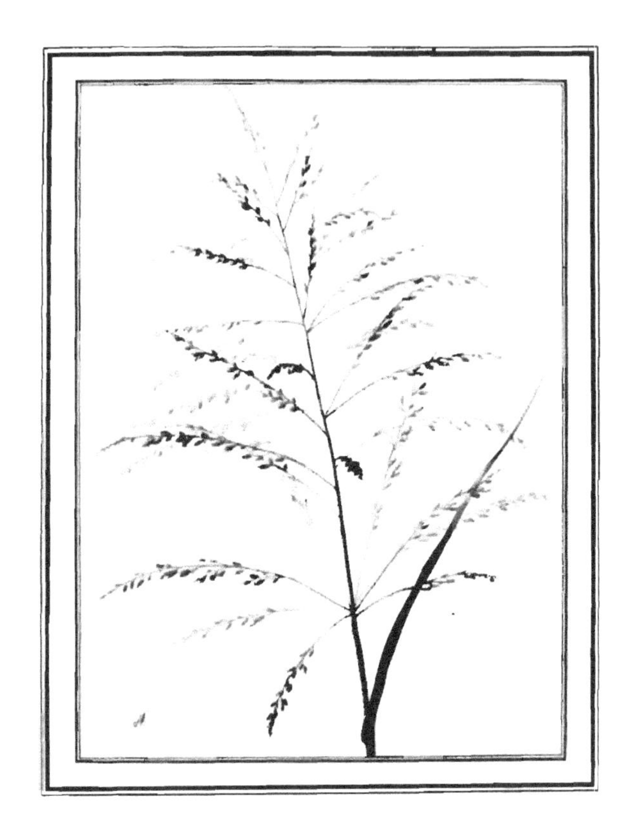

299. *Panicum deustum* (a) p. 110

300. *Panicum deustum* (b) p. 110

301. *Ottelia ulvifolia* (a) p. 110

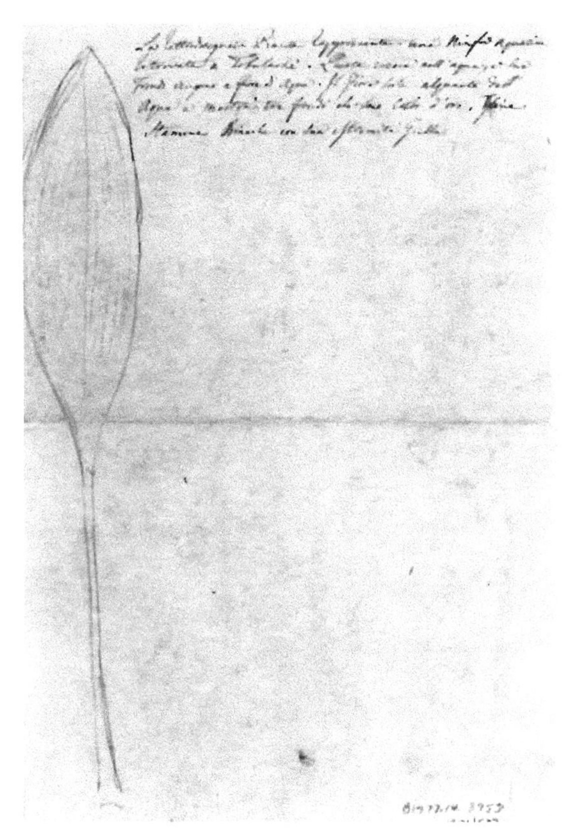

302. *Ottelia ulvifolia* (b) p. 110

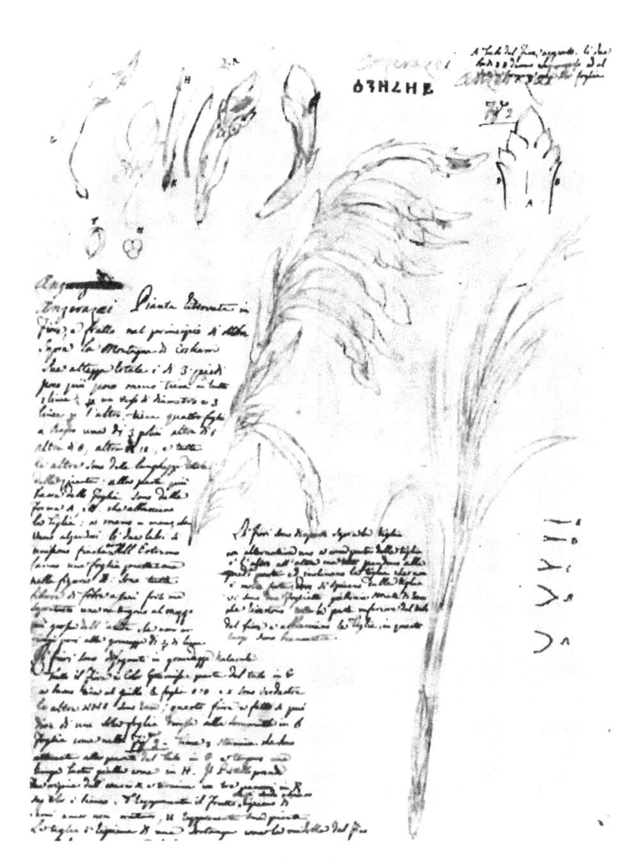

303. *Oenostachys abyssinica* (b) p. 110

304. *Aloe macrocarpa* (a) p. 111

305. *Oenostachys abyssinica* (d) p. 110

306. *Albuca abyssinica* p. 110/111

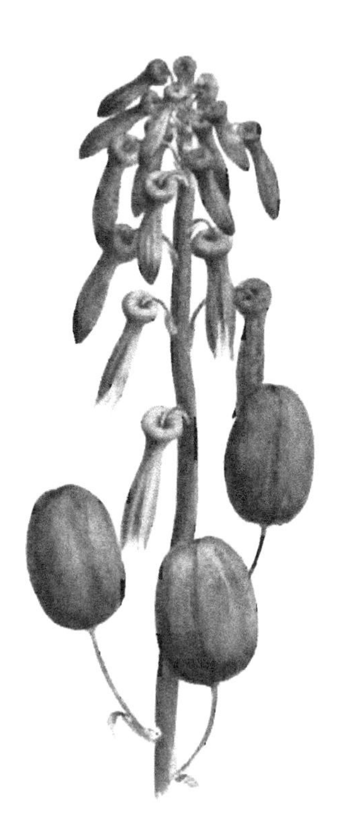

307. *Aloe macrocarpa* (b) p. 111

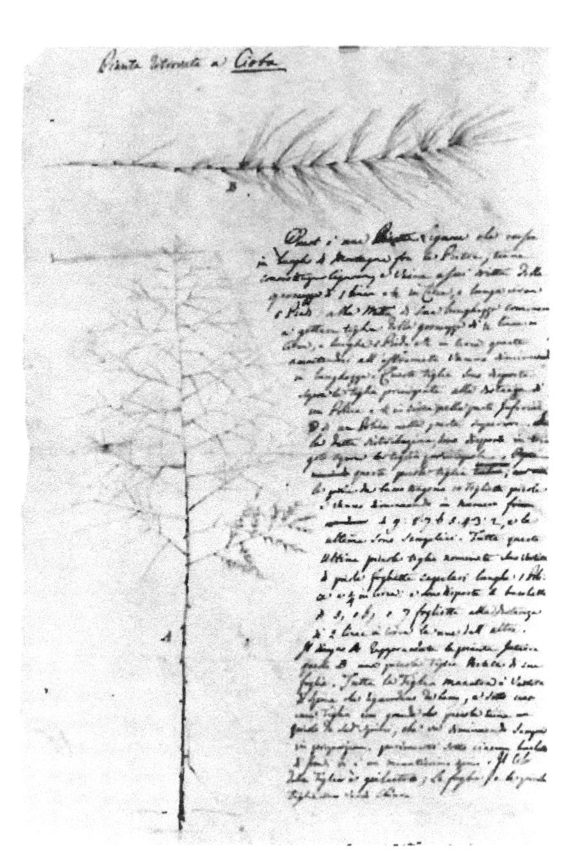

308. *Asparagus* species p. 111

Vi è una sorta di Cantaria grande, e bislunga di color
[illegible] [illegible] che passa in altezza.
Si vende di [illegible] 15 Mazzi per una Pataca.
Un Mazzo contiene 40 fili di 41 grani per ciascuno [illegible]
Un Mazzo tiene 1640 grani; e una Pataca 24600
A Dili un Mercante non voleva darne che 1800 per una tela
la quale si considera una Pataca.

1800 | 24600
13 12/18

100 : 40 : 78
100 | 40

B A

D

C

Quando l'Aloè è fatto [illegible]
[illegible] Natura sua foglia [illegible]
quando [illegible]
l'ovario che [illegible]
Quest'ovario [illegible]
[illegible] la figura [illegible]
tutto d'esso [illegible]
disposte nelle tre case in due lunghe per
ciascuna casa come nel disegno A che
rappresenta una Sezione del frutto per il lungo
e altro disegno B rappresenta una Sezione
trasversale. Quando questo [illegible]
nelle luoghi marcati 1 2 3 [illegible]
d'una sostanza [illegible] e bianche. Il [illegible]
è assai succolenta e [illegible]. Il seme
è racchiuso da due [illegible]
[illegible] al disegno.

Il [illegible] [illegible] D [illegible] [illegible]
[illegible] [illegible] [illegible]
questo è il frutto dell'Aloè che [illegible]
[illegible]; e [illegible] alla foglia [illegible]
[illegible]

309. *Aloe* species 2 p. 111

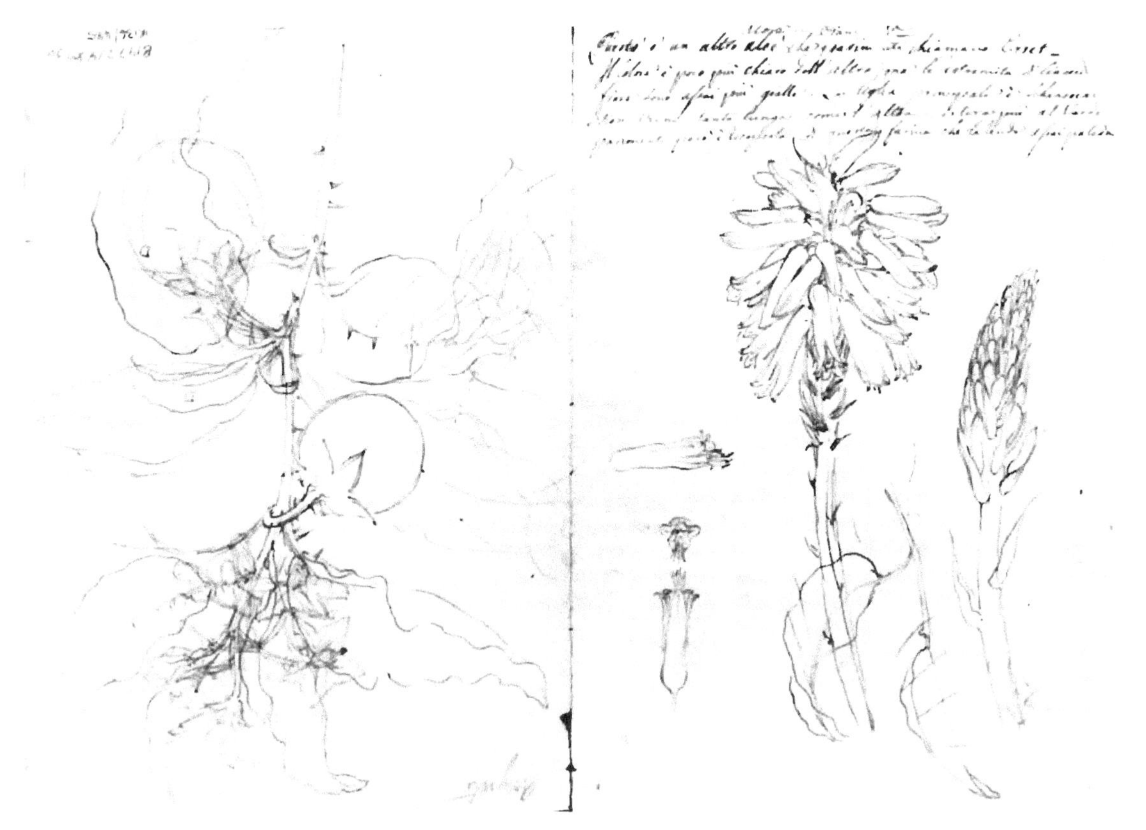

310. (left) *Solanum incanum* (a) p. 102
(right) *Aloe* species 1 p. 111

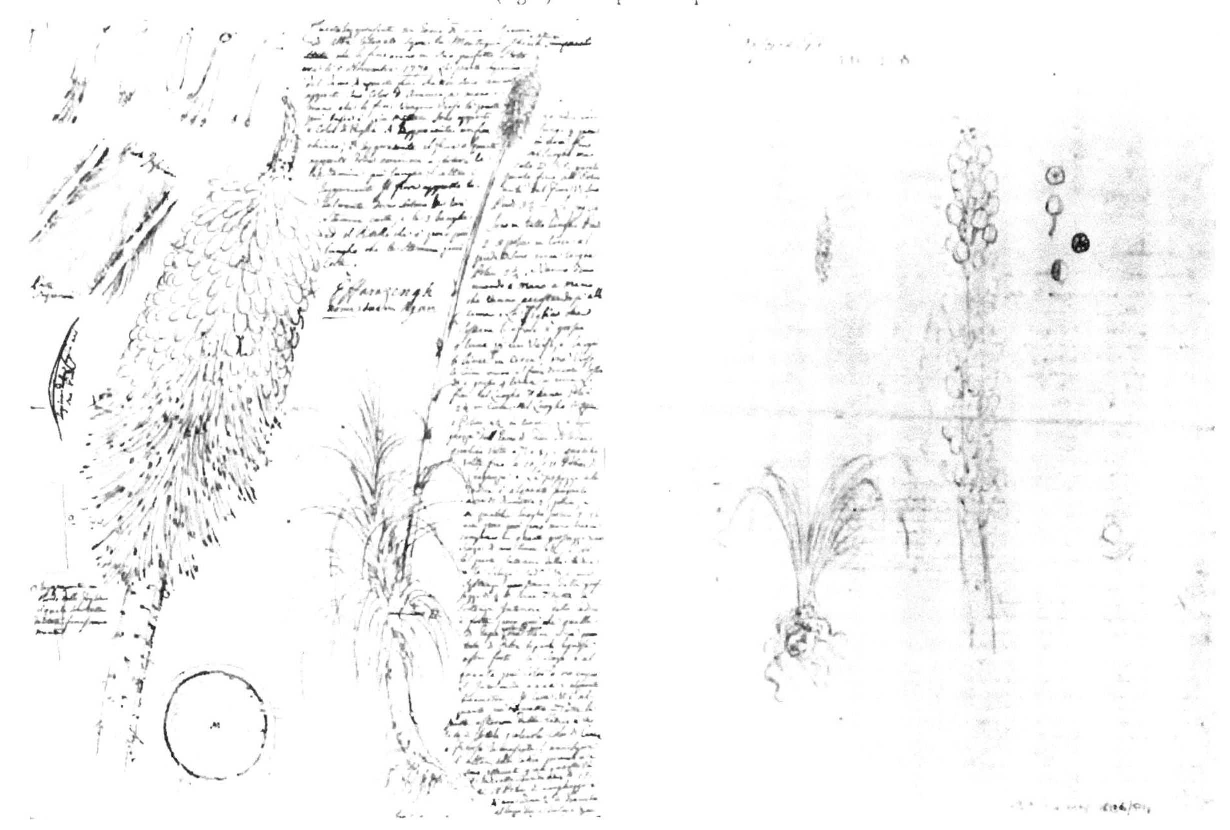

311. *Kniphofia foliosa* p. 111

312. *Kniphofia* prob. *pumila* × *schimperi* p. 112

313. *Kniphofia pumila* (a) p. 111

314. *Kniphofia pumila* (a) p. 111

315. *Sansevieria abyssinica* p. 112

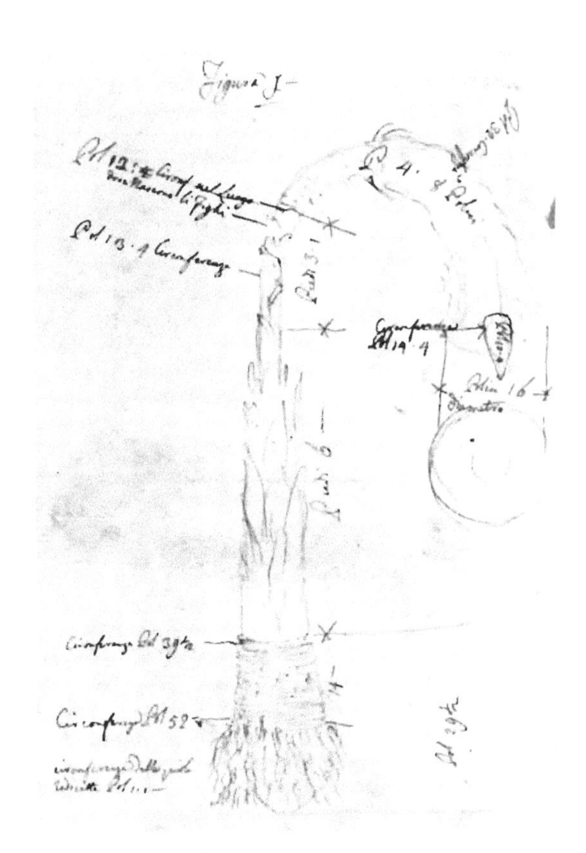

316. *Ensete ventricosum* (a) p. 112

317. *Ensete ventricosum* (b) p. 112 113

318. *Ensete ventricosum* (d) p. 112/113

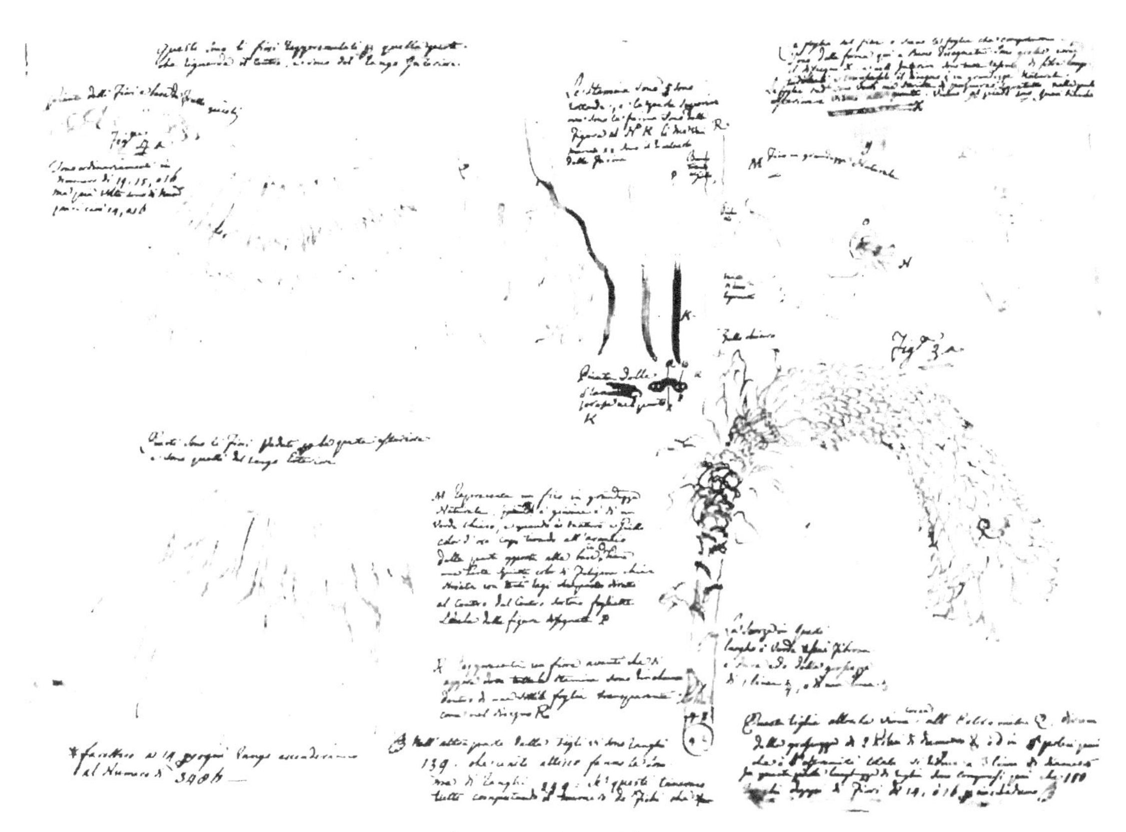

319. *Ensete ventricosum* (e) p. 112–113

320. *Ensete ventricosum* (f) p. 112–113

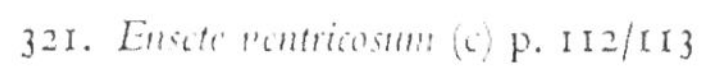

321. *Ensete ventricosum* (c) p. 112/113

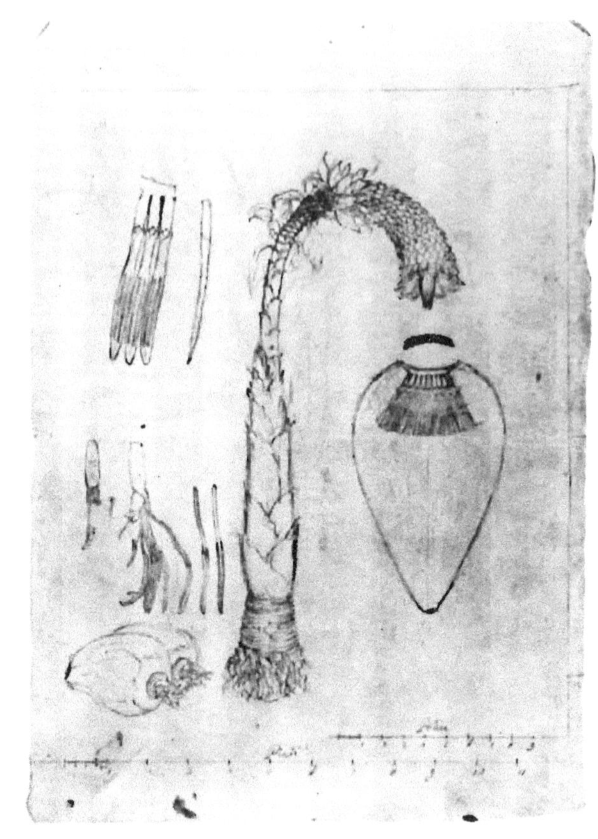

322. *Ensete ventricosum* (i) p. 112/113

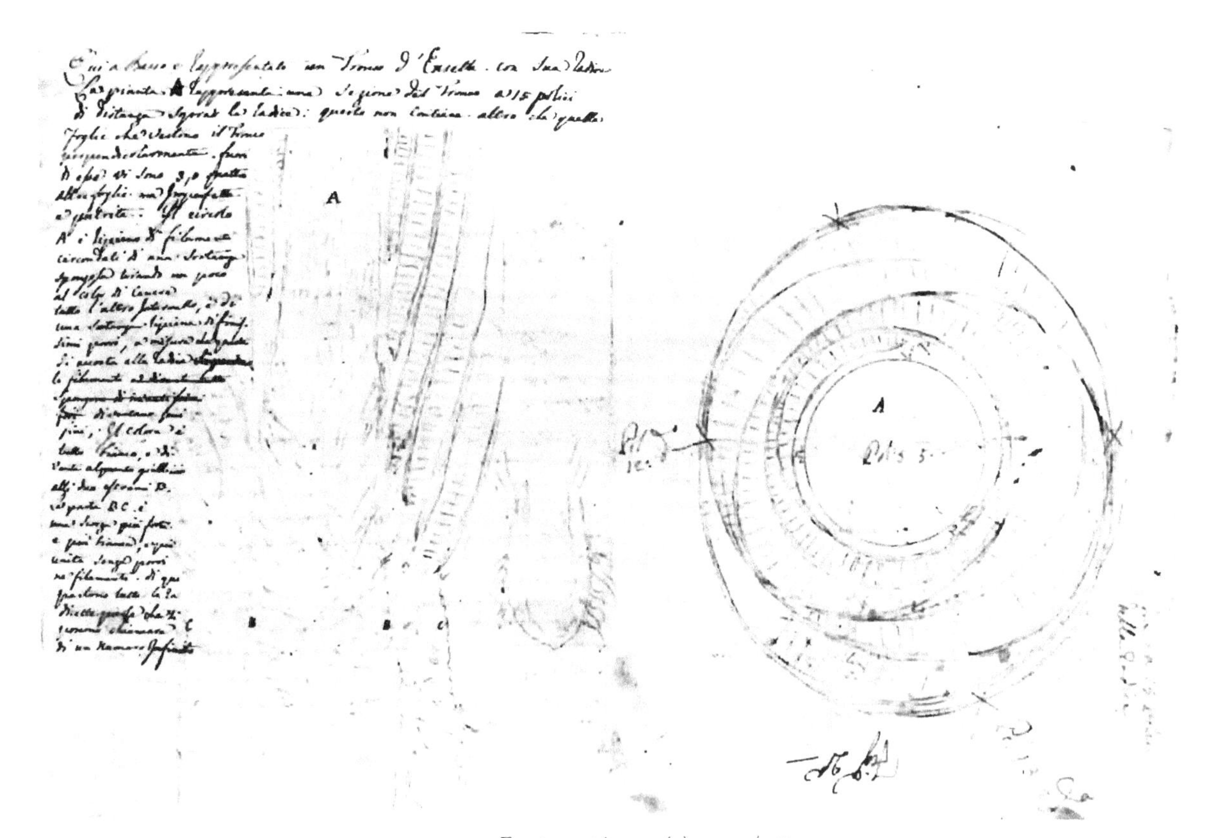

323. *Ensete ventricosum* (g) p. 112/113

Ritrovato a Gondar li 15 Ottobre 1770

L'albero Ensatt è rappresentato nel foglio marcato A: ma il disegno è rilevato da una pianta [illegible] che non tenea nè radice grande nè fiori nè frutti, ma unicamente le foglie in perfezione, le quali sono delle misure descritte e delle forme marcate nel disegno Œ così si fa vedere come sono nel suo piede ed a mano a mano che si vanno allungando come cambiano di figura, sempre conservando però le medesime tessiture di fibre perpendicolari dirette al centro [illegible] tante [illegible] [illegible] disposte alla distanza di ½ linea una dall'altra che forma tante collette. Come dalla figura VS viene rappresentato. Il più basso delle foglie della foglia, interiormente [illegible] [illegible] larghi ⅓ di circonferenza da ambe le parti verde, e le ⅗ che restano nel mezzo sono porporini. Tutta la parte interiore che è cava parimente è colorita della stessa maniera.

Le foglie sono tutte tessute di una quantità di finissime fibre; ed all'intervallo di ¾ di pollice, e di ¼ pollice ancora vi sono fibre più grosse dell'altre che si [illegible] alquanto, e nella parte opposta fanno apparire una cavità appena distinguibile che taglia le foglie di un piano [illegible] ma [illegible]

Un altro albero di Ensett si è ritrovato a Gondar nel med.° tempo il quale dicono alcuni che aveva cinque [illegible] anni questo era cresciuto a sua più gran grandezza. Era in tutto della lunghezza di piedi 16 – compreso la radice ed il fiore, bisogna avvertire che sua lunghezza non apparisce tale perchè oltre 2 piedi ½ di radice che resta seppellita nella terra, allorchè il tronco arriva all'altezza di [illegible] piedi in circa si ricurva facendo un arco [illegible] [illegible] [illegible] rivolti alla terra, il vertice di quest arco [illegible]

in linea perpendicolare sarà lontano dalla terra circa 9 piedi ½ poco più; quando è in questo stato il tronco, e la radice ed il fiore tiene la figura rappresentata alla fig.a 3.a in questo disegno è marcato tutte sue misure ed è spogliato di tutte le frondi e non ha che della scorza [illegible] una [illegible] come viene rappresentato alla fig. [illegible] [illegible] [illegible] è tutta minutamente [illegible] alla fig. [illegible] e le parti delle fiori sono [illegible] marcate all [illegible] foglio

Dove [illegible] a piegare il tronco la comunciano a mettere [illegible] [illegible] de [illegible] che sono tutte doppi [illegible] uno [illegible] come nella fig.a A [illegible] di queste [illegible] dal [illegible] fino all'ultimo [illegible] ne ho contate fino al numero di [illegible]

324. *Ensete ventricosum* (b) p. 112 113

325. *Musa × sapientum* (a) p. 113

326. *Juniperus procera* (a) p. 114

327. *Podocarpus gracilior* p. 114

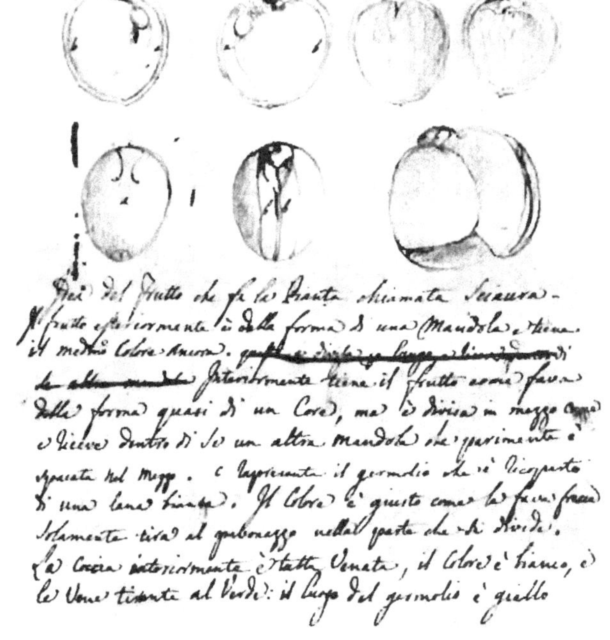

328. Unidentified plant p. 114

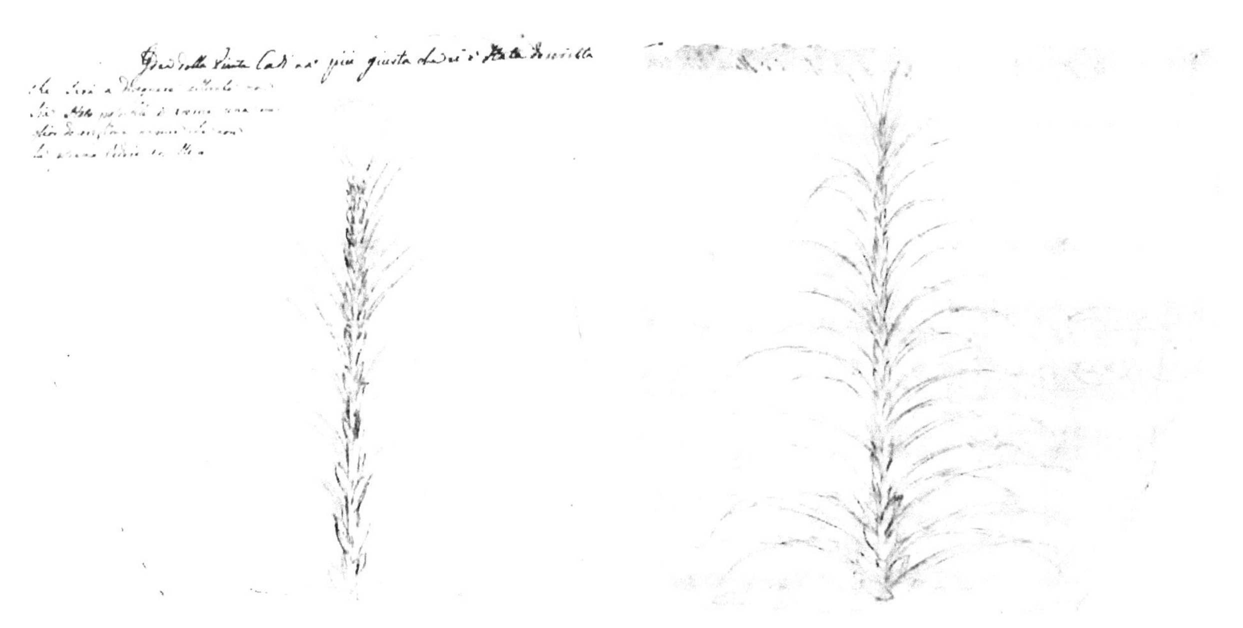

329. Unidentified plant p. 114

330. *Alga* indeterminate species p. 114

· Index I ·

Vernacular Names of Plants with Botanical Equivalents

All the vernacular names, mostly Amharic and Tigrinia, recorded by Bruce and Balugani as transliterations based on the Italian pronunciation, are listed below with their botanical equivalents and page references. Transcriptions of the vernacular names are by Dr. Jacques Mercier, based on the available literature and original field work, using the following abbreviations:

A: Amharic
Ar: Arabic
T: Tigrinia

? before the transcription means that the original name has not been recognized with certainty.
! after the transcription means that the use of this name for such a plant is not common but quite plausible.
? after the transcription means that the use of this name for such a plant is very dubious, but of course not quite impossible at that time and in that place.
( ) Common present use of the vernacular name.

VERNACULAR NAMES *The Bruce/Balugani names are followed by modern transcriptions.*	*SCIENTIFIC NAMES*
Abbeselim, Abbe zelim T [ḥabbi ṣällim]	Jasminum abyssinicum *DC.* 93
Abdeladic, Abdelasis Ar [ḥabb al ʿazīz]	Cyperus esculentus *L.* 107
Agam A [agam] T [ʿagam]	Carissa edulis *(Forssk.) Vahl* 72
Aijè	Diospyros mespiliformis *A. DC.* 80
Aitan balalitti T [ʿəṭan baʿalalti]!	Woodfordia uniflora *(A. Rich.) Koehne* 91
Aja aquareeti A [ahyya. . . .]!	Gardenia ternifolia *Schum. & Thonn.* subsp. jovis-tonantis *(Welw.) Verdc.* 97
Albasis Ar [ḥabb al ʿazīz]	Cyperus esculentus *L.* 107
Allemitth A [almit]!	Astragalus atropilosulus *(Hochst.) Bunge* see Discopodium
Almit A [almit]	Discopodium penninervium *Hochst.* 101
Almol A [aluma]	Discopodium penninervium *Hochst.* 101
Alzazo A [azzo]	Clematis hirsuta *Guill. & Perr.* 95
Amfar A [amfar]	Buddleia polystachya *Fresen.* 90
Angoule, Angule T [ʿangʷəllä]	Solanum incanum *L.* 102
Anguà T [ʿanqʷa] A[anqʷa]!	Boswellia papyrifera *(Del.) Hochst.* 75
Anguah No. 2	Sterculia africana *(Lour.) Fiori* see Boswellia
Anguah, Angurah	Boswellia papyrifera *(Del.) Hochst.* 75
Anseck-due	Kniphofia pumila x schimperi 112
Aquariti ?A ?T	Gardenia ternifolia *Schum. & Thonn.* subsp. jovis-tonantis *(Welw.) Verdc.* 97
Arze	Juniperus procera *Endl.* 114
Atatt A [aṭaṭ] T [ʿaṭʿaṭ]	Maytenus arguta *(Loes.) N. Robson* 77
Atcarò A [atkʷar]	Nuxia oppositifolia *(Hochst.) Benth.* 90
Attath, Attatt A [aṭaṭ] T [ʿaṭaṭ]	Maytenus arguta *(Loes.) N. Robson* 77
Avallò A [abalo]	Brucea antidysenterica *J. Miller* 100

VERNACULAR NAMES	*SCIENTIFIC NAMES*
Ayè T [ʿayä]	Diospyros mespiliformis *A. DC.* 80
Azerazaei A [anzäräzäy]	Oenostachys abyssinica *(A. Rich.) N.E. Br.* 110
Azazo, Azzo A [azzo]	Clematis hirsuta *Guill. & Perr.* 95
Babeer Ar [bardī]	Cyperus papyrus *L.* 107
Balessan, Balissan T [bäläsan] Ar[balasān]	Commiphora gileadensis *(L.) C. Chr.* 75
Berberrà A [bərbərra]	Millettia ferruginea *(Hochst.) Baker* 88
Bohah T [buwah̲]	Rhus ? retinorrhoea *Oliver* 71
Bun AT [bun] Ar[bunn]	Coffea arabica *L.* 97
Burberra Amka A [bərbərra]	Millettia ferruginea *(Hochst.) Baker* 88
Cadi —	Unidentified species 114
Cajem, Cajiem A [k^{w}äšəm]	Myrsine africana *L.* 92
Cangieb, Canjeb AT [qənčəb]	Euphorbia ?tirucalli *L.* 82
Cantaffa, Cantuffa A [qänṭaffa]	Pterolobium stellatum *(Forssk.) Brenan* 89
Carrat, Carratt A [qäräṭ] T [qäräṣ]	Osyris lanceolata *A. DC.* 99
Cashishillo A [k^{w}äšäšəlla]	Acanthus sennii *Chiov.* 70
Chaat AT [čat]	Catha edulis *Forssk.* 76
Chista-cachemantia —	Annona squamosa *L.* 71
Ciath AT [čat]	Catha edulis *Forssk.* 76
Corzama T [kusra]!	Ziziphus abyssinica *A. Rich.* 96
Corzamà, Corzuma —	Bersama abyssinica *Fresen.* 92
Cosciscilla, Coshillilla A [k^{w}äšäšəlla]	{Acanthus sennii *Chiov.* 70 {Echinops macrochaetus *Fresen.* 77
Cottina A [qoṭäṭənna]	Verbascum sinaiticum *Benth.* 100
Cuara A [qwara]	{Erythrina abyssinica *DC.* 87 {Erythrina brucei *Schweinf. emend. Gillett* 87
Cumel T [kumäl]	Mimusops kummel *A. DC.* 99
Cura-magiett A [qura magäṭ]!	Cucumis metuliferus *Naud.* 79
Cusso, Cuzzo AT [koṣo]	Hagenia abyssinica *(Bruce) J.F. Gmel.* 96
Dangheeh, Dangheek, Dangheeli ? A [dängäl]!	Mariscus cyperoides *(L.) O. Kuntze* 108
Daroo T [daʿəro]	Ficus vasta *Forssk.* 92
Deh hack, Dehack —	Mussaenda arcuata *Lam. ex Poir.* 98
Dembelal A [dəmbəlal]	Coriandrum sativum *L.* 104
Denitch, Dinitch A [dənneč]	Coleus edulis *Vatke* 82
Doom Ar [dūm]	Hyphaene thebaica *(L.) Mart.* 113
Dorwan —	Delphinium wellbyi *Hemsley* 95
Effarazengh A [(yä)färäs zäng]!	Kniphofia foliosa *Hochst.* 111
El berdi Ar [bardī]	Cyperus papyrus *L.* 107
Endaud, Endood, Endoud A [əndod]	Phytolacca dodecandra *L'Hérit.* 93
Enset, Ensete, Ensett, Ensette A [ənsät]	Ensete ventricosum *(Welw.) Cheesman* 112
Enzerazai (Enzerazei) A[ənzäräzäy]	Oenostachys abyssinica *(A. Rich.) N.E. Br.* 110
Ergett dimmo Ar [? šagarat al dam]!	Dichrostachys cinerea *(L.) Wight & Arn.* 86
Ergett el krone, Ergett el kroun Ar[?šagarat al qurūn]!	Mimosa pigra *L.* 88
Ergett y dimmo Ar [?šagarat al dam]!	Dichrostachys cinerea *(L.) Wight & Arn.* 86
Erret, Errett A [ərret]	Aloe macrocarpa Todaro 111
Etam balalite, Etan balalli, Etan balabelle T[ʿətan baʿalalti]!	Woodfordia uniflora *(A. Rich.) Koehne* 91

VERNACULAR NAMES	*SCIENTIFIC NAMES*
Farek A [färäq]!	Bauhinia farec *Desv.* 85
Feel fetch, Fiel fetch A [fəyyäläfäğ]?	Hypericum quartinianum *A. Rich.* 82
Gagudei, Gaguedi T	Protea gaguedi *J.F. Gmel.* 94
Gemero A [gəmərro]	Capparis tomentosa *Lam.* 76
Gerget, Gergeth AT [gəršəṭ]	Impatiens rothii *Hook.f.* 74
Ghesh, Geshe A [gešo]	Rhamnus prinoides *L'Hérit.* 96
Geshe el aube Ar	Ischaemum afrum *(J.F. Gmel.) Dandy* 109
Gers seth AT [gəršəṭ]	Impatiens rothii *Hook.f.* 74
Gheccia ?A [gəč̣č̣a]?	Unknown species (mentioned under Trifolium rueppellianum) 89
Ghersehetth, Gherssetth AT [gəršəṭ]	Impatiens rothii *Hook.f.* 74
Ghesh, Gheshe A [gešo]	Rhamnus prinoides *L'Hérit.* 96
Gibara A [ğəbara]	Dracaena steudneri *Engl.* 105
Gir gir Ar [girgīr] ?	Ischaemum afrum *(J.F. Gmel.) Dandy* 109
Goneck, Guoneck T [gonäq]	Dichrostachys cinerea *(L.) Wight & Arn.* 86
Grawa A [grawwa]	Vernonia amygdalina *Del.* 78
Gumarrò A [gumärro]!	Ritchiea albersii *Gilg* 76
Hà A [kaya]	Salix subserrata *Willd.* 98
Habetteri A [abättäre]	Ziziphus abyssinica *A. Rich.* 96
Hatcarù A [atkʷar]	Nuxia oppositifolia *(Hochst.) Benth.* 90
Hecà T [ʿiqqa]	Sansevieria abyssinica *N.E. Br.* 112
Ileef Ar [līf]	Peponium vogelli *(Hook.f.) Engl.* 80
Jadzeferi A [yaṣe. . .]	Desmodium sp. 88
Jajja diitch A [yahyya dənnač]	Coleus species 2 83
Jauira, Javeira A [wäyra]	Olea europaea ssp. africana *(Miller) P.S. Green* 93
Jebaara A [ğəbara]	Dracaena steudneri *Engl.* 105
Jujef —	Myrsine africana *L.* 92
Kantaffa, Kantuffa —	Pterolobium stellatum *(Forssk.) Brenan* 89
Kinchette —	Pycnostachys abyssinica *Fresen.* 84
Kol-qual, Kol-quall A [qulqʷal]-T[qʷälqʷal]	Euphorbia abyssinica *J.F. Gmel.* 81
Korsuma —	Bersama abyssinica *Fresen.* 92
Korzama T [kusra]!	Ziziphus abyssinica *A. Rich.* 96
Krehàhà, Krihaha, Krhaha A [qàrqäha]	Arundinaria alpina *K. Schum.* 108
Kuara A [qʷara]	Erythrina abyssinica *DC.* 87 Erythrina brucei *Schweinf. emend Gillett* 87
Kuaregh —	Cucumis metuliferus *Naud.* 79
Kumel, Kummel, Kummell T [kumäl]	Mimusops kummel *A. DC.* 99
Kuraregh A [(yä)qura aräg]	Momordica foetida *Schum. & Thonn.* 80
Kusso AT [koso]	Hagenia abyssinica *(Bruce) J.F. Gmel.* 96
Lambehk A [ləmbəč̣]	Clausena anisata *(Willd.) Benth.* 98
Lanbutt —	Commelina africana *L.* 107
Leef Ar [līf]	Peponium vogelii *Engl.* 80
Leham T [läḥam]?	Saba comorensis *(Bojer) Pichon* 72
Lembetch A [ləmbəč̣]	Clausena anisata *(Willd.) Benth.* 98
Liff Ar [līf]	Peponium vogelii *Engl.* 80
Lillahoo, Lillahù A [yəllaho]	Pittosporum viridiflorum *Sims* 94
Logheba, Logheta A [länqʷaṭa]	Grewia ferruginea *A. Rich.* 104
Maraqua A [marqʷa]?	Senecio gigas *Vatke* 78
Meim mesalib Ar	?Chloris species 108

VERNACULAR NAMES	*SCIENTIFIC NAMES*
Menghe, Menghi, Mengi T [mängi]	Rhus glutinosa *A. Rich.* 71
Mepta T [mäbṭʿa]	Acokanthera schimperi *(DC.) Schweinf.* 71
Mergiombe, Mergiombei, Mergombey, Mergomeby, Merjombey A[žärč əmbʷay]	Solanum adoense *Hochst. ex A. Rich.* 102 S. piperiferum *A. Rich.* 103
Muem il msalib Ar	?Chloris species 108
Mzenna A [məsanna]	Croton macrostachyus *Hochst. ex A. Rich.* 81
Neccilò, Netchilò, Nev A [näčəllo]	Abutilon longicuspe *A. Rich.* 91
Nook, Nuk A [nug]	Guizotia abyssinica *(Linn.f.) Cass.* 77
Ombuai, Ombuay A [əmbʷay]	Solanum adoense *Hochst. ex A. Rich.* 102
Papyrus —	Cyperus papyrus *L.* 107
Rack Ar [rak] ?	Avicennia marina *(Forssk.) Vierh.* 74
Samf, Sassa A [säsa] ?	Carthamus tinctorius *L.* 77 Millettia ferruginea *(Hochst.) Baker* 88 Albizia gummifera *(J.F. Gmel.) C.A. Sm.* 84
Scangho T	Dregea abyssinica *(Hochst.) K. Schum.* 73
Sciaura —	Indeterminate species 114
Scie A [(ə)še]	Mimusops kummel *A. DC.* 99
Scioneurth —	Crinum zeylanicum *(L.)L.* 106
Scionkust A [šənkurt]	Scadoxus puniceus *L.* 106
Selchienn, Selcienn A [sälčäñ]	Diospyros abyssinica *(Hiern) F. White* 80
Semec, Semeck A [sämäg]	Cardiospermum halicacabum *L.* 99
Semezza, Semezze A [səmizza]-T[səmʿəzza]!	Ruttya speciosa *(Hochst.) Engl.* 70
Seyali Ar [sayāl]	Acacia species 84
Shanfo, Shango T [šanquq]	Dregea abyssinica *(Hochst.) K. Schum.* 73
Shongourt A [šənkurt]	Allium species 111
Somf, Suf, Sussa AT [suf]	Carthamus tinctorius *L.* 77
Tambò T [tambok̲]	Croton macrostachyus *Hochst. ex A. Rich.* 81
Teff A [ṭef]	Eragrostis teff *(Zucc.) Trotter* 109 see Panicum deustum *Thunb.* 110
Terràh —	Jasminum dichotomum *Vahl* 93
Tinguit, Tinguitt A [ṭənğut-ṭəngit]!	Otostegia tomentosa *A. Rich.* subsp. ambigens *(Chiov.) Sebald* 83
Tsadjesar, Tsadjessar, Tsadg-saar, Tzadge saar A [ṭäğğ sar] T?	Cyperus species 1 107
Ueshish A [wəšəš]	Coccinia abyssinica *(Lam.) Cogn.* 79
Umboy A [əmbʷay]	Solanum incanum *L.* 102
Umfar A [amfar]	Buddleia polystachya *Fresen.* 90
Uschis, Usheest, Useish, Ushish A [wəšəš]	Coccinia abyssinica *(Lam.) Cogn.* 79
Valkoffa, Walkaffa, Walkoffa, Walakuffa, Walkuffa A [wəlkəffa]	Dombeya torrida *(J.F. Gmel.) Bamps* 103
Wansey, Wanzey A [wanza]	Cordia abyssinica *R.Br.* 74
Woginoussi, Woginus, Wooginoos, Wryginous A [waginos]	Brucea antidysenterica *J. Miller* 100
Yad zeffere A [yaṣe . . .]	Desmodium species 88
Yesser —	Alga 114
Yeya deetch A [yahyya dənnəč]	Coleus species 2 83
Zatarindi Ar [saʿātar hindī]	Indeterminate Labiatae 84
Zegva AT [zəgba]	Podocarpus gracilior *Pilger* 114
Zuarua T [zəwawʿa]	Erythrina abyssinica *DC.* 87 Erythrina brucei *Schweinf. emend Gillett* 87

· Index II ·

Scientific Names of Plants

Synonyms given in italics

· Index III ·

Names of Places

See also Itinerary pp. 119–20 and Maps 1 & 2 after p. 40

· Index IV ·

Names of Persons, Peoples, Titles etc.

T - #0986 - 101024 - C48 - 276/216/19 [21] - CB - 9781138407657 - Gloss Lamination